Seventh Edition

STUDENT ATLAS OF World Politics

John L. Allen
University of Wyoming

The McGraw-Hill Companies

Book Team

Managing Editor *Larry Loeppke*
Developmental Editor *Nichole Altman*
Developmental Editor *Jill Peter*
Designer *Tara McDermott*
Typesetting Supervisor *Kari Voss*
Typesetter *Jean smith*
Typesetter *Sandy Wille*
Typesetter *Karen Spring*
Cover Design *Maggie Lytle*
Cartography *Carto-Graphics, Oxford, MD*

We would like to thank Digital Wisdom Incorporated for allowing us to use their Mountain High Maps cartography software. This software was used to create maps 74–91.

The credit section for this book begins on page 207 and is considered an extension of the copyright page.

Student atlas of world politics / by John L. Allen.
Dubuque, IA: McGraw-Hill/Contemporary Learning Series, © 2006

7th Edition

208 p.: ill., maps; cm.

I. World politics—1991—Atlases. II. International relations—Atlases.

909.82

ISBN 0-07-352773-4 ISSN: 1524-4556

Printed in the United States of America

123456789QPDQPD5

A Note to the Student

International politics is a drama played out on a world stage. If the events of and subsequent to September 11, 2001–the worldwide war against terrorism, the specific regional military actions in Afghanistan and, more recently, in Iraq–have taught us anything, it is that the drama is very real and the stage is indeed a worldwide one. The maps in this atlas serve as the stage settings for the various scenes in this drama; the data tables are the building materials from which the settings are created. Just as the stage setting helps bring to life and give meaning to the actions and words of a play, so can these maps and tables enhance your understanding of the vast and complex drama of global politics, including the emergence of global terrorism as a political instrument. Use this atlas in conjunction with your text on international politics or international affairs. It will help you become more knowledgeable about this international stage as well as the actors.

The maps and data sets in the *Student Atlas of World Politics*, seventh edition, are designed to introduce you to the importance of the connections between geography and world politics. The maps are not perfect representations of reality–no maps ever are–but they do represent "models," or approximations of the real world, that should aid in your understanding of the world drama.

You will find your study of this atlas more productive in relation to your study of international politics if you examine the maps on the following pages in the context of five distinct analytical themes:

1. *Location: Where Is It?* This involves a focus on the precise location of places in both absolute terms (the latitude and longitude of a place) and in relative terms (the location of a place in relation to the location of other places). When you think of location, you should automatically think of both forms. Knowing something about absolute location will help you to understand a variety of features of physical geography, since such key elements are often so closely related to their position on the earth. But it is equally important to think of location in relative terms. The location of places in relation to other places is often more important in influencing social, economic, and cultural characteristics than are the factors of physical geography. Certainly the relative location of the World Trade Center was crucial to the identification of the WTC as both a symbolic and actual center of economic activity–and this identification played an important role in its selection as a target for terrorist action. Equally, both the relative and absolute location of Baghdad play an important role in the ongoing struggle to create a politically stable Iraq.
2. *Place: What Is It Like?* This encompasses the political, economic, cultural, environmental, and other characteristics that give a place its identity. You should seek to understand the similarities and differences of places by exploring their basic characteristics. Why are some places with similar environmental characteristics so very different in economic, cultural, social, and political ways? Why are other places with such different environmental characteristics so seemingly alike in terms of their institutions, their economies, and their cultures? The place characteristics of parts of the world American students have known little about (like Afghanistan and Iraq) have now emerged as vital components of our necessary understanding of the implementation of military and political strategies.
3. *Human/Environment Interactions: How Is the Landscape Shaped?* This theme focuses on the ways in which people respond to and modify their environments. On the world stage, humans are not the only part of the action. The environment also plays a role in the drama of international politics. But the characteristics of the environment do not exert a controlling influence over human activities; they only provide a set of alternatives from which different cultures, in different times, make their choices. Observe the relationship between the basic elements of physical geography such as climate and terrain and the host of ways in which humans have used the land surfaces of the world. To know something of the relationship between people and the environment in the arid parts of the Old World is to begin to understand the nature of political, economic, and even religious conflicts between the inhabitants of those regions and others. The ongoing unrest in the Dafur region of Sudan or military conflict in the Congo Basin are at least partly attributable to the interaction between people and their environment.
4. *Movement: How Do People Stay in Touch?* This examines the transportation and communication systems that link people and places. Movement or "spatial interaction" is

the chief mechanism for the spread of ideas and innovations from one place to another. It is spatial interaction that validates the old cliché, "the world is getting smaller." We find McDonald's restaurants in Tokyo and Honda automobiles in New York City because of spatial interaction. And the spread of global terrorism is, first and foremost, a process of spatial interaction. Advanced transportation and communication systems have made possible such events as transpired in New York City in September, 2001, or in Iraq in May, 2003, and have transformed the world into which your parents were born. And the world that greets your children will be very different from your world. None of this would happen without the force of movement or spatial interaction.

5. *Regions: Worlds Within a World.* This theme, perhaps the most important for this atlas, helps to organize knowledge about the land and its people. The world consists of a mosaic of "regions" or areas that are somehow different and distinctive from other areas. The region of Anglo-America (the United States and Canada) is, for example, different enough from the region of Western Europe that geographers clearly identify them as two unique and separate areas. Yet despite their differences, Anglo-Americans and Europeans share a number of similarities: common cultural backgrounds, comparable economic patterns, shared religious traditions, and even some shared physical environmental characteristics. Conversely, although the regions of Anglo-America and Southwestern Asia (the "Middle East") are also easily distinguished as distinctive units of the Earth's surface with some shared physical environmental characteristics, the inhabitants of these two regions have fewer similarities and more differences between them than is the case with Anglo-America and Western Europe: different cultural traditions, different institutions, different linguistic and religious patterns. An understanding of both the differences and similarities between regions like Anglo-America and Europe on the one hand, or Anglo-America and Southwest Asia on the other, will help you to understand much that has happened in the human past or that is currently transpiring in the world around you. At the very least, an understanding of regional similarities and differences will help you to interpret what you read on the front page of your daily newspaper or view on the evening news report on your television set.

Not all of these themes will be immediately apparent on each of the 105 maps and 14 tables in this atlas. But if you study the contents of *Student Atlas of World Politics*, seventh edition, along with the reading of your text and think about the five themes, maps and tables and text will complement one another and improve your understanding of global politics. As Shakespeare said, "All the world's a stage." Your challenge is now to understand both the stage and the drama being played on it.

A Word about Data Sources

At the very outset of your study of this atlas, you should be aware of some limitations of the maps and data tables. In some instances, a map or a table may have missing data. This may be the result of the failure of a country to report information to a central international body (like the United Nations or the World Bank). Alternatively, it may reflect shifts in political boundaries, internal or external conflicts, or changes in responsibility for reporting data have caused certain countries (for example, those countries that made up the former Yugoslavia) to delay their reports. It is always our wish to be as up-to-date as is possible; earlier editions of this atlas were lacking more data than this one and subsequent versions will have still more data, particularly on the southeastern European countries, the independent countries formerly part of the now-defunct Soviet Union, or on African and Asian nations that are just beginning to reach a point in their economic and political development where they can consistently report reliable information. In the meantime, as events continue to restructure our world, it's an exciting time to be a student of international events!

John L. Allen
University of Wyoming

What's New In This Edition

The Student Atlas of World Politics, Seventh Edition, reflects current political, economic, demographic, and environmental changes in every world region. This edition is substantially enlarged from its predecessor and is the most comprehensive version yet of a book that has long been a standard in the field. Here, in one volume are 105 carefully selected full-color thematic maps (11 more than in the Fifth Edition), 18 reference maps rich in details of physical and political geography (14 more than in the Sixth Edition), and 114 tables of valuable and current data relevant to international political issues and affairs. A number new thematic maps highlight global and regional geopolitical events and processes (including recent and ongoing terrorist and military action in several regions of the world). Other new thematic maps on land use ("Environment and Economy") have been added to the reference maps in the concluding section of the Atlas. Current hotspots or "flashpoints" are identified in their own section and include maps of:

- Macedonia
- Chechnya
- Israel and Its Neighbors
- Jammu and Kashmir
- Kurdistan
- Central Africa/Congo
- Sri Lanka
- Sudan/Darfur
- Somalia
- Afghanistan
- Iraq
- South Central Eurasia

The tables include the latest available country-by-country data on a wide array of political, military, economic, demographic, and environmental issues.

This unique combination of maps and data makes the atlas an invaluable pedagogical tool. It also serves to introduce students to the five basic themes of spatial analysis:

- Location: Where is it?
- Place: What is it like?
- Human-Environment Interaction: How is the landscape shaped?
- Movement: How do people stay in touch?
- Regions: How is the surface of the world arranged and organized?

Concise and affordable, this up-to-date *Student Atlas of World Politics* is suitable for any course dealing in current world affairs.

About the Authors

John L. Allen is professor and chair of the Department of Geography at the University of Wyoming and professor emeritus of geography at the University of Connecticut, where he taught from 1967 to 2000. He is a native of Wyoming and received his bachelor's degree in 1963 and his M.A. in 1964 from the University of Wyoming, and his Ph.D. in 1969 from Clark University. His areas of special interest are perceptions of the environment and the impact of human societies on environmental systems. Dr. Allen is the author and editor of many books and articles as well as several other student atlases, including the best-selling *Student Atlas of World Geography*.

Acknowledgments

The authors wish to recognize with gratitude the advice, suggestions, and general assistance of the following reviewers:

Robert Bednarz
Texas A & M University

Gerald E. Beller
West Virginia State College

Kenneth L. Conca
University of Maryland

Femi Ferreira
Hutchinson Community College

Paul B. Frederic
University of Maine at Farmington

James F. Fryman
University of Northern Iowa

Michael Gold-Biss
St. Cloud State University

Herbert E. Gooch III
California Lutheran University

Lloyd E. Hudman
Brigham Young University

Edward L. Jackiewicz
Miami University of Ohio

Artimus Keiffer
Indiana University–Purdue University at Indianapolis

Richard L. Krol
Kean College of New Jersey

Jeffrey S. Lantis
The College of Wooster

Robert Larson
Indiana State University

Mark Lowry II
United States Military Academy at West Point

Max Lu
Kansas State University

Taylor E. Mack
Mississippi State University

Kenneth C. Martis
West Virginia University

Calvin O. Masilela
West Virginia University

Patrick McGreevy
Clarion University

Tyrel G. Moore
University of North Carolina at Charlotte

David J. Nemeth
The University of Toledo

Emmett Panzella
Point Park College

Daniel S. Papp
University System of Georgia

Lance Robinson
United States Air Force Academy

Jefferson S. Rogers
University of Tennessee at Martin

Barbara J. Rusnak
United States Air Force Academy

Mark Simpson
University of Tennessee at Martin

Jutta Weldes
Kent State University

Table of Contents

Part III The Global Economy 59

Part IV Population and Human Development 71

Part V Food, Energy, and Materials 84

Part VI Environmental Conditions 94

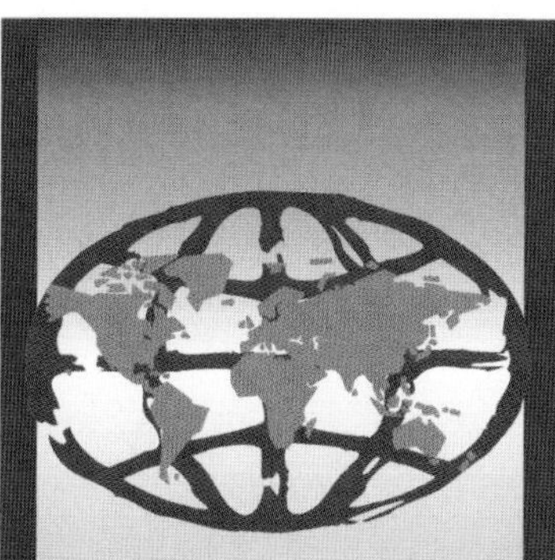

Part I

The Contemporary World

Map 1 World Political Boundaries

The international system includes states (countries) as the most important component. The boundaries of countries are the primary source of political division in the world, and for most people nationalism is the strongest source of political identification.

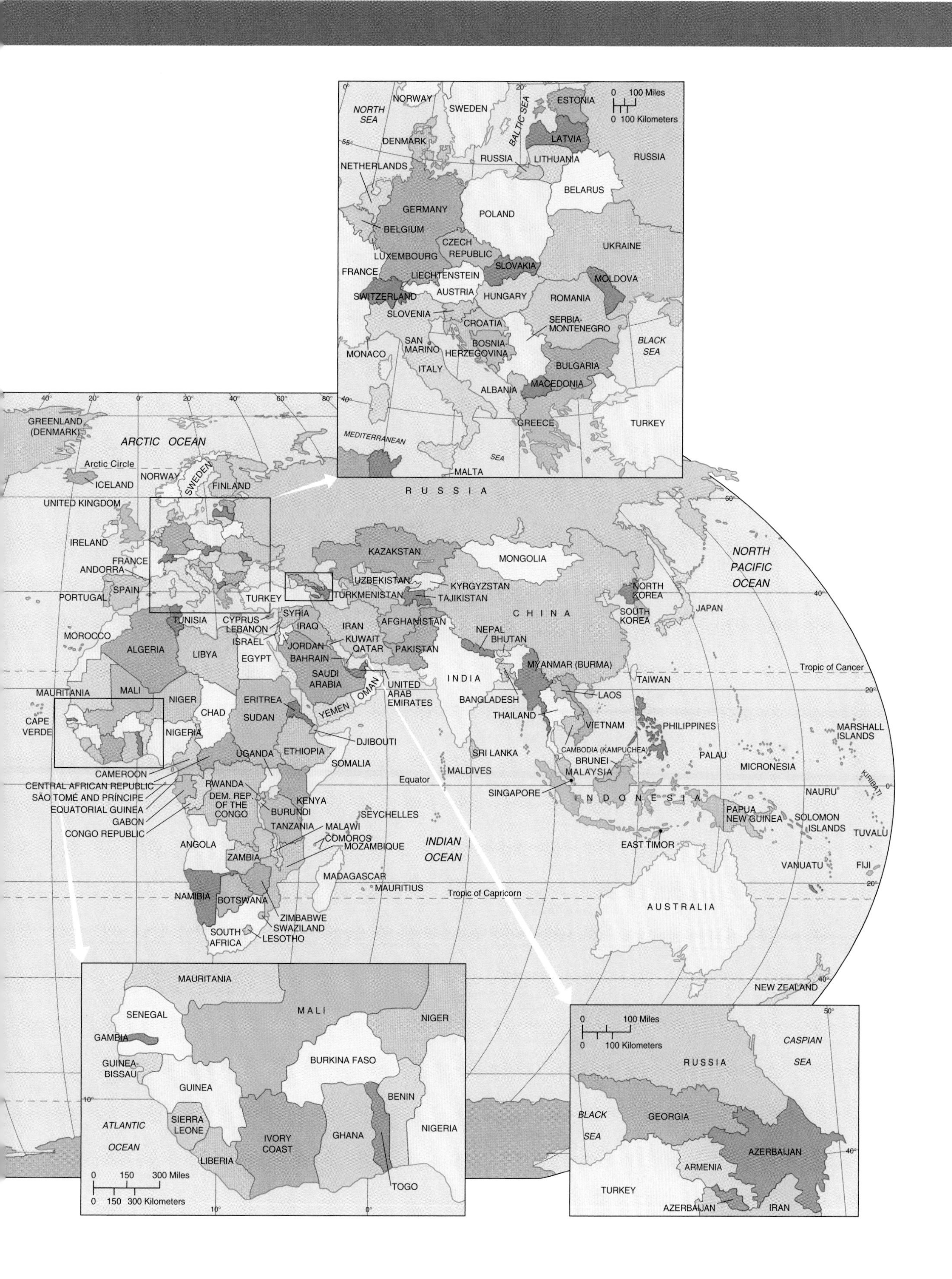

Map 2 World Climate Regions

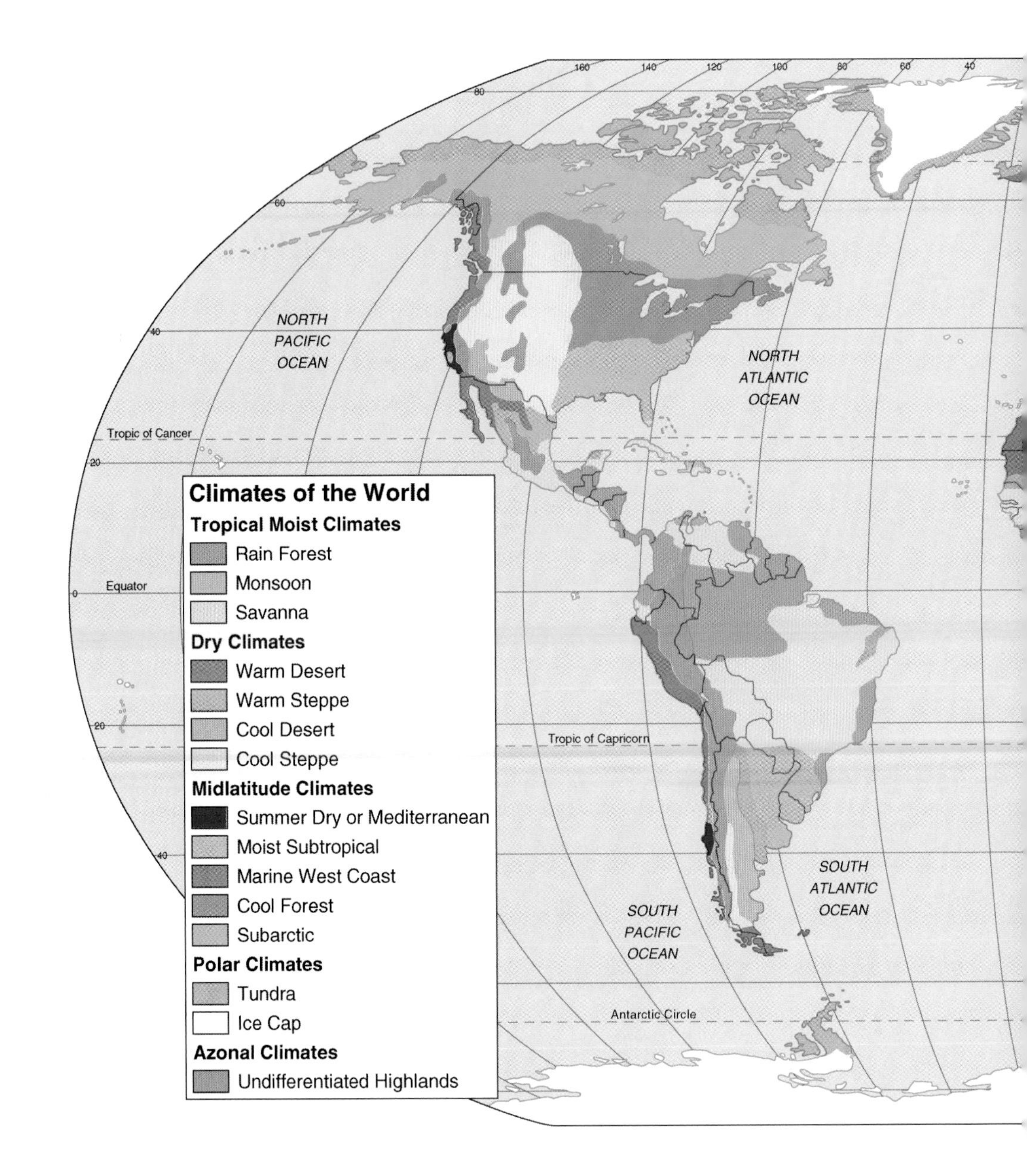

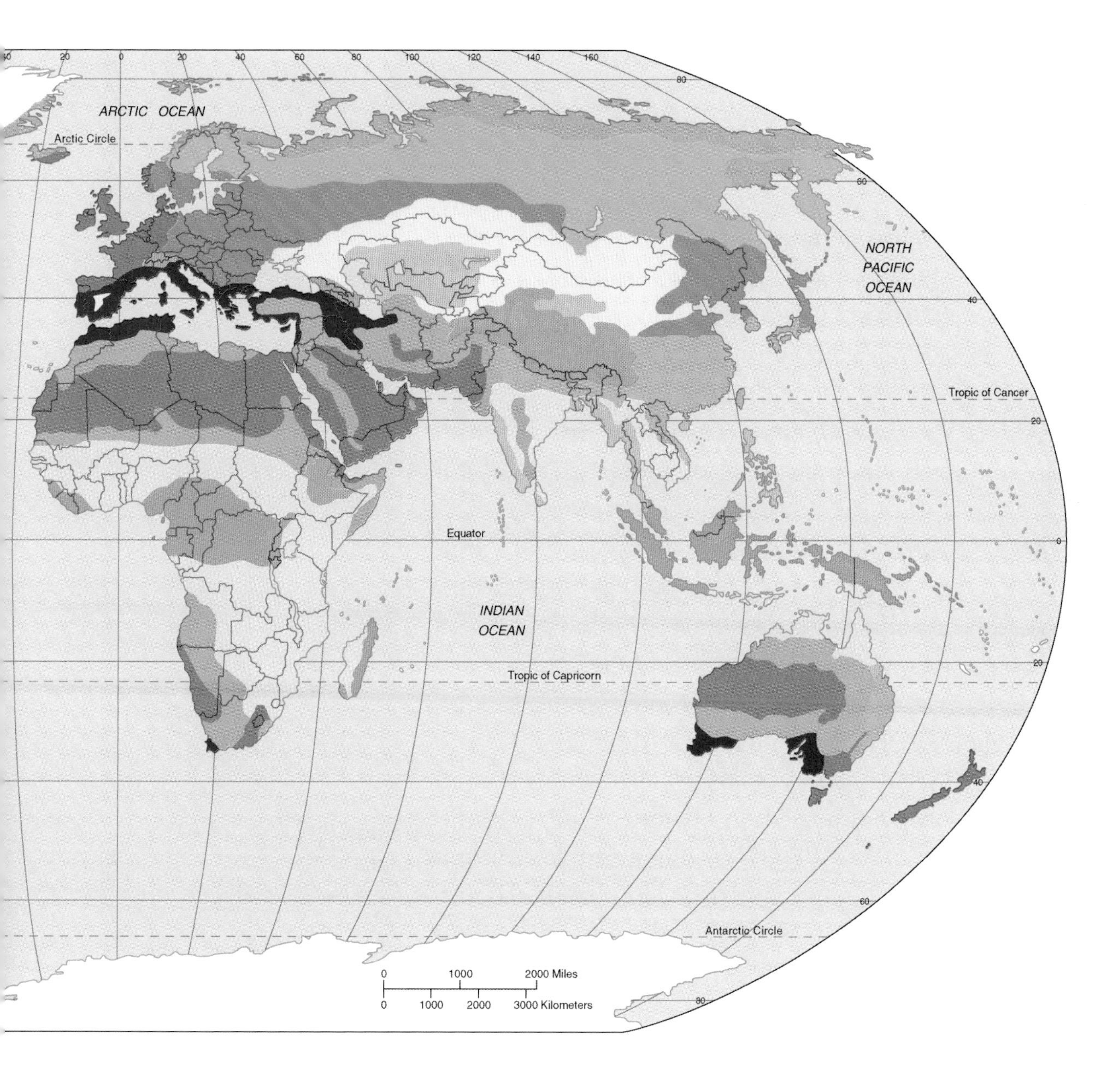

Of the world's many physical geographic features, climate (the long-term average of such weather conditions as temperature and precipitation) is the most important. It is climate that conditions the types of natural vegetation patterns and the types of soil that will exist in an area. It is also climate that determines the availability of our most precious resource: water. From an economic standpoint, the world's most important activity is agriculture; no other element of physical geography is more important for agriculture than climate.

Map 3 World Topography

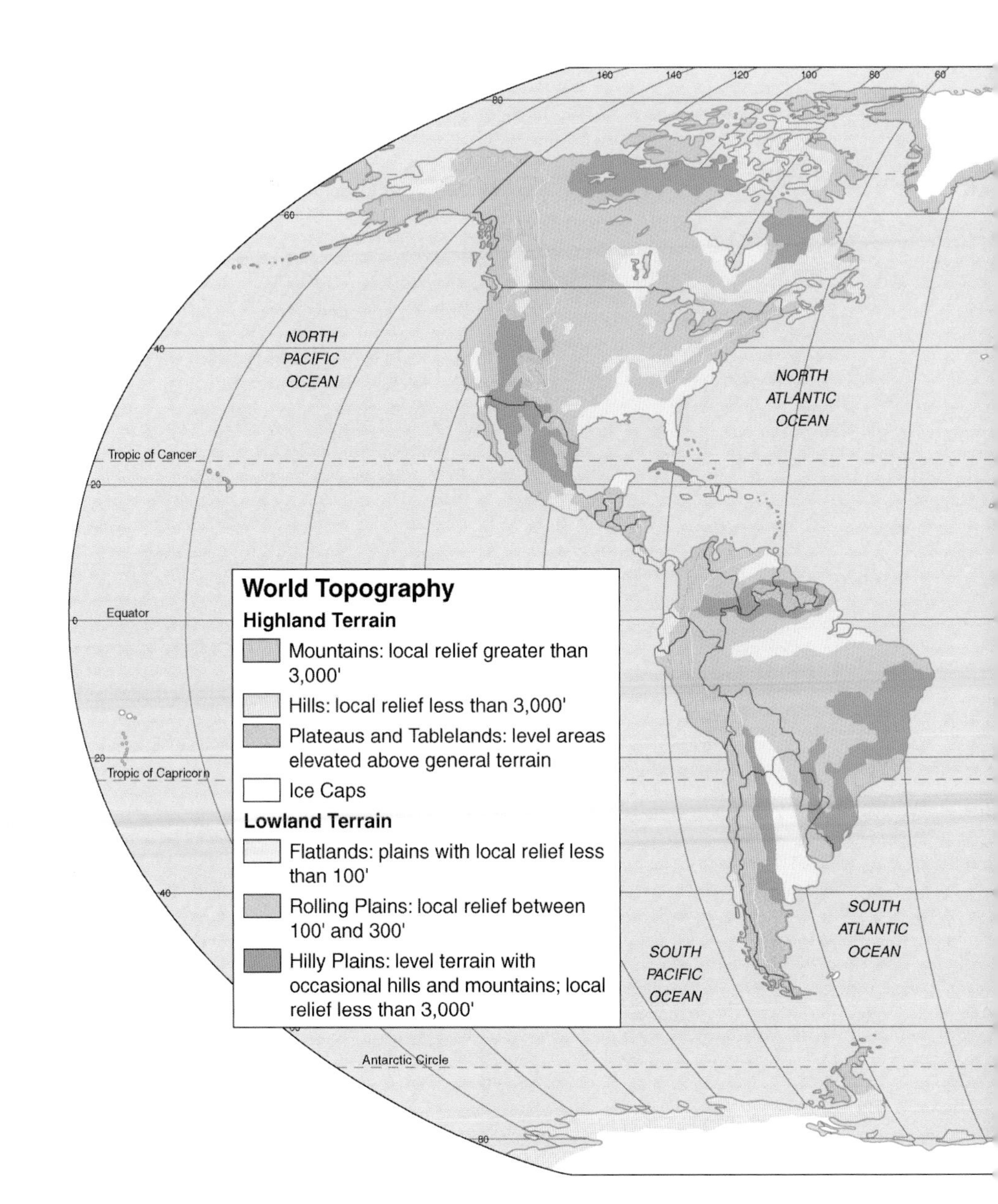

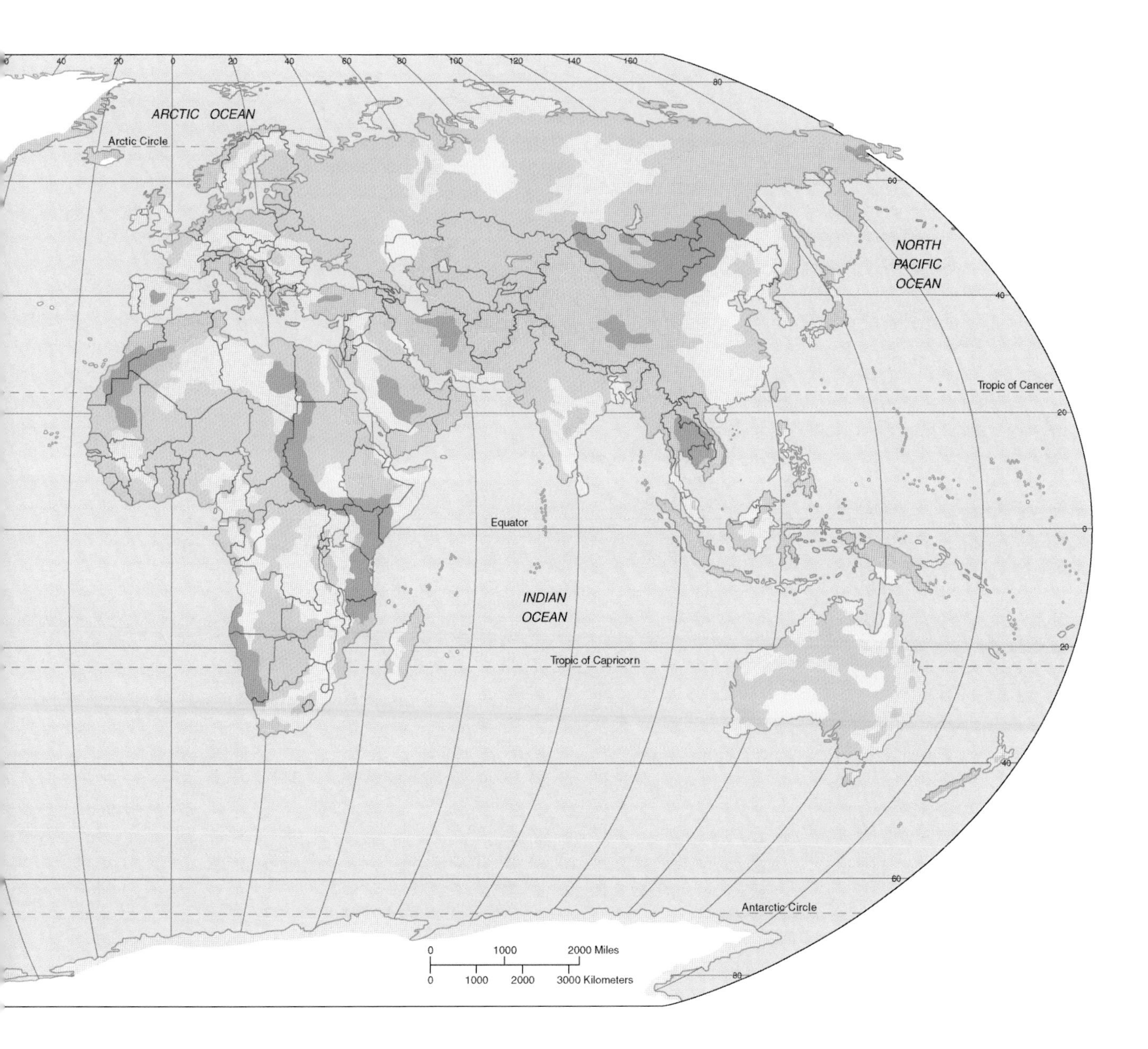

Second only to climate as a conditioner of human activity—particularly in agriculture and in the location of cities and industry—is topography or terrain. It is what we often call *landforms*. A comparison of this map with the map of land use (Map 6) will show that most of the world's productive agricultural zones are located in lowland regions. Where large regions of agricultural productivity are found, we tend to find urban concentrations and, with cities, industry. There is also a good spatial correlation between the map of landforms and the map showing the distribution and density of the human population (Map 7). Normally, the world's landforms shown on this map are the result of extremely gradual primary geologic activity, such as the long-term movement of crustal plates (sometimes called continental drift). This activity occurs over hundreds of millions of years. Also important is the more rapid (but still slow by human standards) geomorphological or erosional activity of water, wind, and glacial ice; and waves, tides, and currents. Some landforms may be produced by abrupt or cataclysmic events, such as a major volcanic eruption or a meteor strike, but these are relatively rare and their effects are usually too minor to show up on a map of this scale.

Map 4 World Ecological Regions

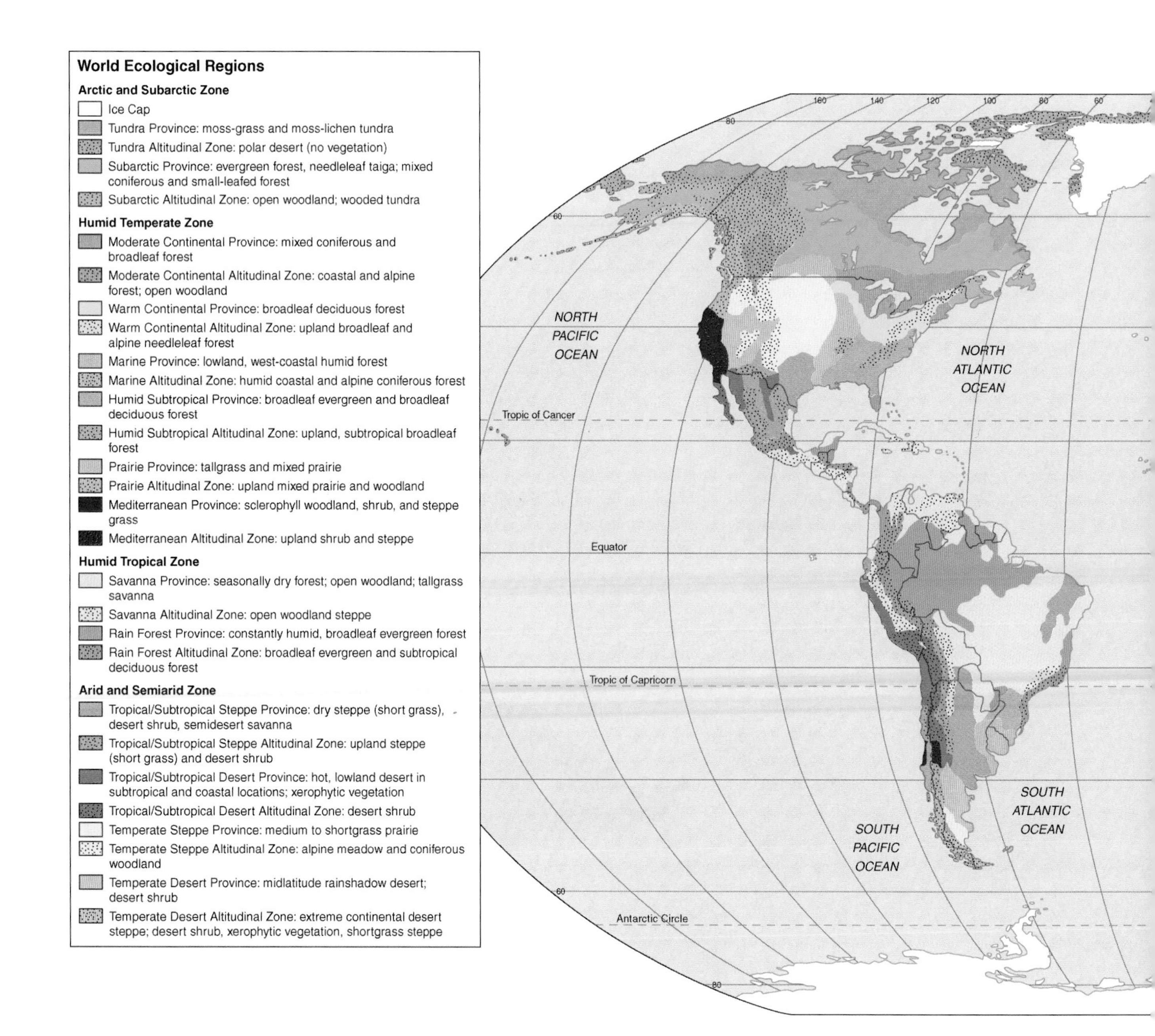

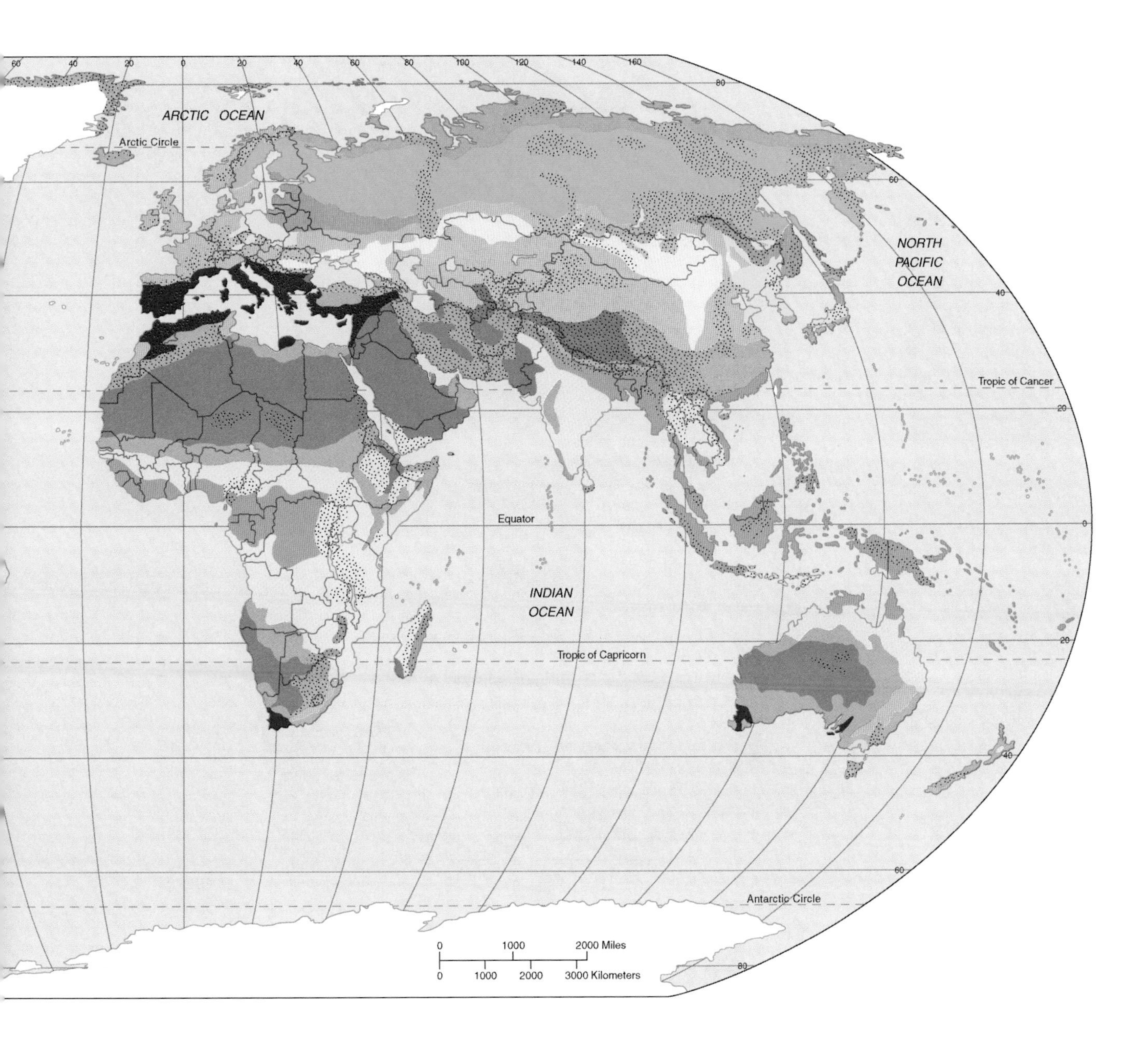

Ecology is the study of the relationships between living organisms and their environmental surroundings. Ecological regions are distinctive areas within which unique sets of organisms and environments are found. Within each ecological region, a particular combination of vegetation, wildlife, soil, water, climate, and terrain defines that region's habitability, or ability to support life, including human life. Like climate and landforms, ecological relationships are crucial to the existence of agriculture, the most basic of our economic activities, and important for many other kinds of economic activity as well.

Map 5 World Natural Hazards

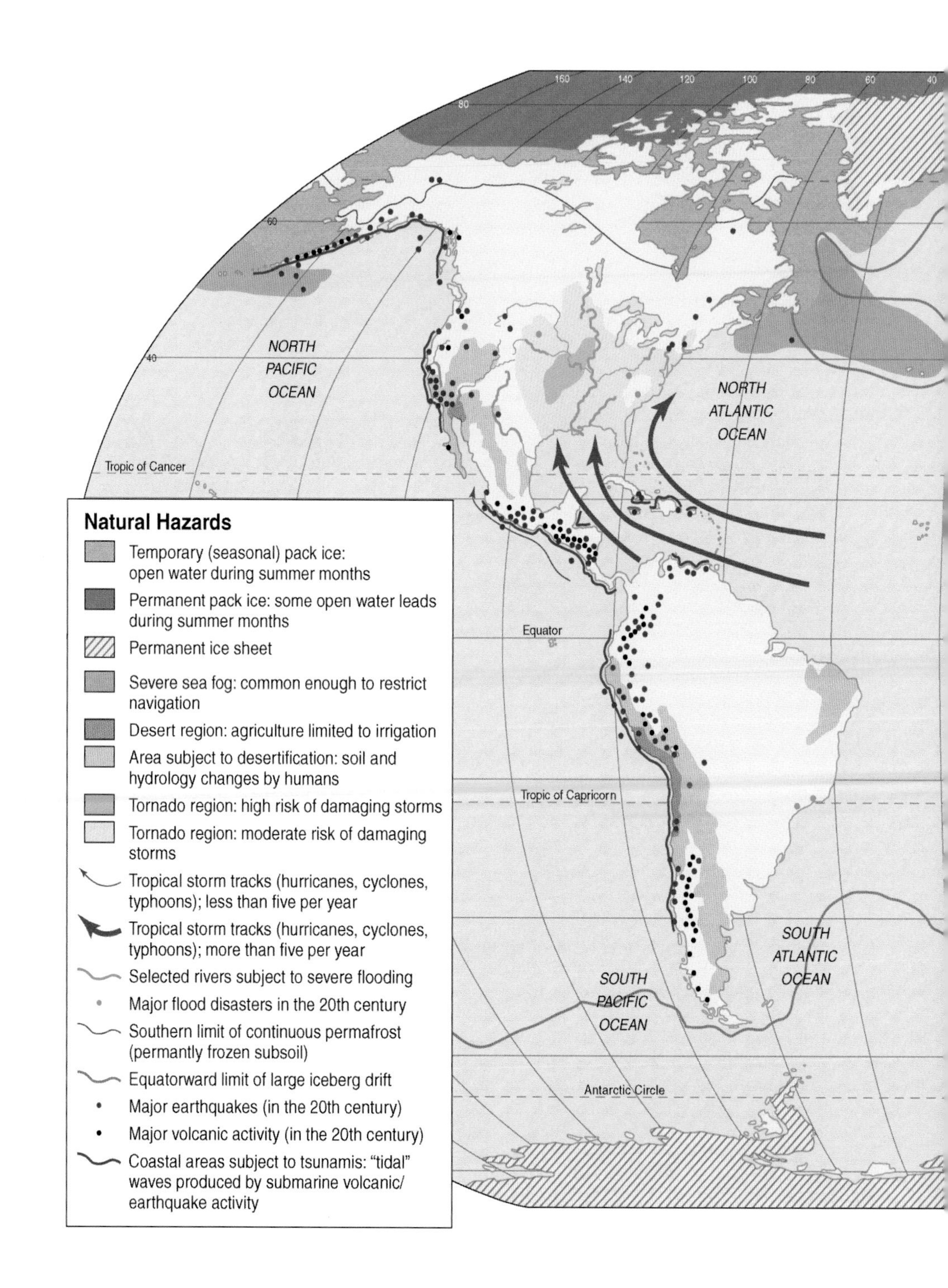

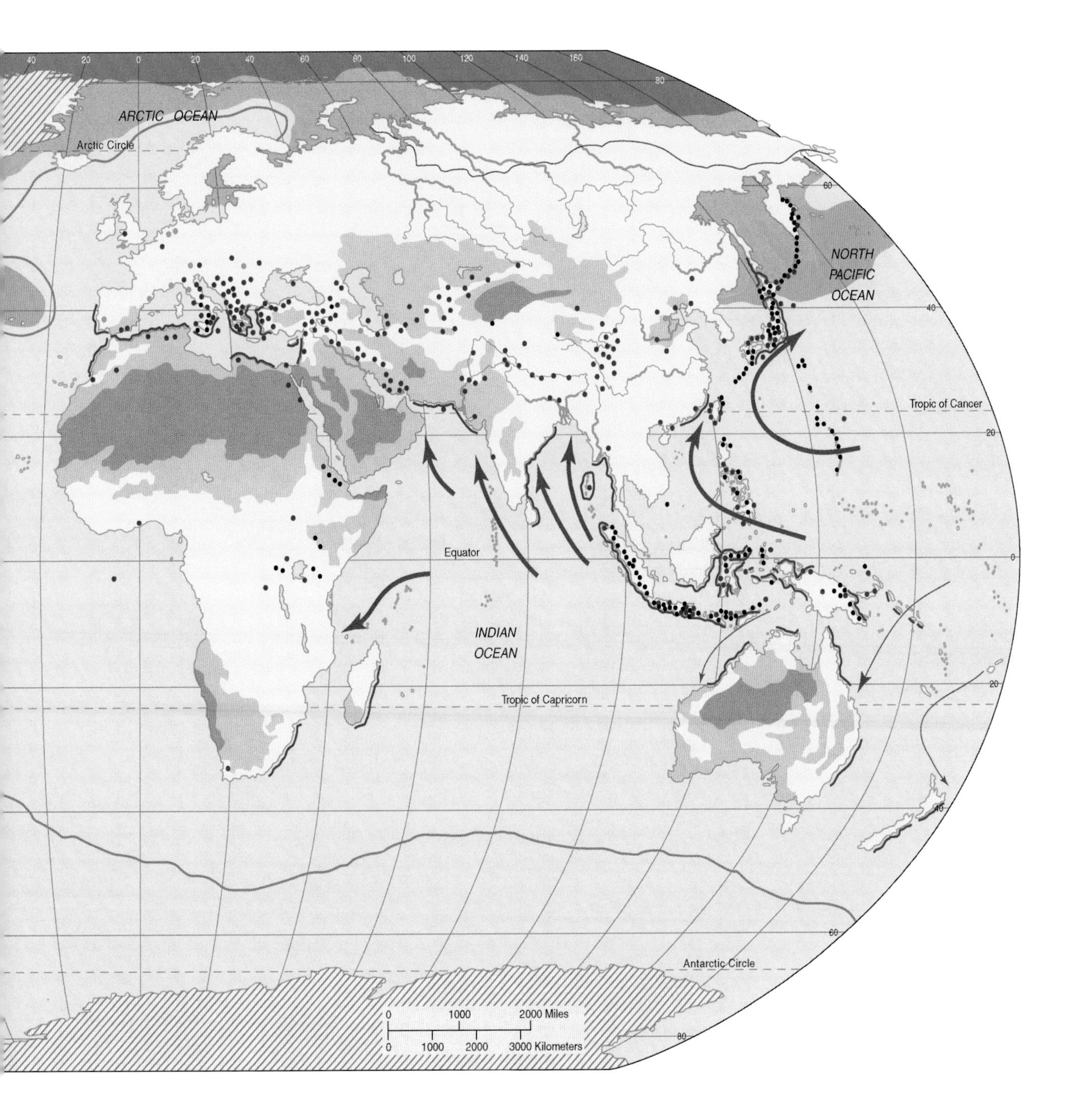

Unlike other elements of physical geography, natural hazards are unpredictable. There are certain regions, however, where the probability of the occurrence of a particular natural hazard is high. This map shows regions affected by major natural hazards at rates that are higher than the global norm. Persistent natural hazards may undermine the utility of an area for economic purposes. Some scholars suggest that regions of environmental instability may be regions of political instability as well.

Map 6 Land Use Patterns of the World

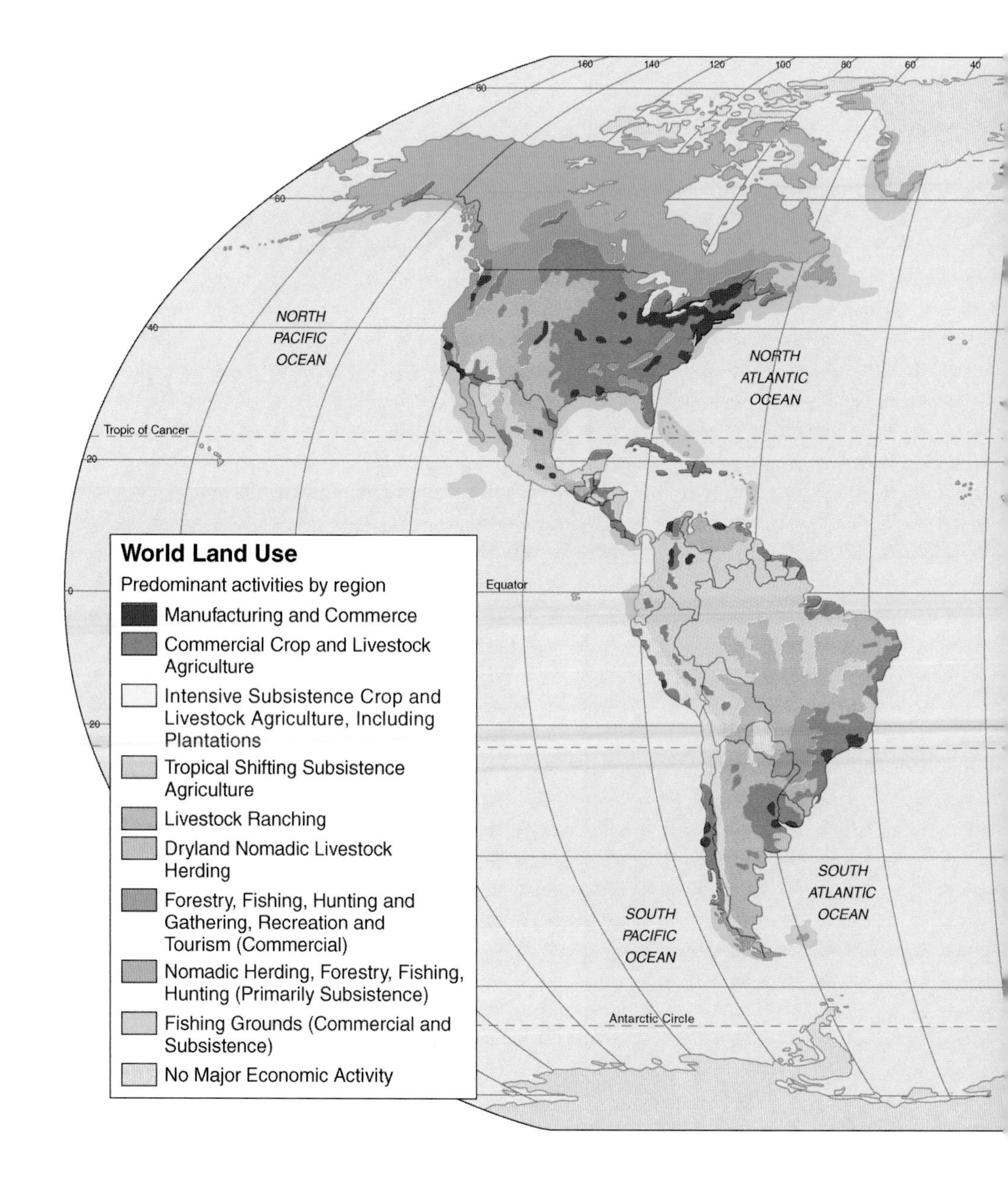

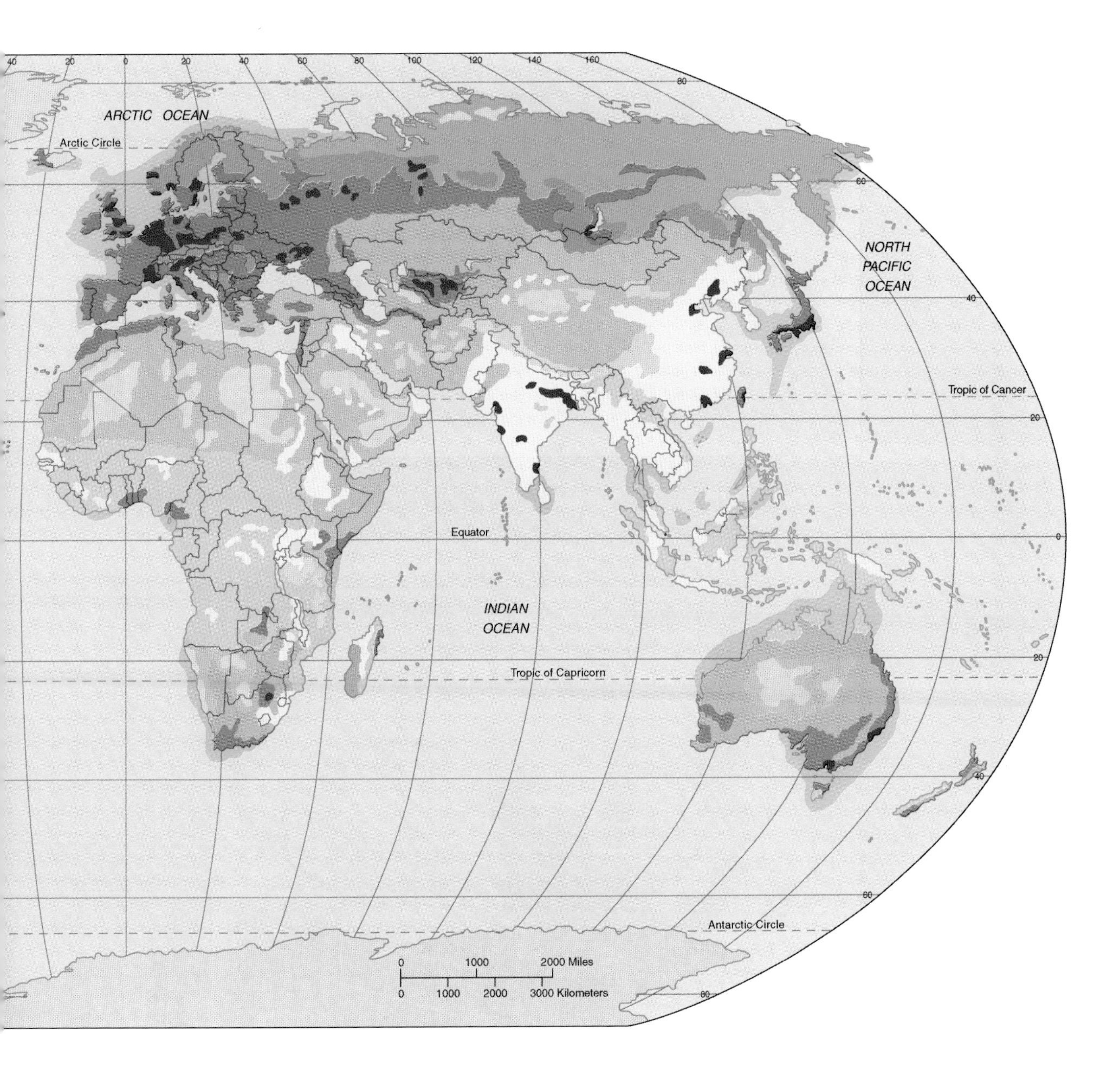

Many of the major land use patterns of the world (such as urbanization, industry, and transportation) are relatively small in area and are not easily seen on maps, but the most important uses people make of the earth's surface have more far-reaching effects. This map illustrates, in particular, the variations in primary land uses (such as agriculture) for the entire world. Note the differences between land use patterns in the more developed countries of the middle latitude zones and the less developed countries of the tropics.

Map 7 World Population Density

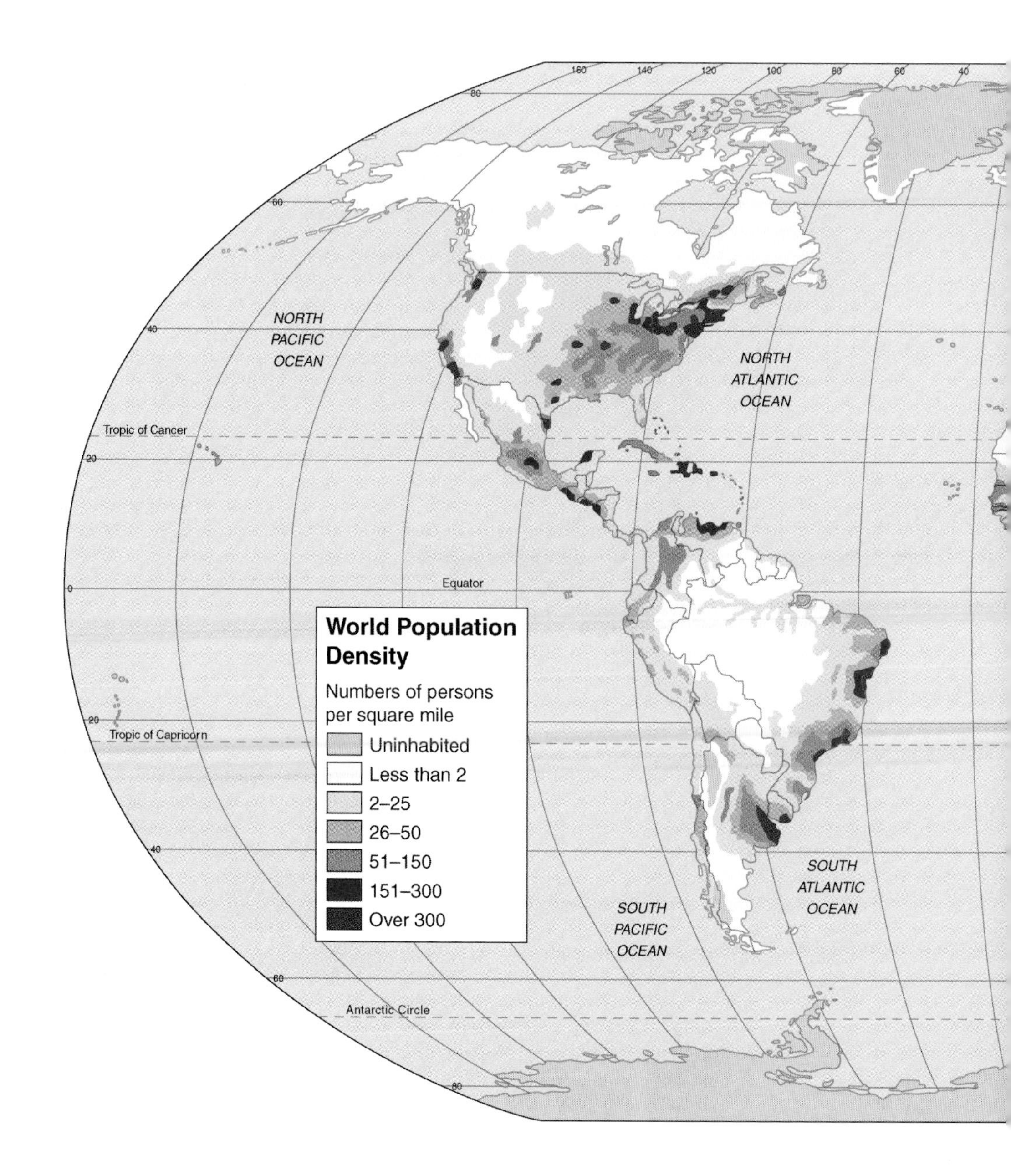

No feature of human activity is more reflective of environmental conditions than where people live. In the areas of densest populations, a mixture of natural and human factors has combined to allow maximum food production, maximum urbanization, and maximum centralization of economic activities. Three great concentrations of human population appear on the map—East Asia, South Asia, and Europe—with a fourth, lesser concentration in eastern North America (the "Megalopolis" region of the United States and Canada). One of these great population clusters—South Asia—is still growing rapidly and is expected to become even more densely populated during the twenty-first century. The

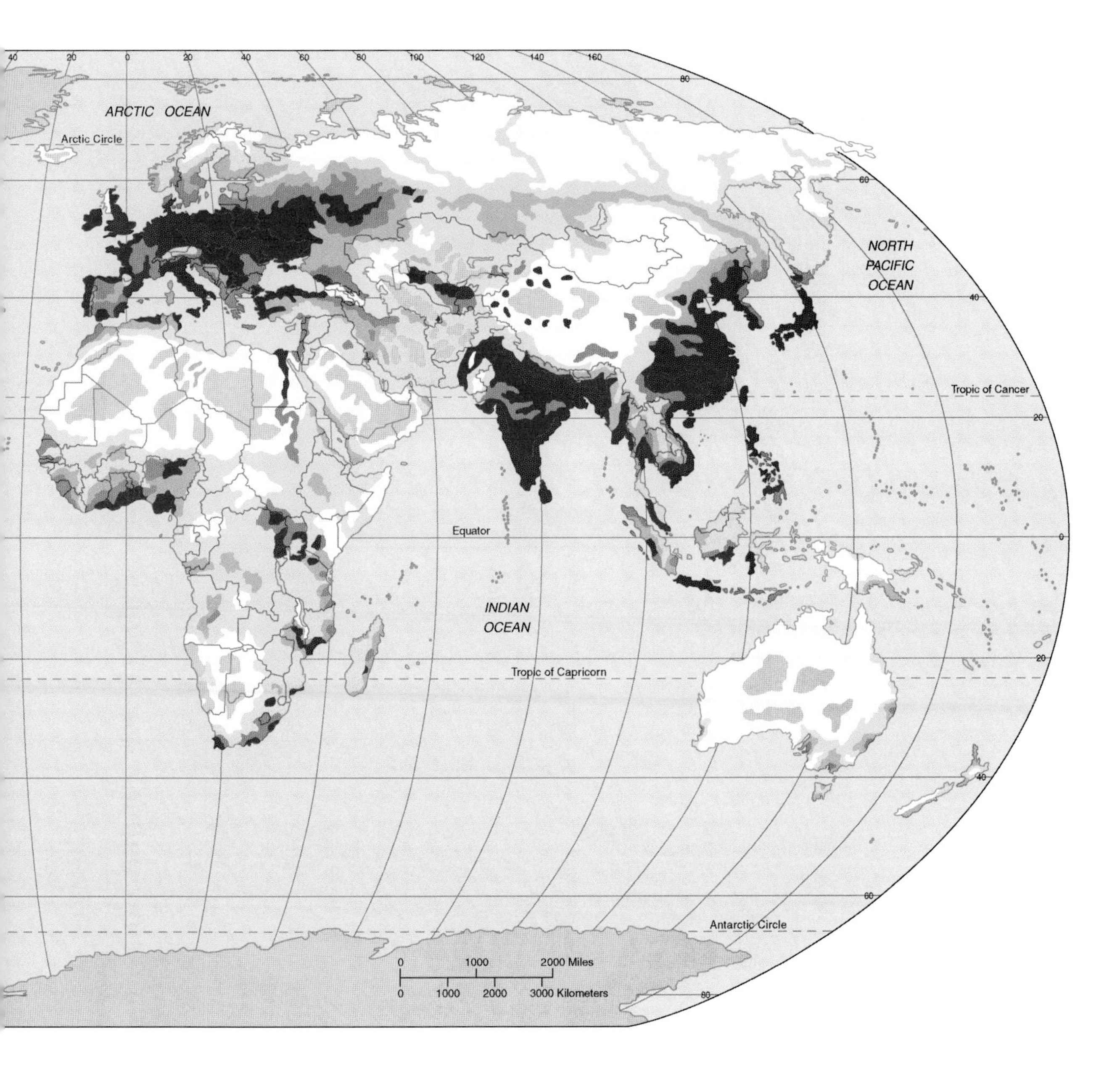

other concentrations are likely to remain about as they now appear. In Europe and North America, this is the result of economic development that has caused population growth to level off during the last century. In East Asia, population has also begun to grow more slowly. In the case of Japan and the Koreas, this is the consequence of economic development; in the case of China, it is the consequence of government intervention in the form of strict family planning. The areas of future high density (in addition to those already existing) are likely to be in Middle and South America and Africa, where population growth rates are well above the world average.

Map 8 World Religions

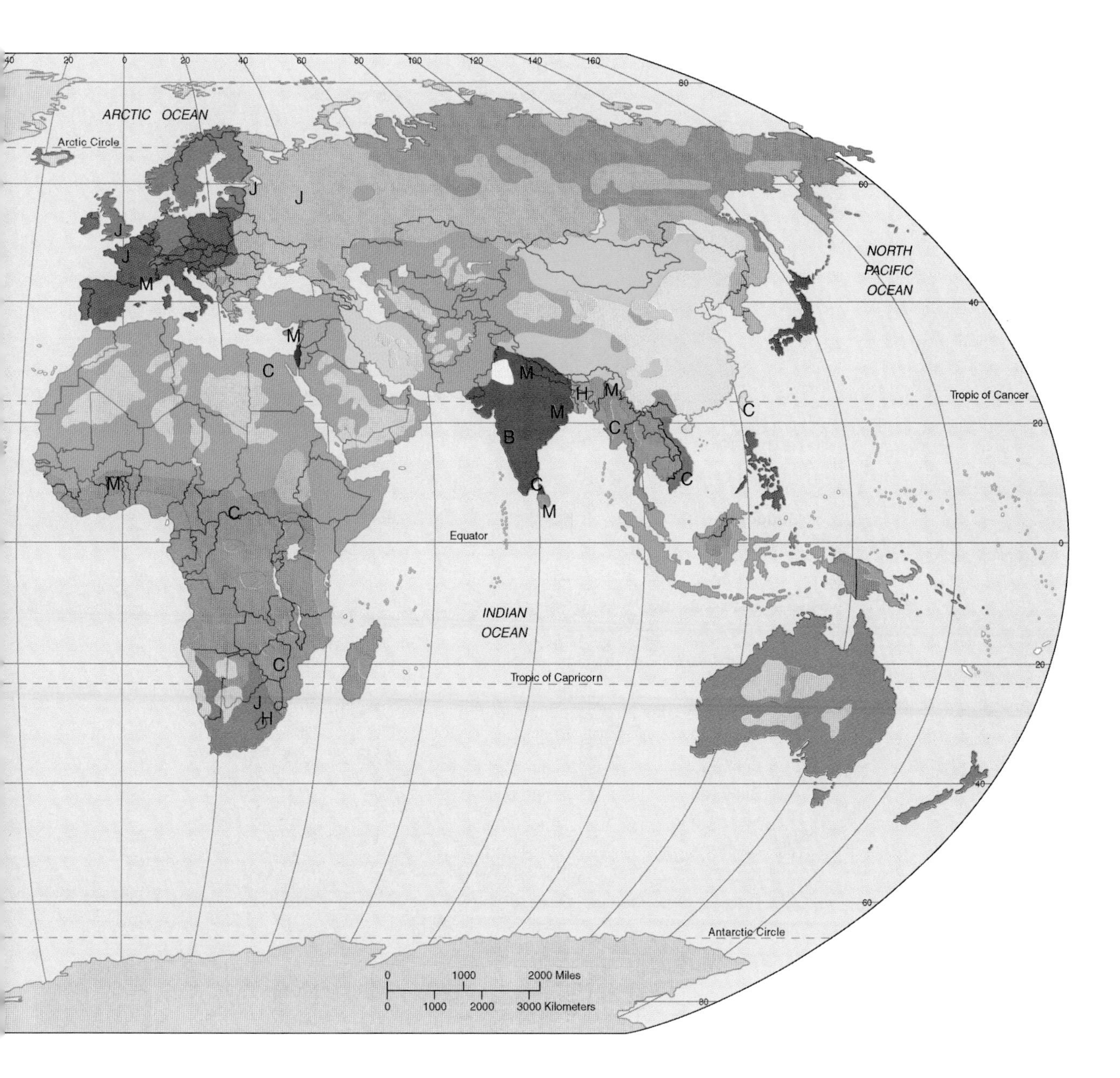

Religious adherence is one of the fundamental defining characteristics of culture. A depiction of the spatial distribution of religions is, therefore, as close as we can come to a map of cultural patterns. More than just a set of behavioral patterns having to do with worship and ceremony, religion is an important conditioner of how people treat one another and the environments that they occupy. In many areas of the world, the ways in which people make a living, the patterns of occupation that they create on the land, and the impacts that they make on ecosystems are the direct consequence of their adherence to a religious faith.

Map 9 World Languages

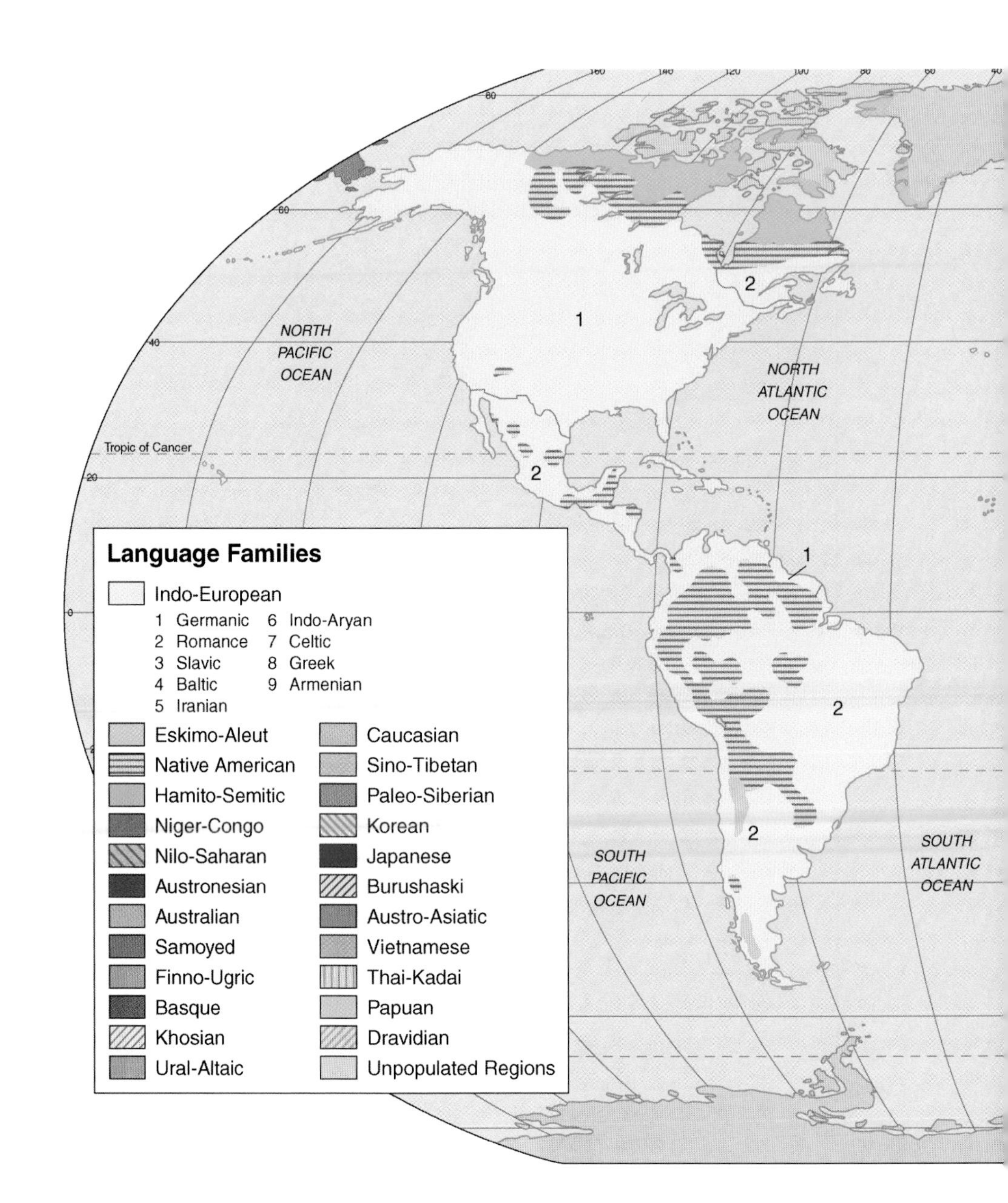

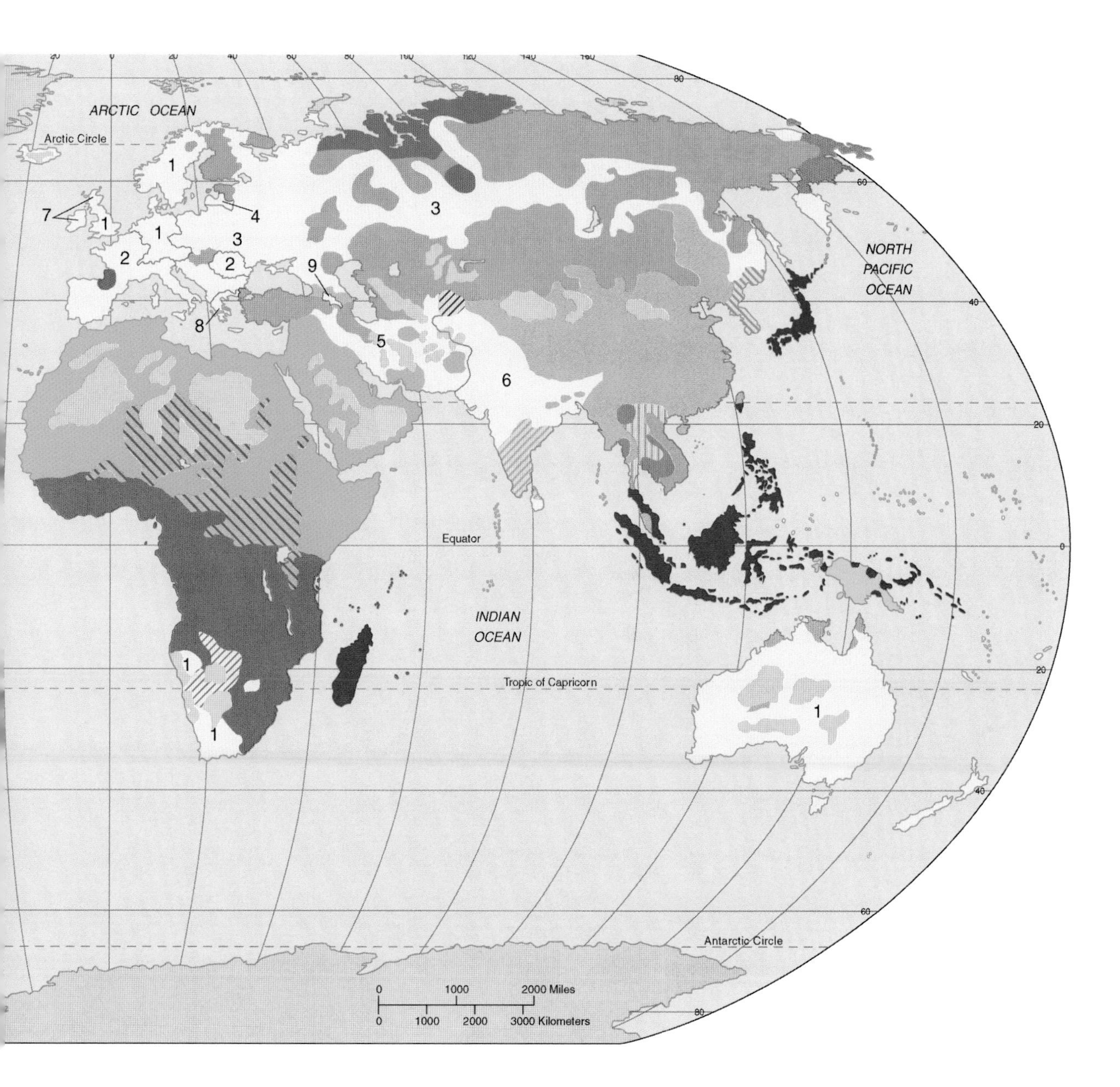

Like religion, language is an important defining characteristic of culture. It is perhaps the most durable of all cultural traits. Even after centuries of exposure to other languages or of conquest by speakers of other languages, the speakers of a specific tongue will often retain their own linguistic identity. As a geographic element, language helps us to locate areas of potential conflict, particularly in regions where two or more languages overlap. Many, if not most, of the world's conflict zones are areas of linguistic diversity. Language also provides clues that enable us to chart the course of human migrations, as shown in the distribution of Indo-European languages. And it helps us to understand some of the reasons behind important historical events; linguistic identity differences played an important part in the disintegration of the Soviet Union.

Map 10 World External Migrations in Modern Times

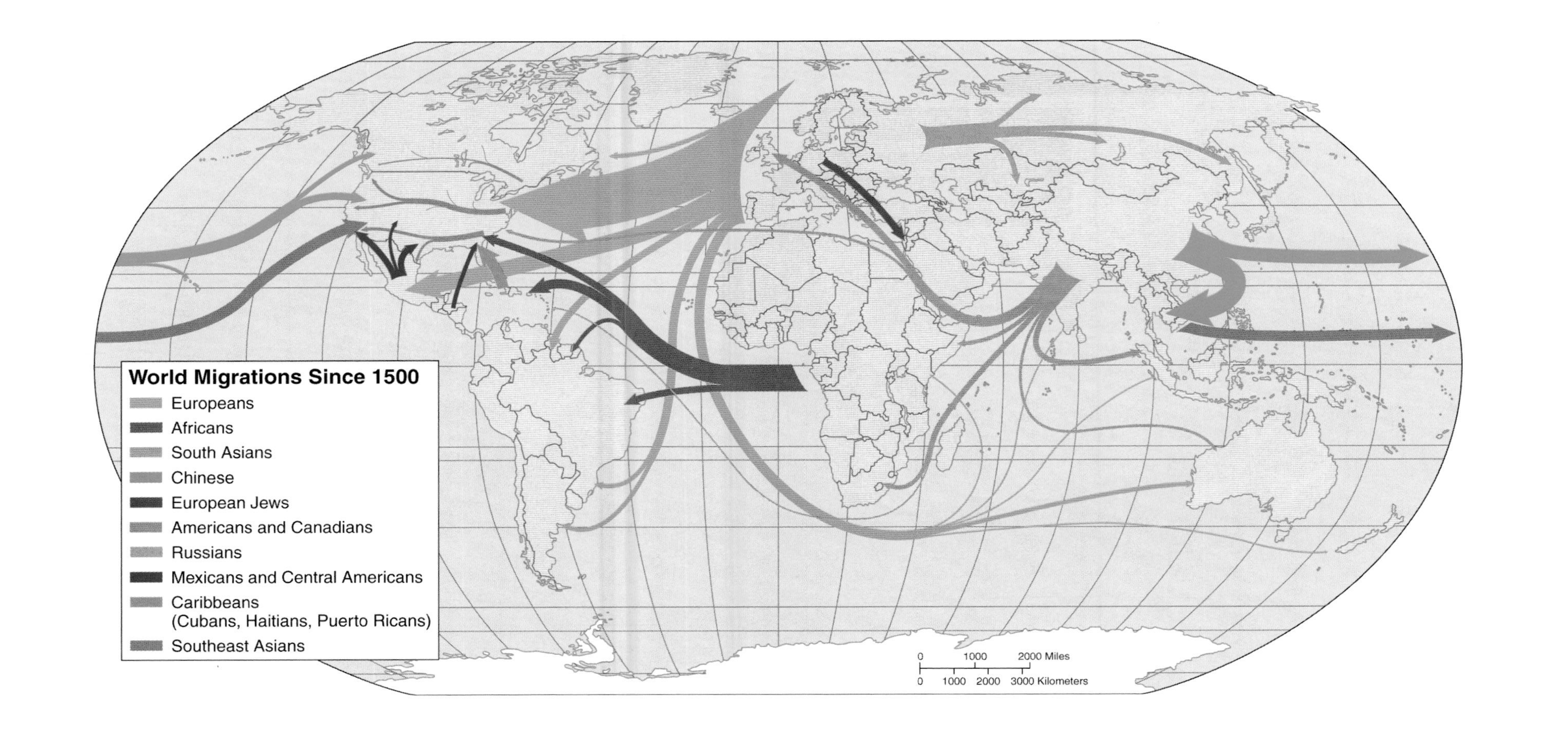

Migration has had a significant effect on world geography, contributing to cultural change and development, to the diffusion of ideas and innovations, and to the complex mixture of people and cultures found in the world today. Internal migration occurs within the boundaries of a country; external migration is movement from one country or region to another. Over the last 50 years, the most important migrations in the world have been internal, largely the rural-to-urban migration that has been responsible for the recent rise of global urbanization. Prior to the mid-twentieth century, three types of external migrations were most important: voluntary, most often in search of better economic conditions and opportunities; involuntary or forced, involving people who have been driven from their homelands by war, political unrest, or environmental disasters, or who have been transported as slaves or prisoners; and imposed, not entirely forced but which conditions make highly advisable. Human migrations in recorded history have been responsible for major changes in the patterns of languages, religions, ethnic composition, and economies. Particularly during the last 500 years, migrations of both the voluntary and involuntary or forced type have literally reshaped the human face of the earth.

Part II

States: The Geography of Politics and Political Systems

Map 11 Political Boundary Types

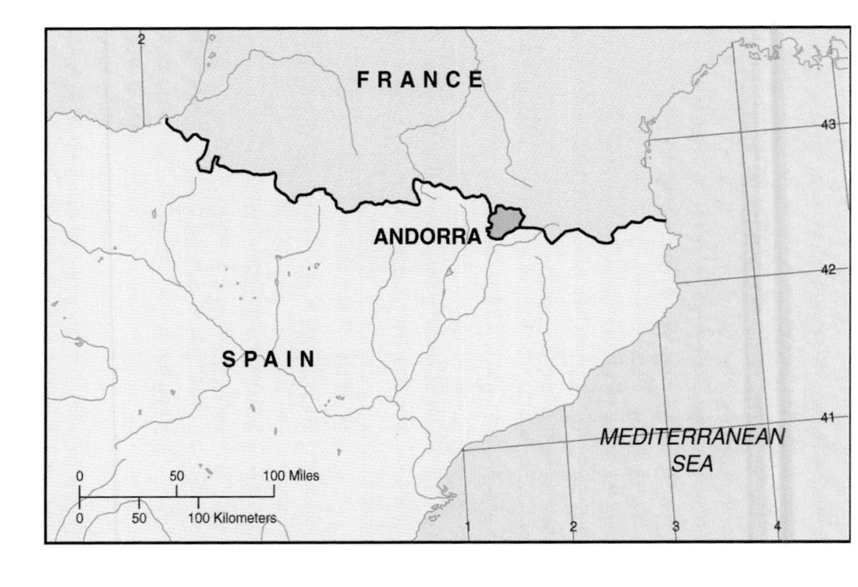

Antecedent: Antecedent boundaries are those that existed as part of the cultural landscape before the establishment of political territories. The boundary between Spain and France is the crest of the Pyrenees Mountains, long a cultural and linguistic barrier and a region of sparse population that is reflected on population density maps even at the world scale.

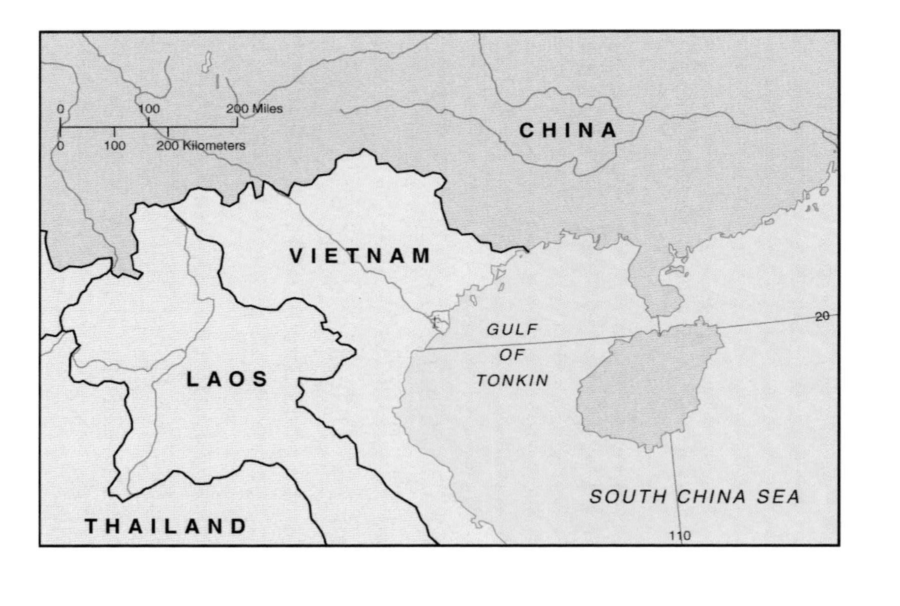

Subsequent: Subsequent boundaries are those that develop along with the cultural landscape of a region, part of a continuing evolution of political territory to match cultural region. The border region between Vietnam and China has developed over thousands of years of adjustment of territory between the two different cultural realms. Following the end of the Vietnam War, a lengthy border conflict between Vietnam and China suggests that the process is not yet completed.

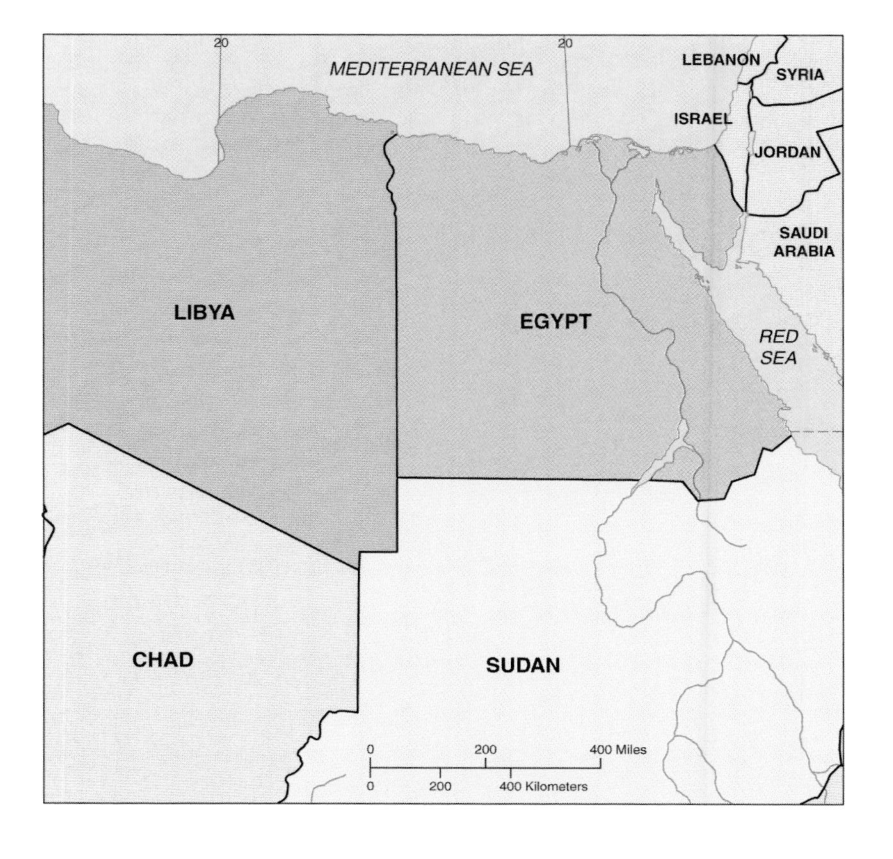

Superimposed: Superimposed boundaries are drawn arbitrarily across a uniform or homogenous cultural landscape. These boundaries often result from the occupation of territory by an expansive settlement process (see, for example, many of the boundaries of the western states in the United States) or from the process whereby colonial powers divided territory to suit their own needs rather than those of the indigenous population. The borders of Egypt, Libya, and Sudan meet in the center of a uniform cultural and physical region, artificially dividing what from a natural and human perspective is unified.

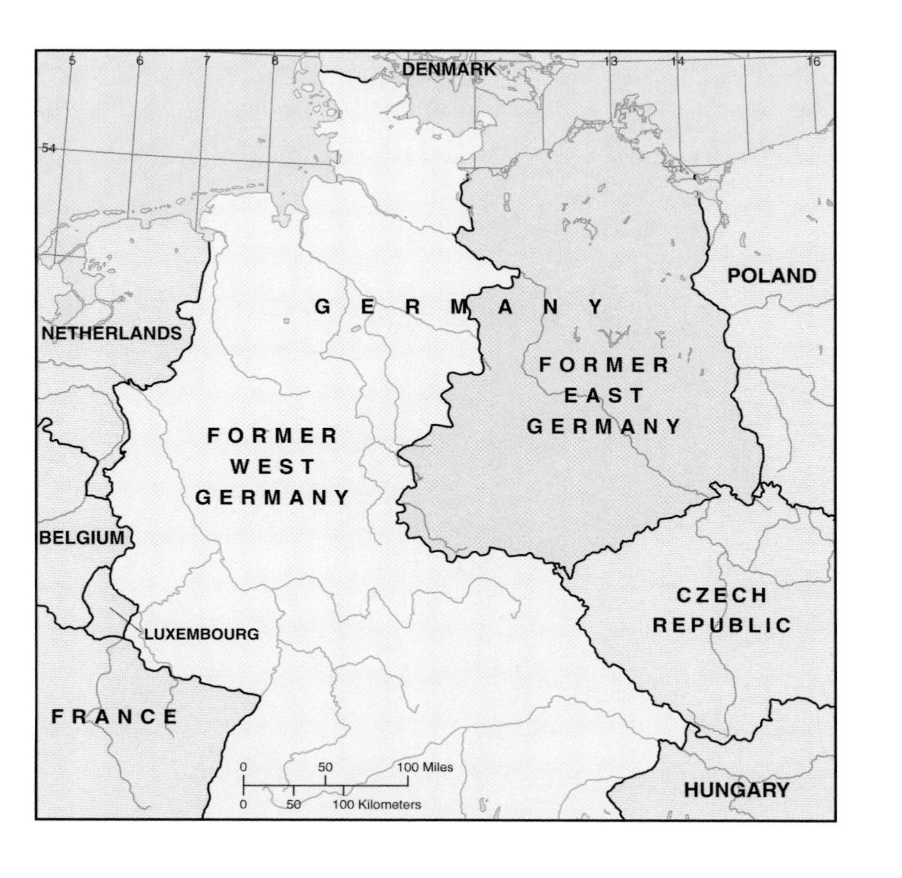

Relict: A relict boundary is like a relict landscape. The boundary between the former North and South Vietnam, along the Ben Hai River, is an example of a relict boundary. So too is the dividing line between the former Federal Republic of Germany (West Germany) and the German Democratic Republic (East Germany). Germany has been unified since 1990, with reintegration of the former Communist East into the West German economy happening progressively and more rapidly than expected. Nevertheless, there are still significant and visible differences between the urban German west and the rural east, between a progressive and modern economic landscape and a deteriorating one.

Map 12 Political Systems

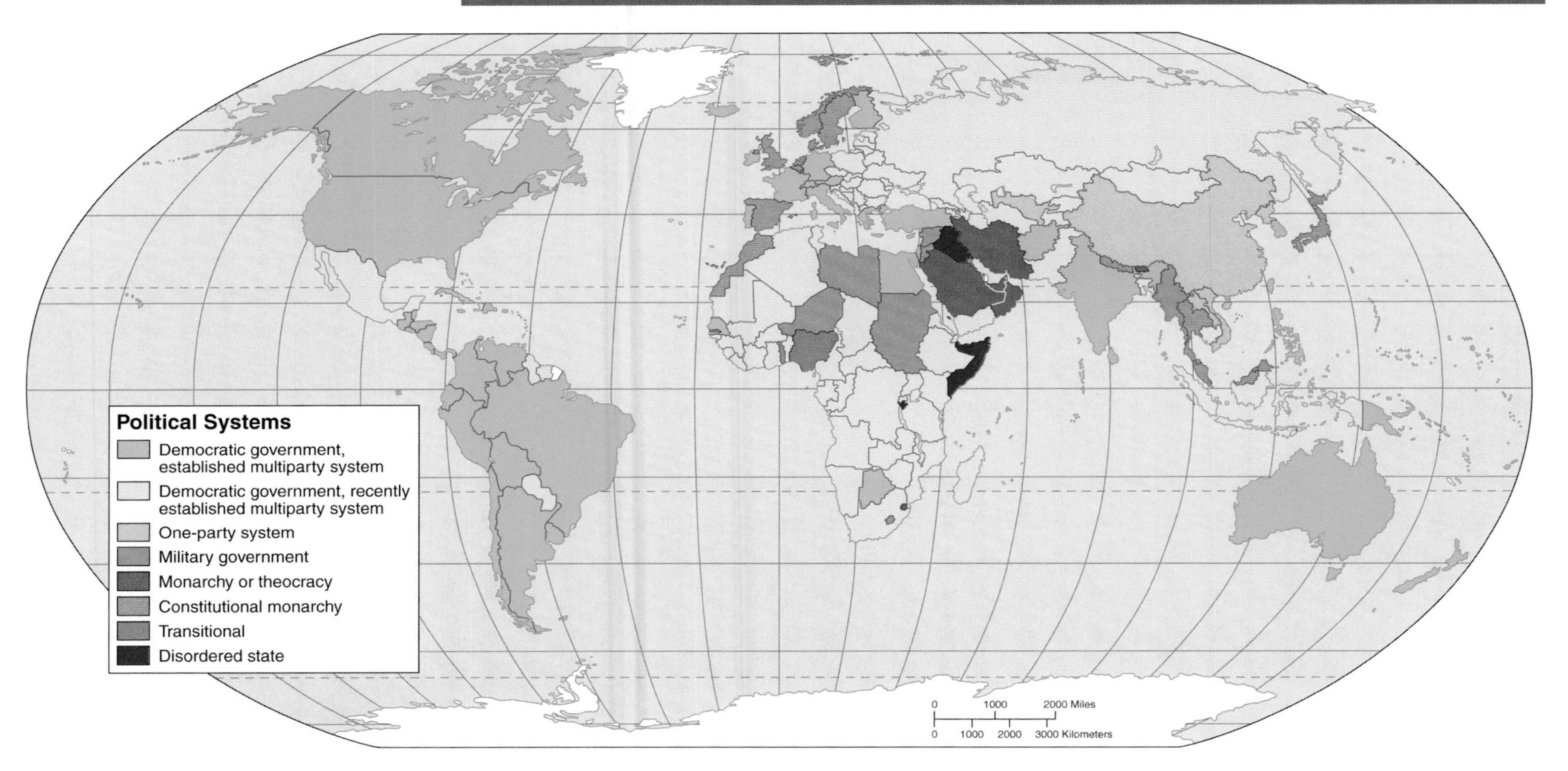

World political systems have changed dramatically during the last decade and may change even more in the future. The categories of political systems shown on the map are subject to some interpretation: established multiparty democracies are those in which elections by secret ballot with adult suffrage are and have been long-term features of the political landscape; recently established multiparty democracies are those in which the characteristic features of multiparty democracies have only recently emerged. The former Soviet satellites of eastern Europe and the republics that formerly constituted the USSR are in this category; so are states in emerging regions that are beginning to throw off the single-party rule that often followed the violent upheavals of the immediate postcolonial governmental transitions. The other categories are more or less obvious. One-party systems are states where single-party rule is constitutionally guaranteed or where a one-party regime is a fact of political life. Monarchies are countries with heads of state who are members of a royal family. In a constitutional monarchy, such as the U.K. and the Netherlands, the monarchs are titular heads of state only. Theocracies are countries in which rule is within the hands of a priestly or clerical class; today, this means primarily fundamentalist Islamic countries such as Iran. Military governments are frequently organized around a junta that has seized control of the government from civil authority, usually with the promise to return the reins of government to civil authority once order has been restored. Transitional governments are those in which there is a shift of authority taking place from a military ruling junta or from autocratic civil authority to a freely-elected governemnt. Finally, disordered states are countries so beset by civil war, ethnic conflict, or other insurgency that no organized government can be said to exist throughout the country's leagal boundaries.

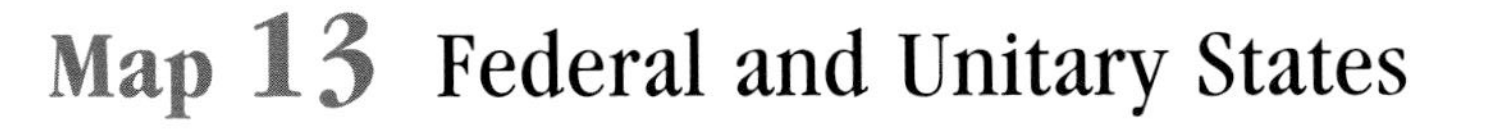

Map 13 Federal and Unitary States

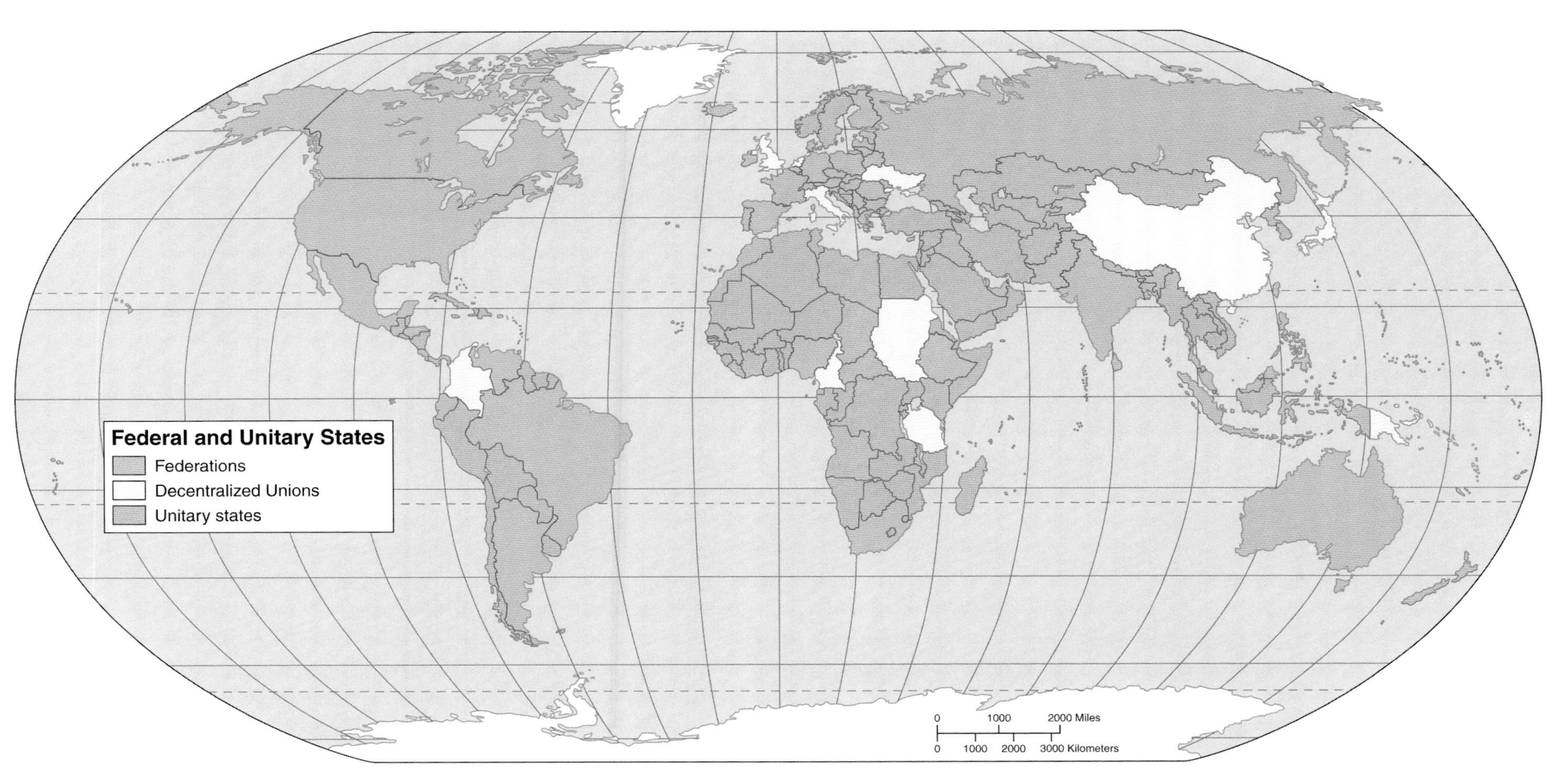

Countries vary greatly in the degree to which the national government makes all decisions or certain matters are left to governments of subdivisions. In federal states, each subdivision (state, province, and so on) has its own capital, and certain aspects of life are left to the subdivisions to manage. For example, in the United States, motor vehicle and marriage/divorce laws are powers for states rather than the federal government. Federal systems are suited to large countries with a diversity of cultures. They easily accommodate new territorial subdivisions, which can take their places among those already a part of the country. In unitary states, major decisions come from the national capital, and subdivisions administer those decisions. Unitary systems are best suited to small states with relatively homogeneous ethnic and cultural makeup. Some countries, the decentralized unions, have a mix of systems—for example, provinces that follow the dictates of the central government but other subdivisions that have more autonomy.

Map 14 Sovereign States: Duration of Independence

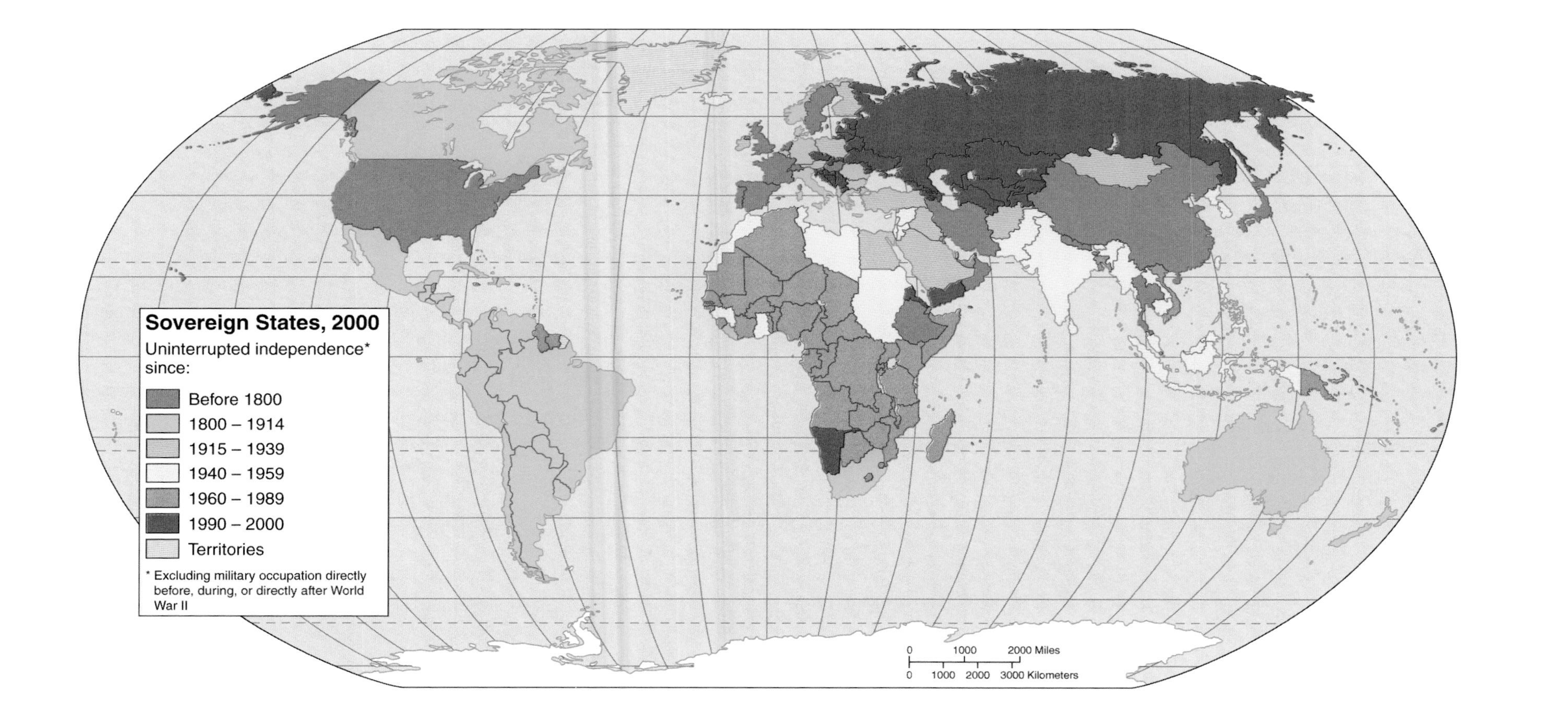

Most countries of the modern world, including such major states as Germany and Italy, became independent after the beginning of the nineteenth century. Of the world's current countries, only 27 were independent in 1800. (Ten of the 27 were in Europe; the others were Afghanistan, China, Colombia, Ethiopia, Haiti, Iran, Japan, Mexico, Nepal, Oman, Paraguay, Russia, Taiwan, Thailand, Turkey, the United States, and Venezuela.) Following 1800, there have been four great periods of national independence. During the first of these (1800–1914), most of the mainland countries of the Americas achieved independence. During the second period (1915–1939), the countries of Eastern Europe emerged as independent entities. The third period (1940–1959) includes World War II and the years that followed, when independence for African and Asian nations that had been under control of colonial powers first began to occur. During the fourth period (1960–1989), independence came to the remainder of the colonial African and Asian nations, as well as to former colonies in the Caribbean and the South Pacific. More than half of the world's countries came into being as independent political entities during this period. Finally, in the last decade (1990–2000), the breakup of the existing states of the Soviet Union, Yugoslavia, and Czechoslovakia created 22 countries where only 3 had existed before.

Map 15 Political Realms: Regional Changes, 1945–2003

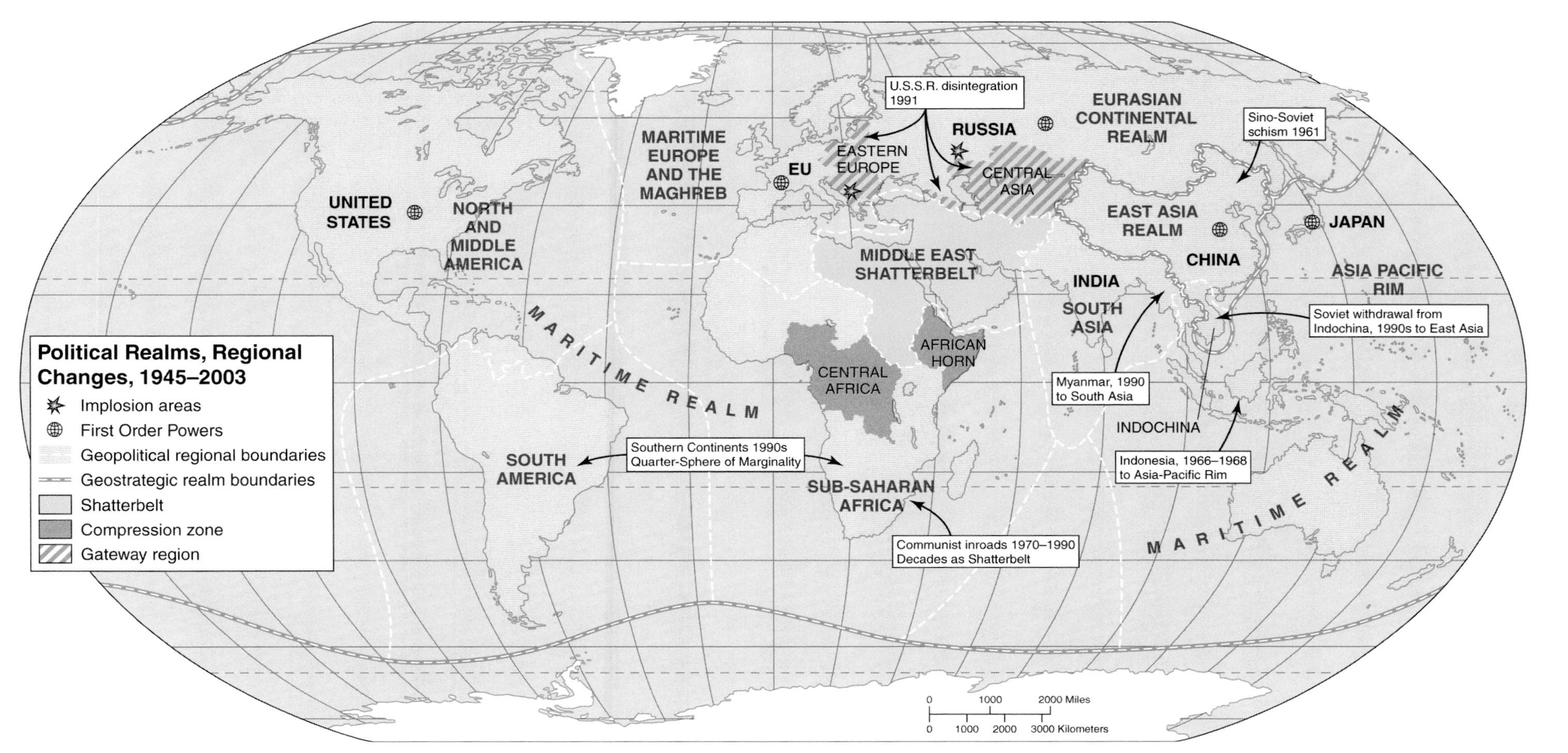

The Cold War following World War II shaped the major outlines of today's geopolitical relations. The Cold War included three phases. In the first, from 1945–1956, the Maritime Realm established a ring around the Continental Eurasian Realm in order to prevent its expansion. This phase included the Korean War (1950–1953), the Berlin Blockade (1948), the Truman doctrine and Marshall Plan (1947), and the founding of NATO (1949) and the Warsaw Pact (1955). Most of the world fell within one of the two realms: the Maritime (dominated by the United States) or the Eurasian Continental Realm (dominated by the Soviet Union). The Soviet Union sought to establish a ring of satellite states to protect it from a repeat of the invasions of World War II. The United States and other Maritime Realm states, in turn, sought to establish a ring of allies around the Continental Realm to prevent its expansion. South Asia was politically independent, but under pressure from both realms. During the second phase (1957–1979), Communist forces from the Continental Eurasian Realm penetrated deeply into the Maritime Realm. The Berlin Wall went up in 1961, Soviet missiles in Cuba ignited a crisis in 1962, and the United States became increasingly involved in the war in Vietnam (late 1960s). The Soviet Union sought increased political and military presence along important waterways including those in the Middle East, Southeast Asia, and the Caribbean. These regions became especially dangerous shatterbelts. The third phase (1980–1989) saw the retreat of Communist power from the Maritime Realm. China, after ten years of radical Communism and chaos of the Cultural Revolution (1966–1976), broke away from the Continental Eurasian Realm to establish a new East Asian realm. Soviet influence declined in the Middle East, Sub-Saharan Africa, and Latin America. In 1989 the Berlin Wall fell, and Eastern Europe began to establish democratic governments. In 1991 the Soviet Union broke apart as its constituent republics became independent states.

Map 16 European Boundaries, 1914–1948

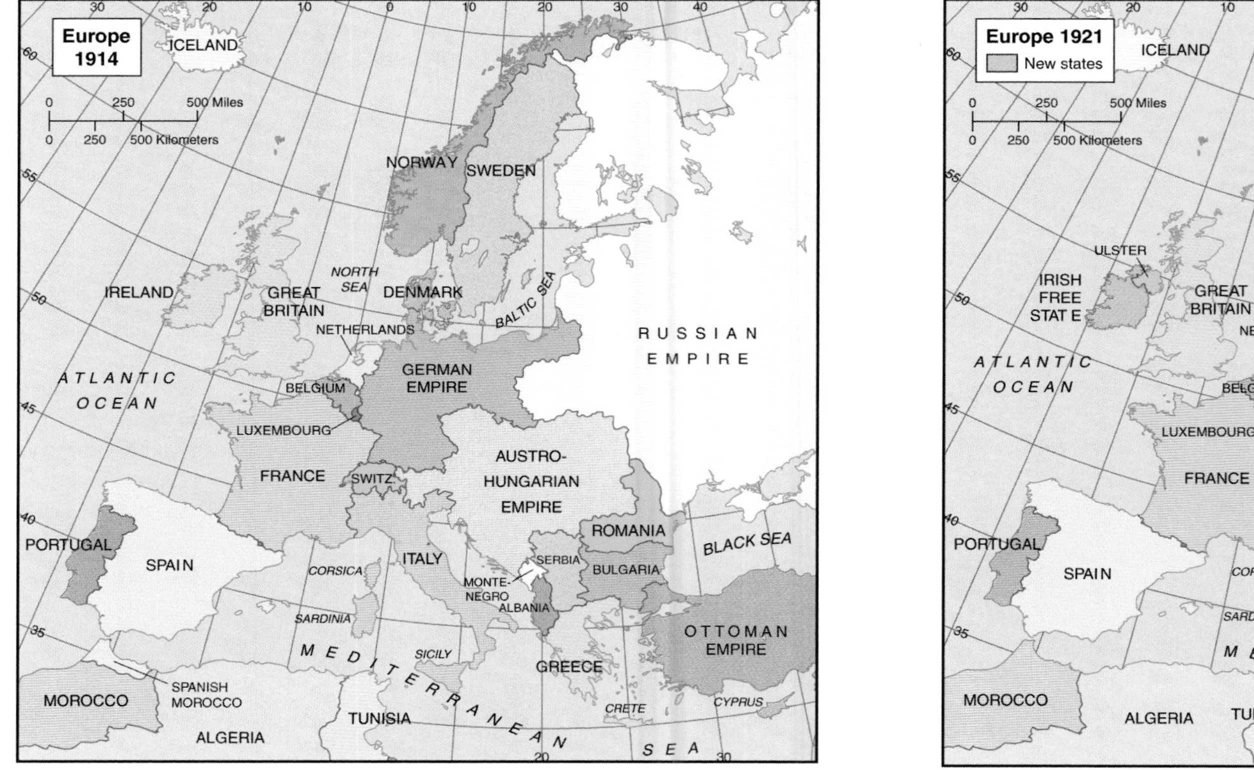

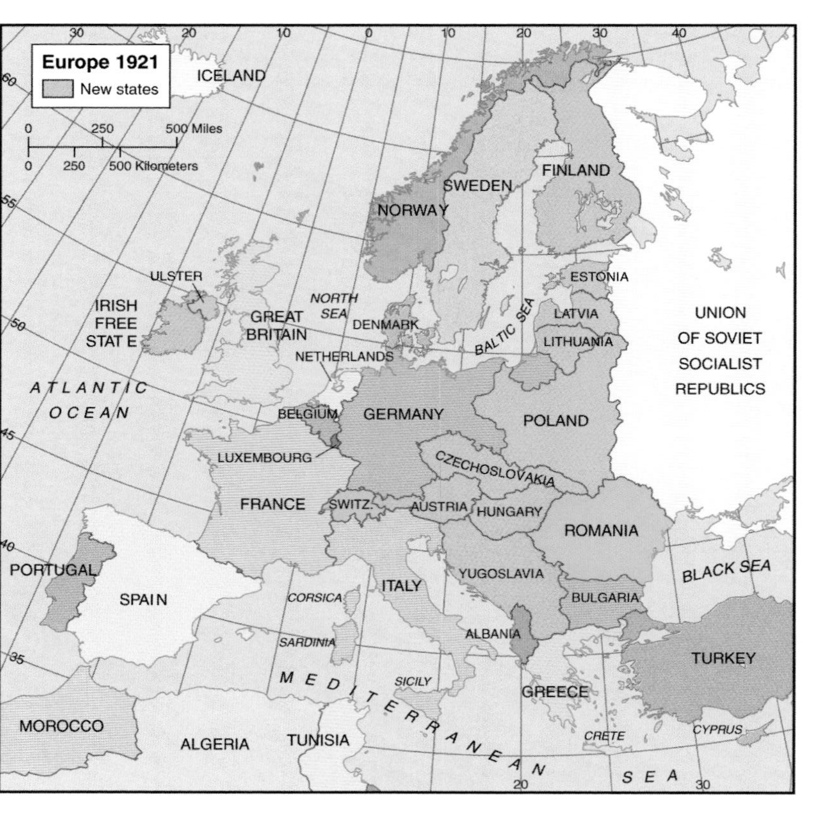

In 1914, on the eve of the First World War, Europe was dominated by the United Kingdom and France in the west, the German Empire and the Austro-Hungarian Empire in central Europe, and the Russian Empire in the east. Battle lines for the conflict that began in 1914 were drawn when the United Kingdom, France, and the Russian Empire joined together as the Triple Entente. In the view of the Germans, this coalition was designed to encircle Germany and its Austrian ally, which, along with Italy, made up the Triple Alliance. The German and Austrian fears were heightened in 1912–14 when a Russian-sponsored "Balkan League" pushed the Ottoman Turkish Empire from Europe, leaving behind the weak and mutually antagonistic Balkan states Serbia and Montenegro. In August 1914, Germany and Austria-Hungary attacked in several directions and World War I began. Four years later, after massive loss of life and destruction, the central European empires were defeated. The victorious French, English, and Americans (who had entered the war in 1917) restructured the map of Europe in 1919, carving nine new states out of the remains of the German and Austro-Hungarian empires and the westernmost portions of the Russian Empire which, by the end of the war, was deep in the Revolution that deposed the czar and brought the Communists to power in a new Union of Soviet Socialist Republics.

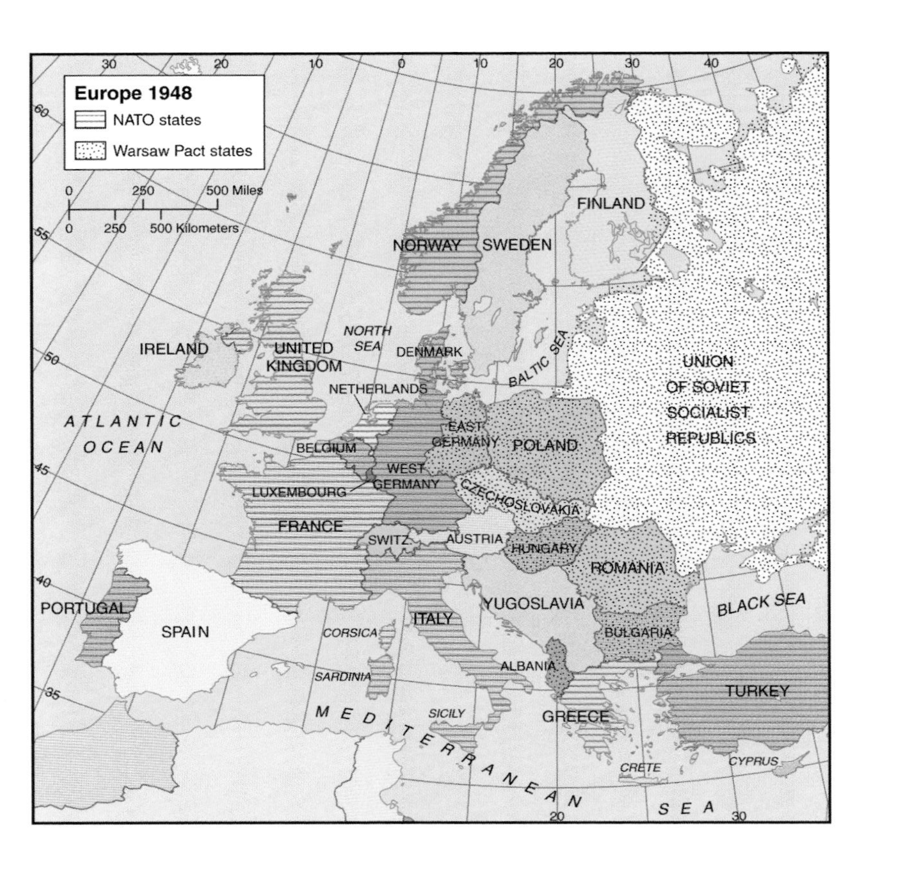

When the victorious Allies redrew the map of central and eastern Europe in 1919, they caused as many problems as they were trying to solve. The interval between the First and Second World Wars was really just a lull in a long war that halted temporarily in 1918 and erupted once again in 1939. Defeated Germany, resentful of the terms of the 1918 armistice and 1919 Treaty of Versailles and beset by massive inflation and unemployment at home, overthrew the Weimar republican government in 1933 and installed the National Socialist (Nazi) party led by Adolf Hitler in Berlin. Hitler quickly began making good on his promises to create a "thousand year realm" of German influence by annexing Austria and the Czech region of Czechoslovakia and allying Germany with a fellow fascist state in Mussolini's Italy. In September 1939 Germany launched the lightning-quick combined infantry, artillery, and armor attack known as *der Blitzkrieg* and took Poland to the east and, in quick succession, the Netherlands, Belgium, and France to the west. By 1943 the greater German Reich extended from the Russian Plain to the Atlantic and from the Black Sea to the Baltic. But the Axis powers of Germany and Italy could not withstand the greater resources and manpower of the combined United Kingdom-United States-USSR–led Allies and, in 1945, Allied armies occupied Germany. Once again, the lines of the central and eastern European map were redrawn. This time, a strengthened Soviet Union took back most of the territory the Russian Empire had lost at the end of the First World War. Germany was partitioned into four occupied sectors (English, French, American, and Russian) and later into two independent countries, the Federal Republic of Germany (West Germany) and the German Democratic Republic (East Germany). Although the Soviet Union's territory stopped at the Polish, Hungarian, Czechoslovakian, and Romanian borders, the eastern European countries (Poland, East Germany, Czechoslovakia, Hungary, Romania, Yugoslavia, Albania, and Bulgaria) became Communist between 1945 and 1948 and were separated from the West by the Iron Curtain.

Map 17 An Age of Bipolarity: The Cold War ca. 1970

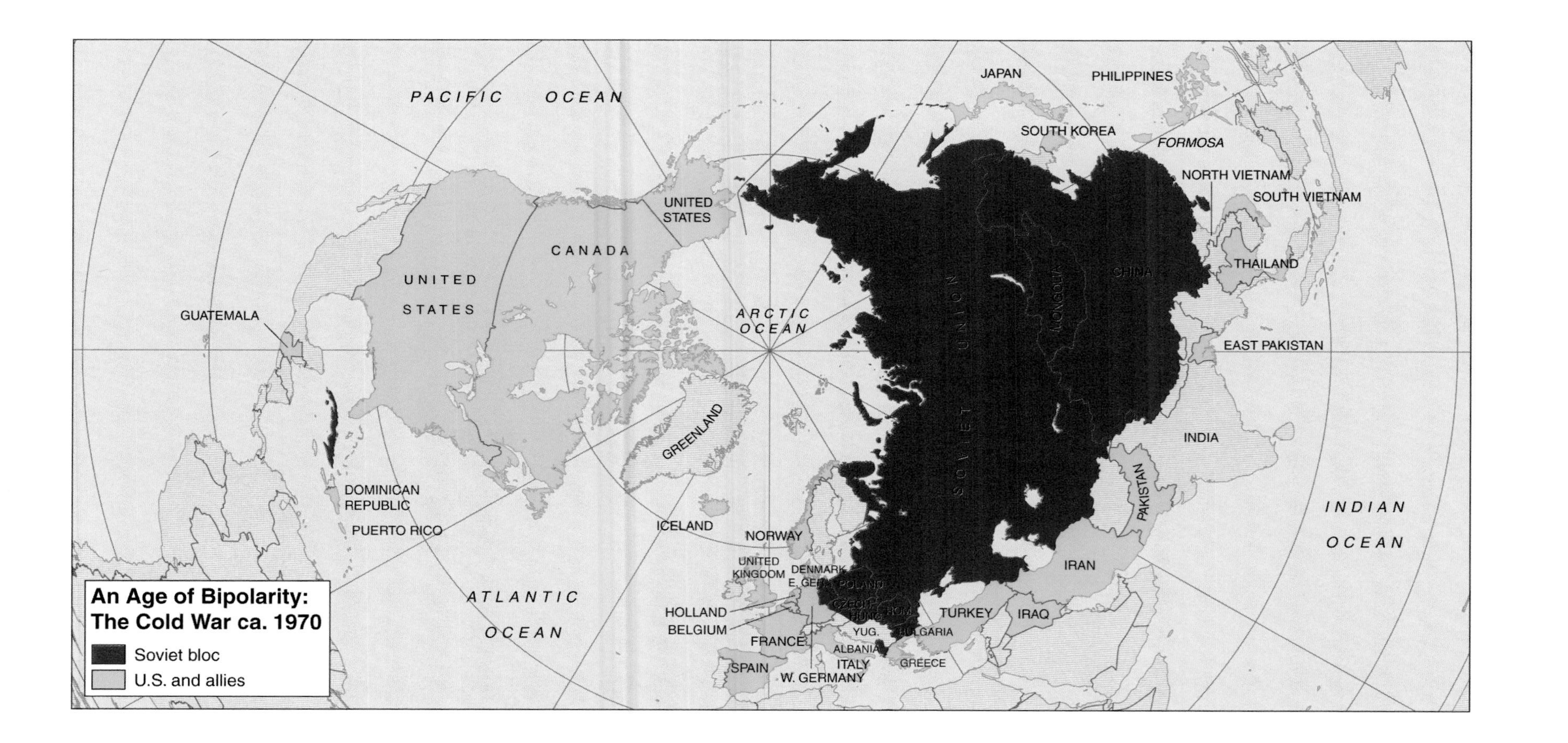

Following the Second World War, the world was divided into two armed camps led by the United States and the Soviet Union. The Soviet Union and its allies, the Warsaw Pact countries, feared a U.S.-led takeover of the eastern European countries that became Soviet satellites after the war and the replacing of a socialist political and economic system with a liberal one. The United States and its allies, the NATO (North Atlantic Treaty Organization) countries, equally feared that the USSR would overrun western Europe. Both sides sought to defend themselves by building up massive military arsenals. The United States, adopting an international geopolitical strategy of containment, sought to ring the Soviet Union with a string of allied countries and military bases that would prevent Soviet expansion in any direction. The levels of spending on military hardware contributed to the devolution of the Soviet Union, and the obsolescence of alliances and military bases in an age of advanced guidance and delivery systems made the U.S. military containment less necessary. Following a peak in the early 1960s, the cold war gradually became less significant and the age of bipolar international power essentially ended with the dissolution of the USSR in 1991.

Map 18 Europe: Political Changes, 1989–2005

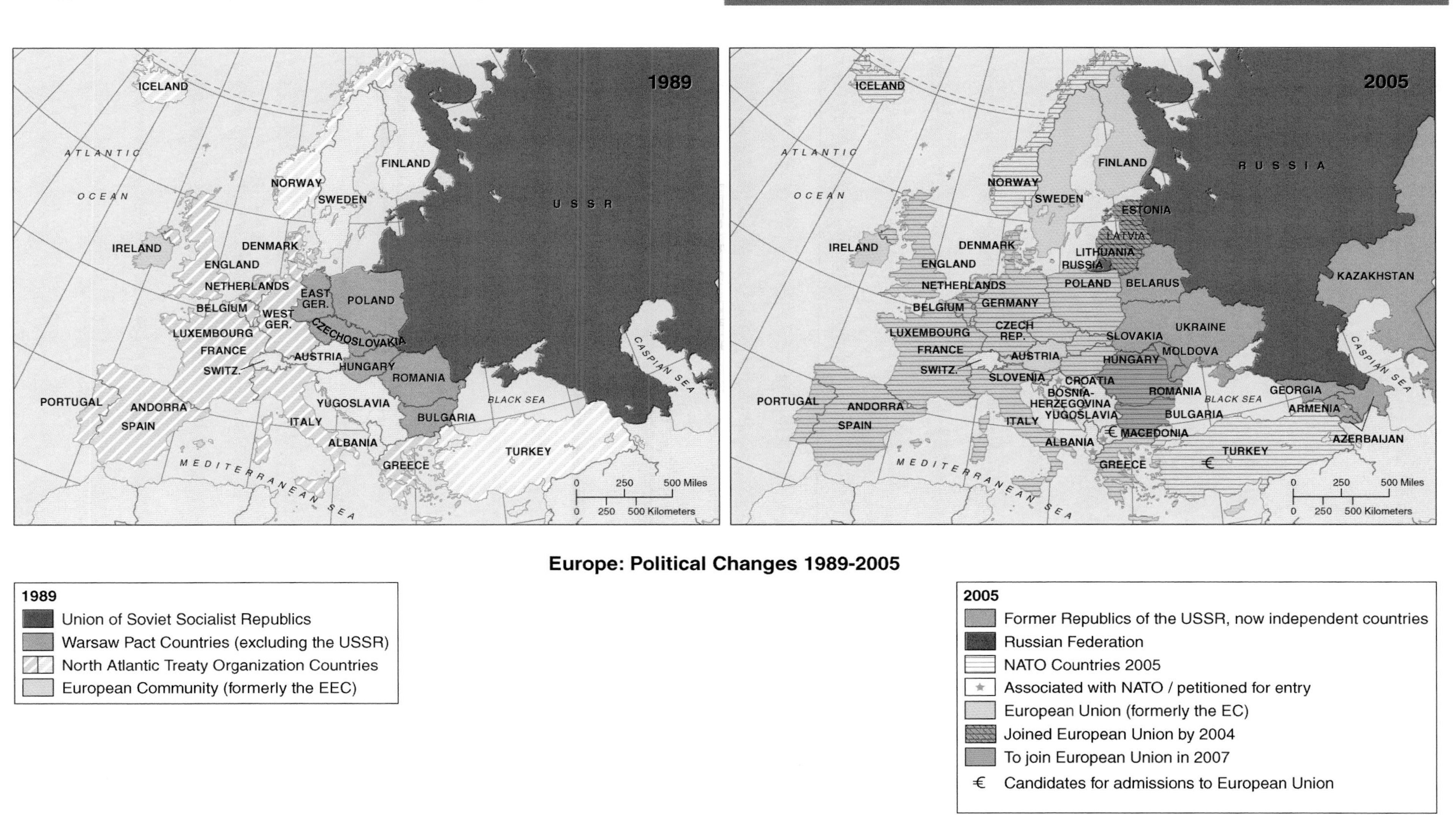

Europe: Political Changes 1989-2005

During the last decade of the twentieth century, one of the most remarkable series of political geographic changes of the last 500 years took place. The bipolar East-West structure that had characterized Europe's political geography since the end of the Second World War altered in the space of a very few years. In the mid-1980s, as Soviet influence over eastern and central Europe weakened, those countries began to turn to the capitalist West. Between 1989, when the country of Hungary was the first Soviet satellite to open its borders to travel, and 1991, when the Soviet Union dissolved into 15 independent countries, abrupt change in political systems occurred. The result is a new map of Europe that includes a number of countries not present on the map of 1989. These countries have emerged as the result of reunification, separation, or independence from the former Soviet Union. The new political structure has been accompanied by growing economic cooperation.

Map 19 The European Union

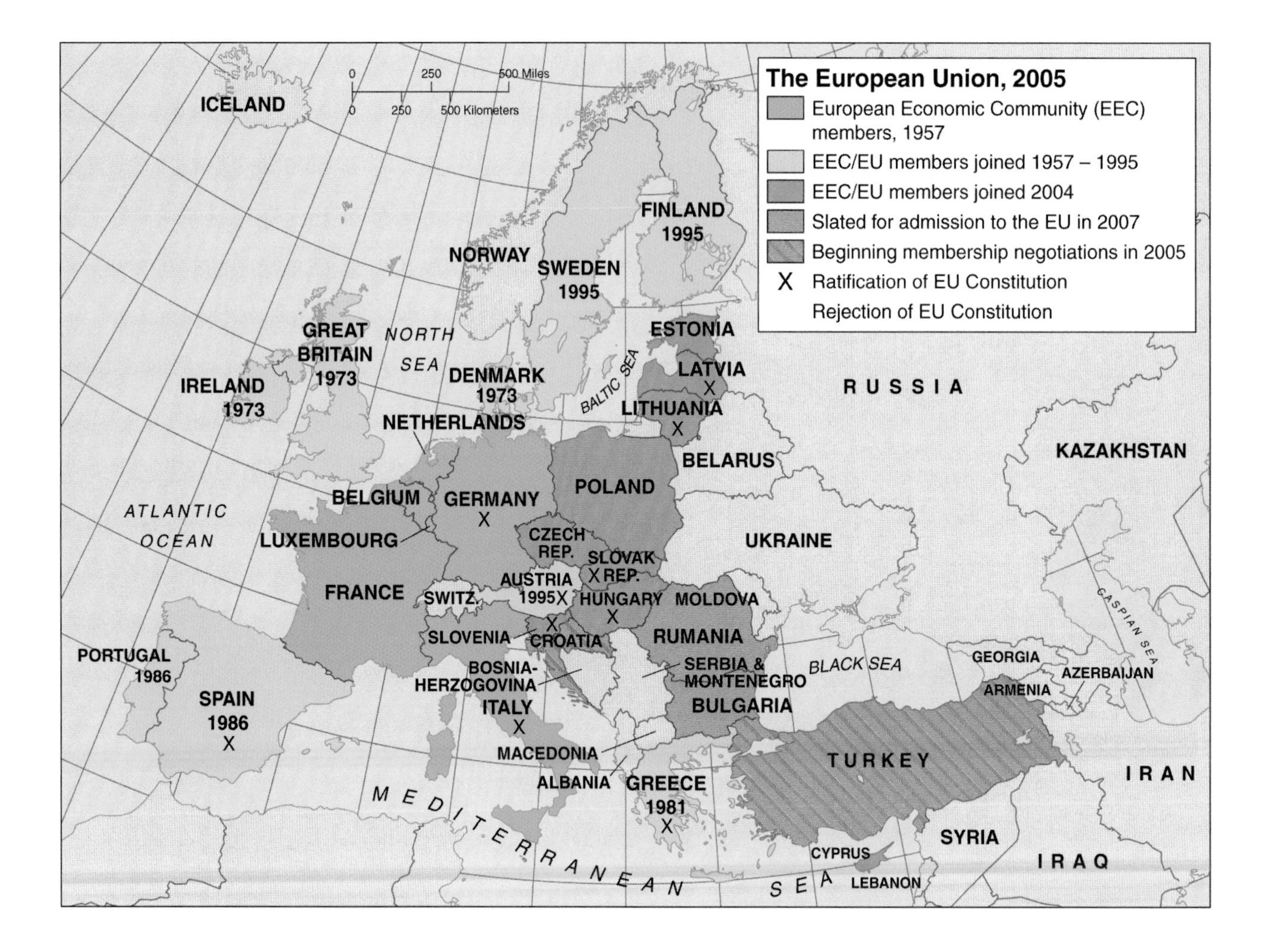

After World War II, a number of European leaders became convinced that the only way to secure a lasting peace between their countries was to unite them economically and politically. The first attempts at this were made in 1951, when the European Coal and Steel Community (ECSC) was set up, with six members: Belgium, West Germany, Luxembourg, France, Italy and the Netherlands. The ECSC was such a success that, within a few years, these same six countries decided to go further and integrate other sectors of their economies. In 1957 they signed the Treaties of Rome, creating the European Atomic Energy Community (EURATOM) and the European Economic Community (EEC). The member states set about removing trade barriers between them and forming a "common market." The original six member countries were joined in the common market of the EEC by Denmark, Ireland and the United Kingdom in 1973, followed by Greece in 1981, and Spain and Protugal in 1986. In 1992 the 12 countries of the EEC signed The Treaty of Maastricht, which introduced new forms of co-operation between the member state governments—particularly in defense and legal systems—and created the European Union (EU). The original 12 EU members were joined by Austria, Finland and Sweden in 1995. In 2004, 10 new members joined the EU: Cyprus, the Czech Republic, Estonia, Hungary, Latvia, Lithuania, Malta, Poland, Slovakia and Slovenia. Bulgaria and Romania are expected to be admitted to membership in 2007; Croatia and Turkey are beginning membership negotiations in 2005. The EU has meant the dropping of trade barriers and labor migration barriers among member countries, along with economic and political cooperation in a number of areas. The adoption of the Euro in 2002 as the common currency of the majority of EU countries is expected to aid in the integration of the European economy. The major issue for the European Union is the ratification of a constitution, which must be unanimously adopted. As of June 2005, although 12 countries had ratified the constitution, two (France and The Netherlands) had rejected it in popular referendums, making the expansion of political integration of the EU questionable. However, the economic components of the EU will remain unchanged.

Map 20 The Geopolitical World at the Beginning of the Twenty-First Century

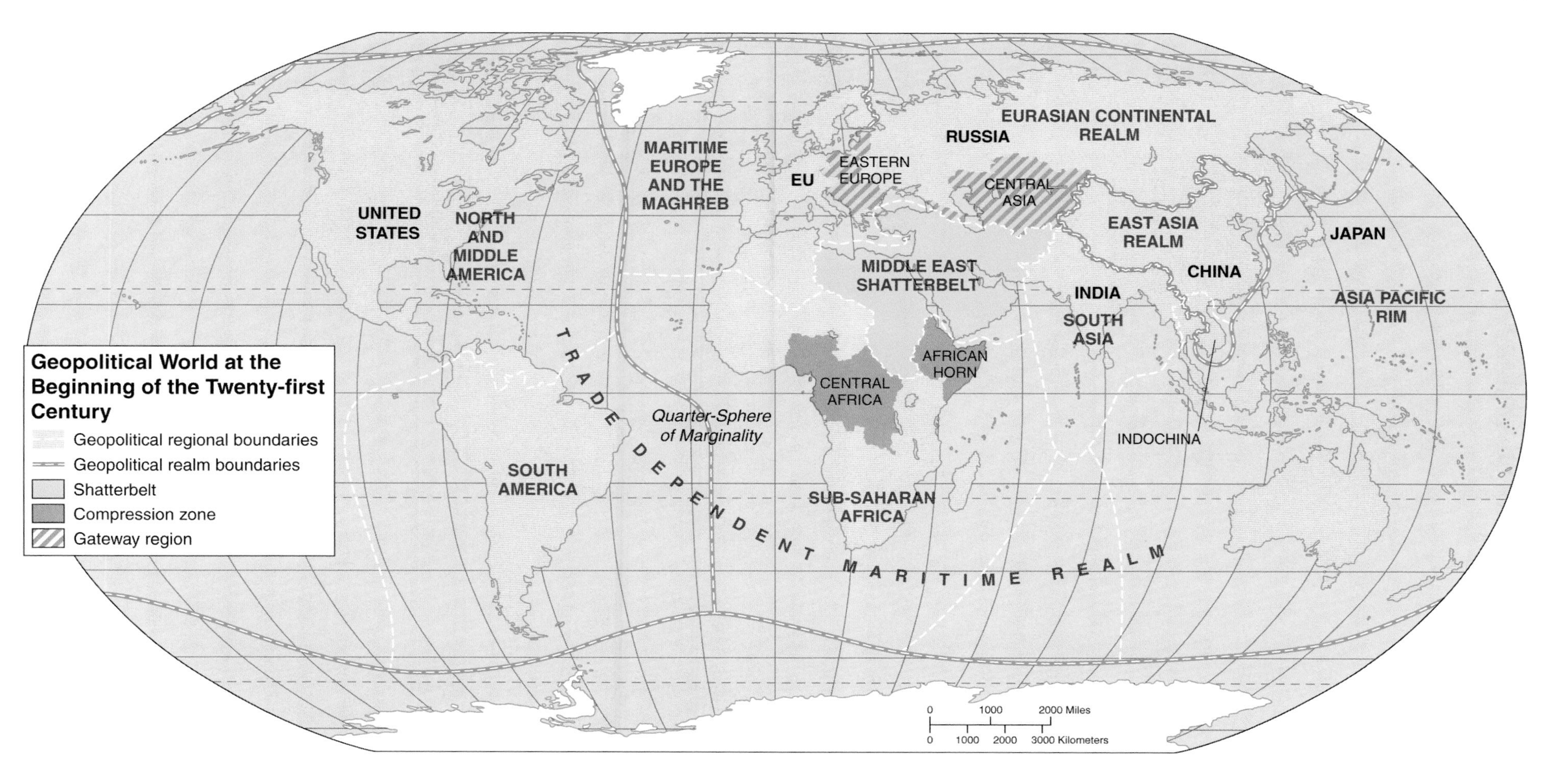

In the geostrategic structure of the world, the largest territorial units are realms. They are shaped by circulation patterns that link people, goods, and ideas. Realms are shaped by maritime and continental influences. Today's Atlantic and Pacific Trade-Dependent Maritime Realm has been shaped by international exchange over the oceans and their interior seas as mercantilism, capitalism, and industrialization gave rise to maritime-oriented states and to economic and political colonialism. The world's leading trading and economic powers are part of this realm. The Eurasian Continental Realm, centered around Russia, is inner-oriented, less influenced by outside economic or cultural forces, and politically closed, even after the fall of Communism. Expansion of NATO in Europe has increased its feeling of being "hemmed in." East Asia has mixed Maritime and Continental influences. China has traditionally been continental, but reforms that began in the late 1970s increased the importance of its maritime-oriented southern coasts. Even so, its trade volume is still low, and it maintains a hold on inland areas like Tibet and Xinjiang. Realms are subdivided into regions, some dependent on others, as South America is on North America. Regions located between powerful realms or regions may be shatterbelts (internally divided and caught up in competition between Great Powers) or gateways (facilitating the flow of ideas, goods, and people between regions). Compression zones are areas of conflict, but they are not contested by major powers.

Map 21 The Middle East: Territorial Changes, 1918–Present

The Middle East, encompassing the northeastern part of Africa and southwestern Asia, has experienced a turbulent history. In the last century alone, many of the region's countries have gone from being ruled by the Turkish Ottoman Empire, to being dependencies of Great Britain or France, to being independent. Having experienced the Crusades and colonial domination by European powers, the region's predominantly Islamic countries are now resentful of interference in the region's affairs by countries with a European and/or Christian heritage. The tension between Israel (settled largely in the late nineteenth and twentieth centuries by Jews of predominantly European background) and its neighbors is a matter of European–Middle Eastern cultural stress as well as a religious conflict between Islamic Arab culture and Judaism.

Map 22 Africa: Colonialism to Independence, 1910–2002

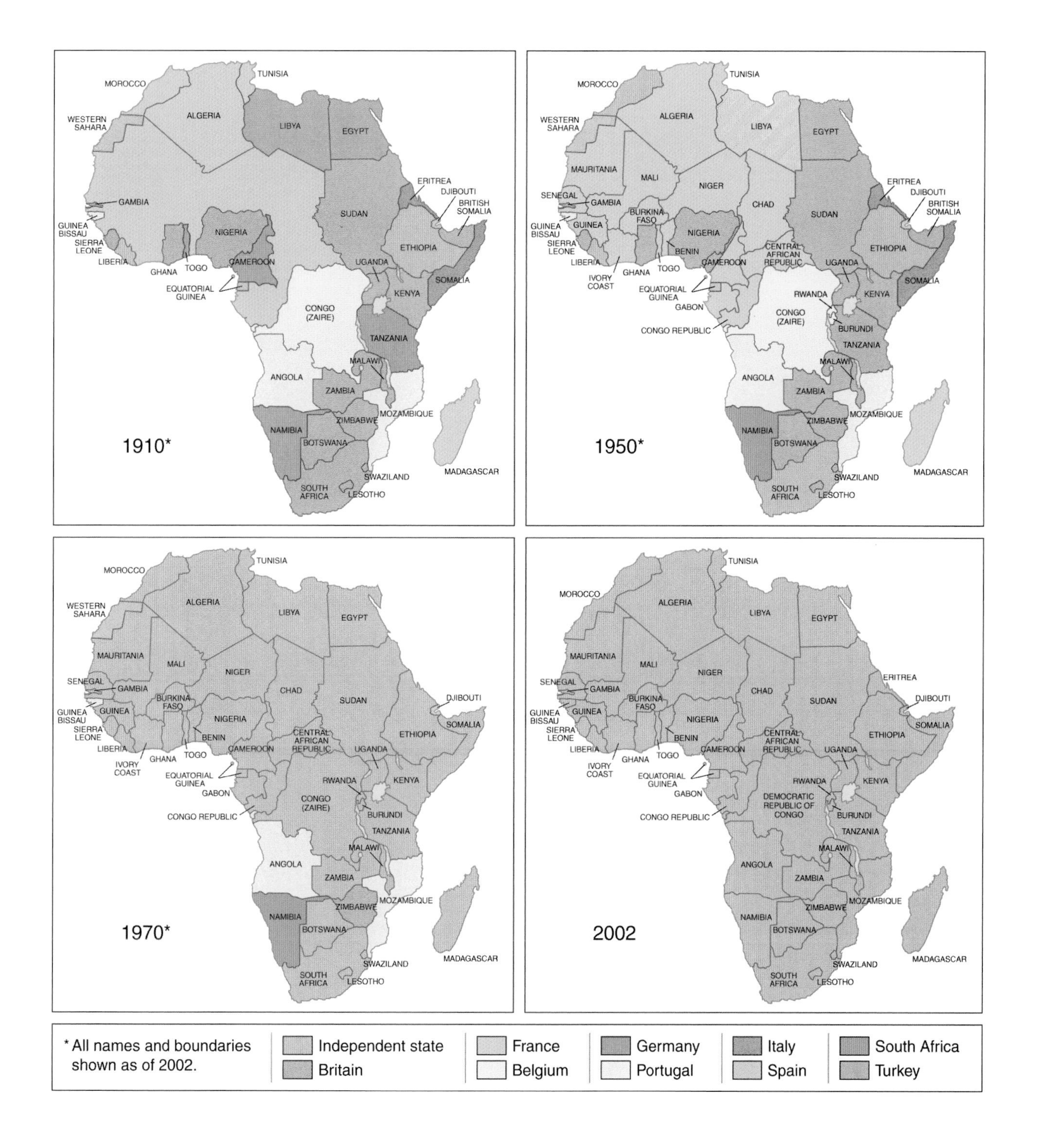

In few parts of the world has the transition from colonialism to independence been as abrupt as on the African continent. Unlike the states of Middle and South America, which generally achieved independence from their colonial masters in the early nineteenth century, most African states did not become independent until the twentieth century, often not until after World War II. In part because they retain borders that are legacies of their former colonial status, many of these recently created African states are beset by internal problems related to tribal and ethnic conflicts.

Map 23 South Africa: Black Homelands and Post-Apartheid Provinces

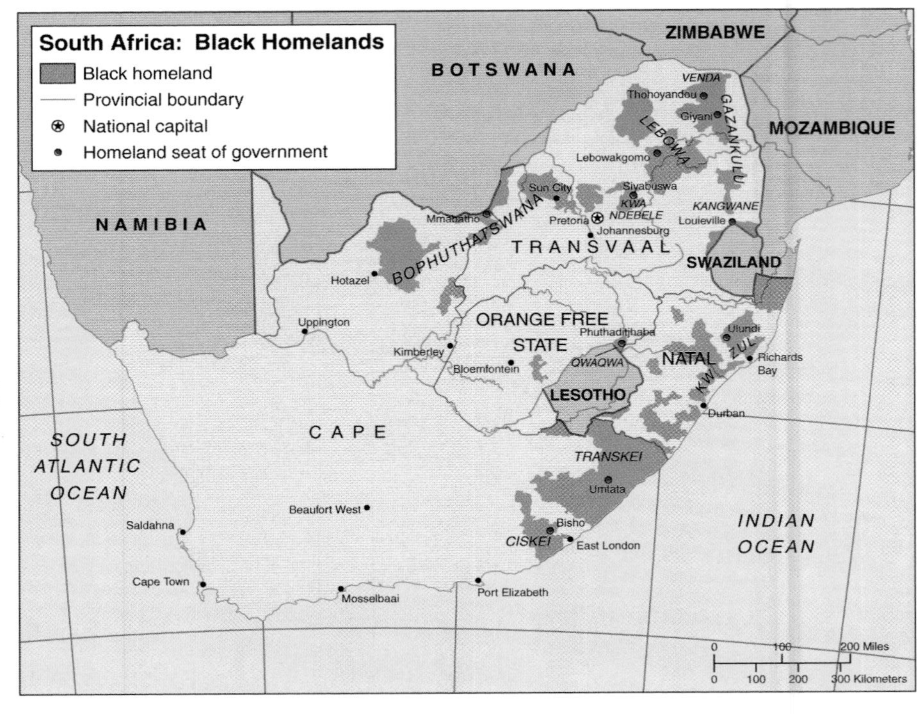

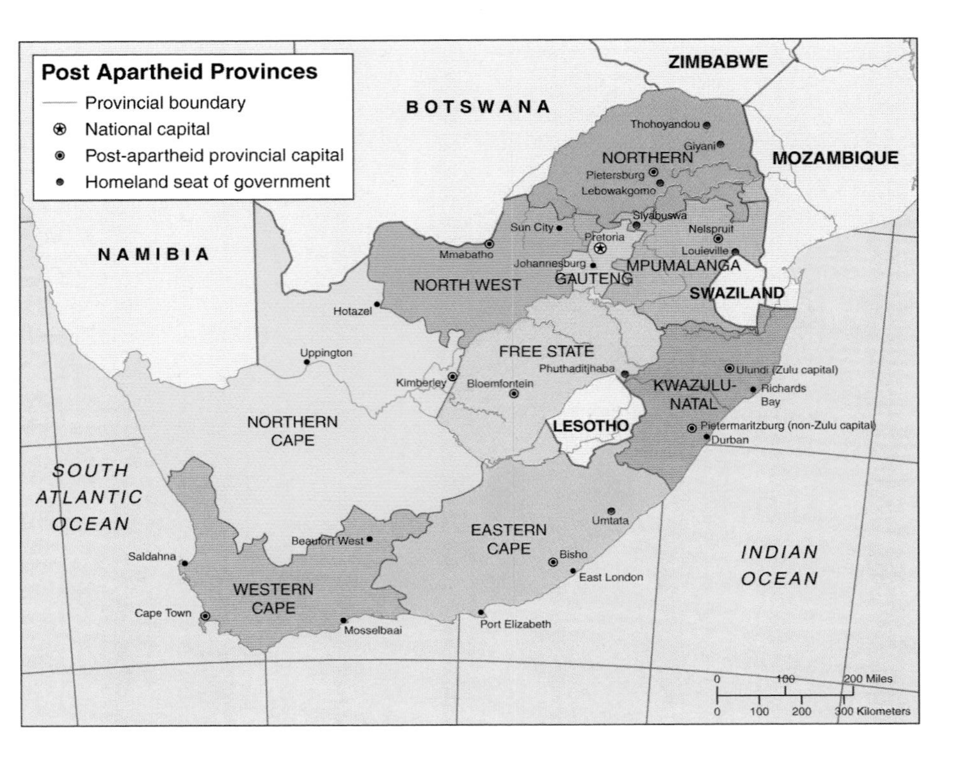

After their defeat by the British in the Boer War (1899–1902), the Dutch-descended Afrikaners negotiated with the British for greater powers in South Africa. Eventually, they became the most powerful group and imposed "separate development" or *apartheid* on the country. African (and other minority) populations would live completely separated from white South Africans. Millions were forced to relocate to the ancestral areas, where "homelands" that would be declared "independent" were set up for them. The amounts and quality of the land were completely insufficient to support the populations assigned to them, and many of the "homelands" were fragmented as well. Thousands of black Africans flocked to the black "townships" around major cities, looking for work. Here, they were foreigners in their own land. After the fall of *apartheid* in 1994, the "homelands" were abolished, and South Africa's political geography was reorganized, with each new province centered around its dominant ethnic group. These provinces now serve as subdivisions within the country without restrictions by race or ethnicity on where people can live. This organization was important in the peaceful transfer to majority rule.

Map 24 Asia: Colonialism to Independence, 1930–2002

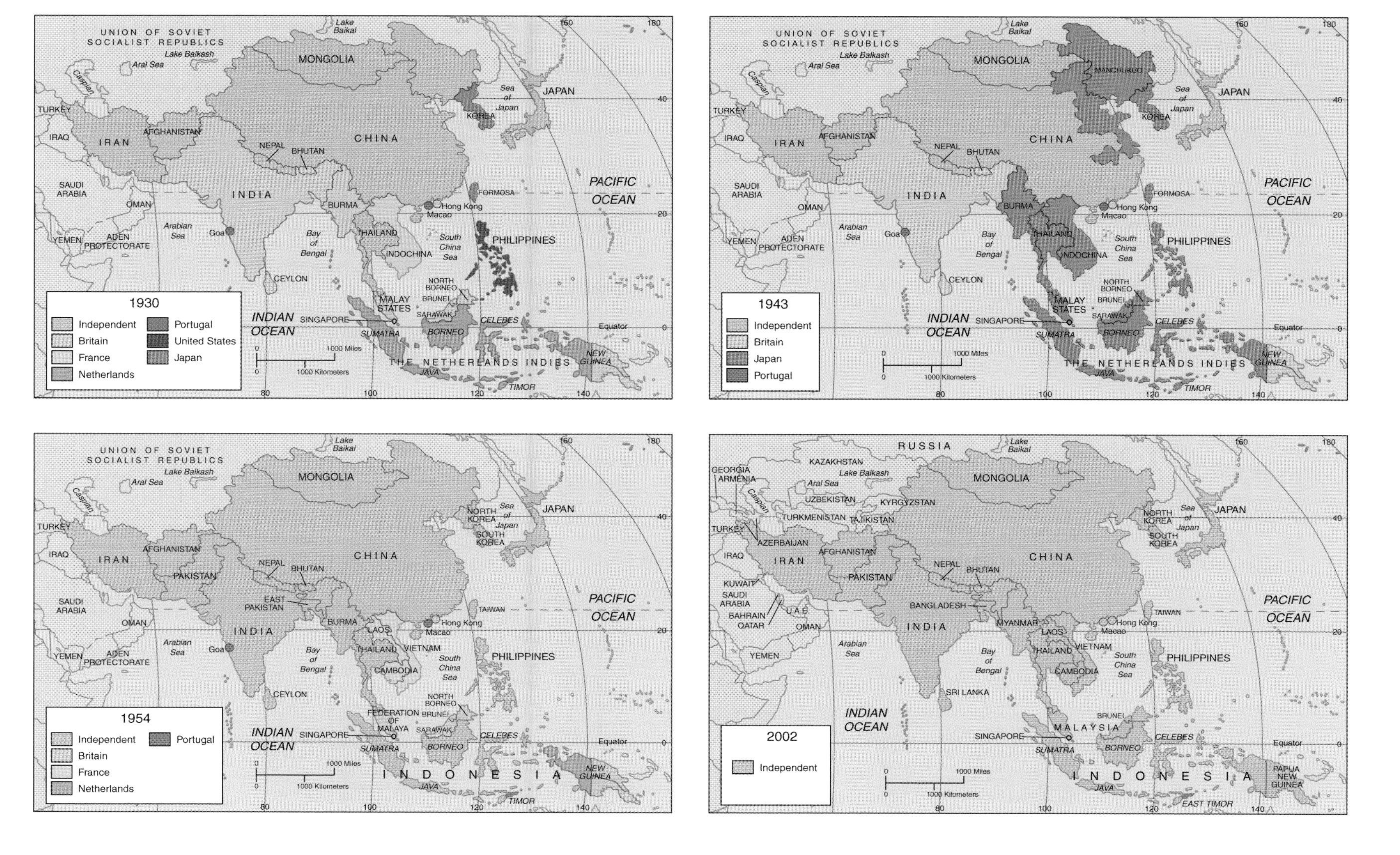

Asian countries, like those in Africa, have recently emerged from a colonial past. With the exception of China, Japan, and Thailand, virtually all Asian nations were until not long ago under the colonial control of Great Britain, France, Spain, the Netherlands, or the United States. For a short period of time between 1930 and 1945, Japan itself was a colonial power with considerable territories on the Asian mainland. The unraveling of colonial control in Asia, particularly in South and Southeast Asia, has precipitated internal conflicts in the newly independent states that make up a significant part of the political geography of the region. The last vestiges of European colonialism in Asia disappeared with the cession of Hong Kong (1997) and Macau (1999) to China.

Map 25 Global Distribution of Minority Groups

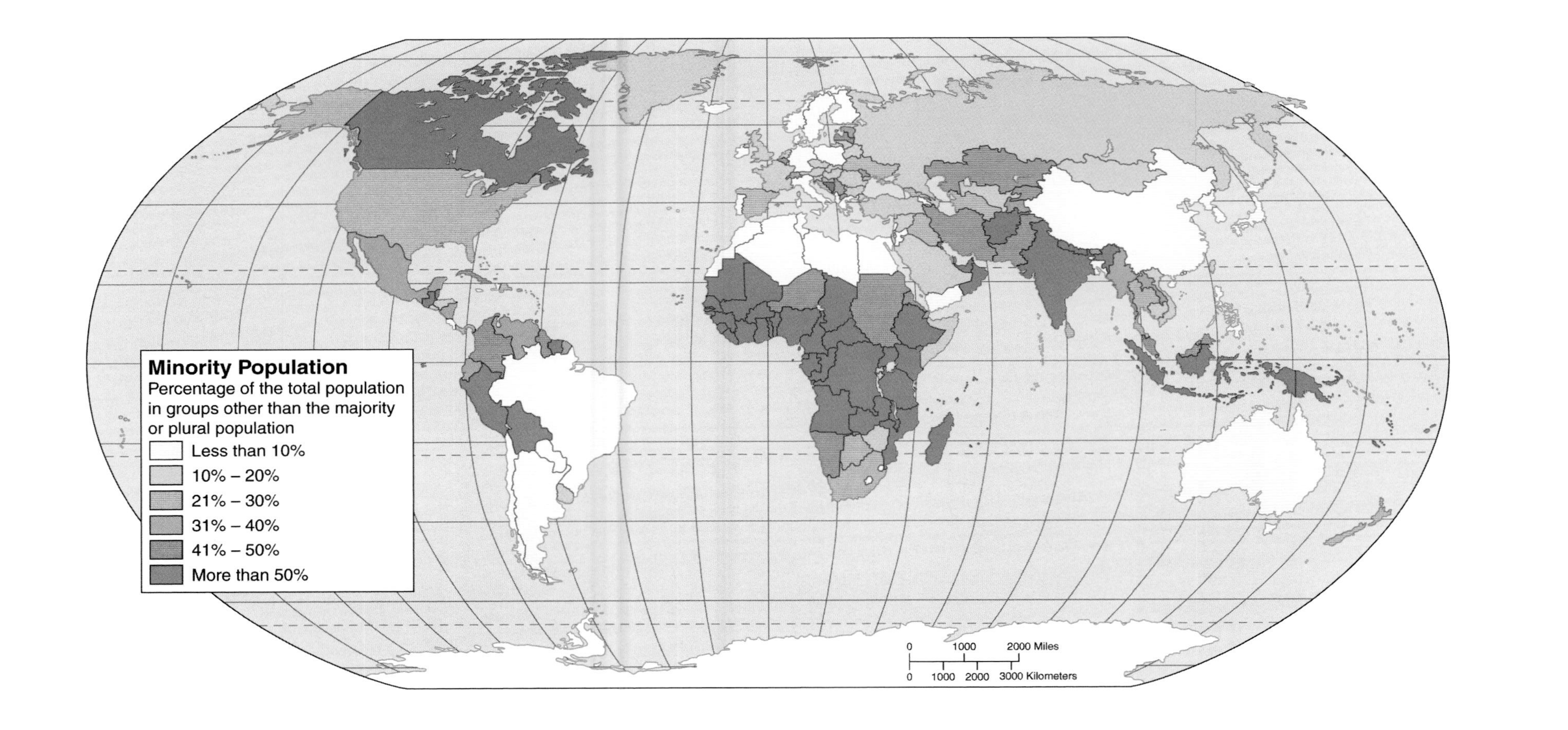

The presence of minority ethnic, national, or racial groups within a country's population can add a vibrant and dynamic mix to the whole. Plural societies with a high degree of cultural and ethnic diversity should, according to some social theorists, be among the world's most healthy. Unfortunately, the reality of the situation is quite different from theory or expectation. The presence of significant minority populations played an important role in the disintegration of the Soviet Union; the continuing existence of minority populations within the new states formed from former Soviet republics threatens the viability and stability of those young political units. In Africa, national boundaries were drawn by colonial powers without regard for the geographical distribution of ethnic groups, and the continuing tribal conflicts that have resulted hamper both economic and political development. Even in the most highly developed regions of the world, the presence of minority ethnic populations poses significant problems: witness the separatist movement in Canada, driven by the desire of some French-Canadians to be independent of the English majority, and the continuing ethnic conflict between Flemish-speaking and Walloon-speaking Belgians. This map, by arraying states on a scale of homogeneity to heterogeneity, indicates areas of existing and potential social and political strife.

Map 26 Linguistic Diversity

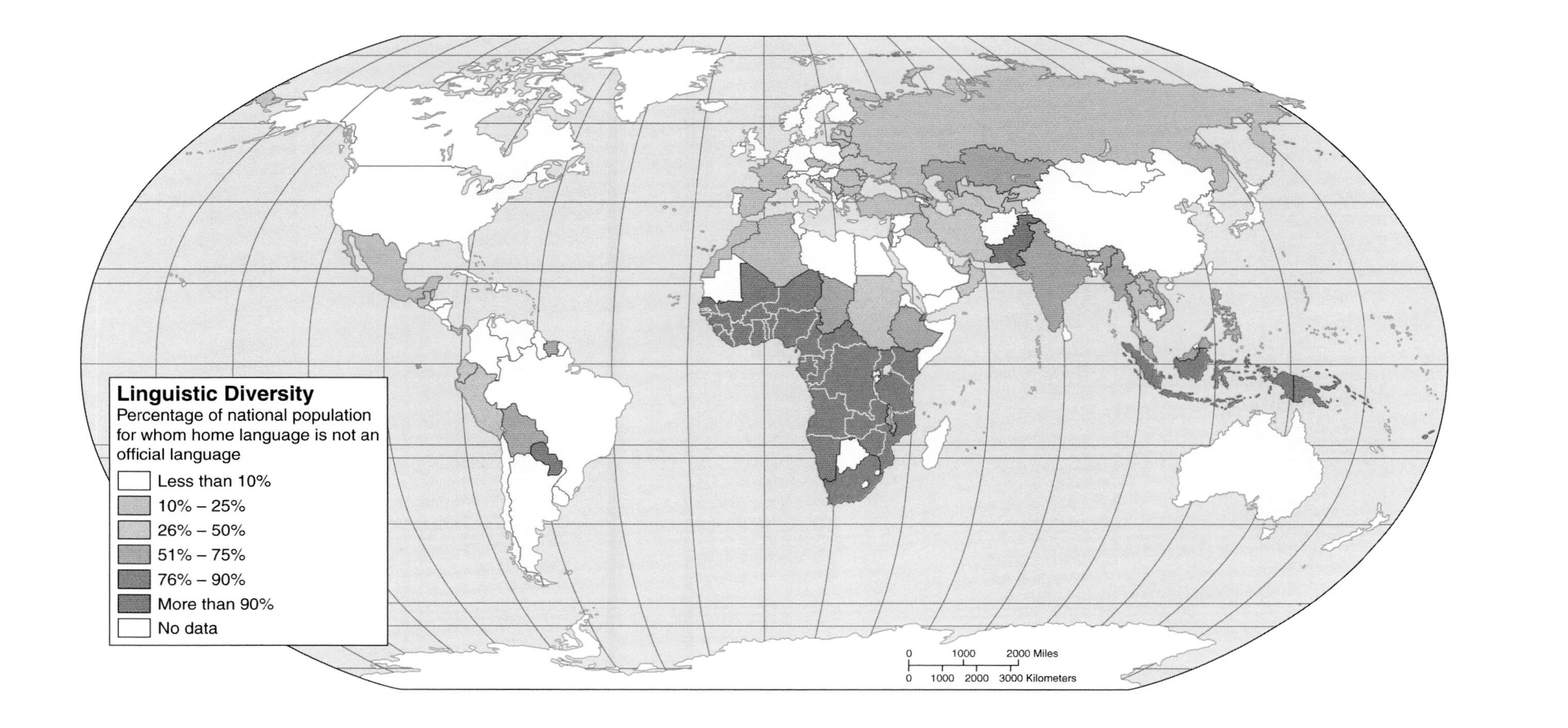

Of the world's approximately 5,300 languages, fewer than 100 are official languages, those designated by a country as the language of government, commerce, education, and information. This means that for much of the world's population the language that is spoken in the home is different from the official language of the country of residence. The world's former colonial areas in Middle and South America, Africa, and South and Southeast Asia stand out on the map as regions in which there is significant disparity between home languages and official languages. To complicate matters further, for most of the world's population, the primary international languages of trade and tourism (French and English) are neither home nor official languages.

Map 27 International Conflicts in the Post–World War II World

	Conflict[1]	Start Date	Major Belligerent Countries[2] (in alphabetical order)	
1	Palestine	1948	Egypt Iraq Israel	Jordan Lebanon Syria
2	Korean	1950	China North Korea South Korea United Nations: United States and 11 other countries	
3	Soviet-Hungarian	1956	Hungary	Soviet Union
4	Sinai	1956	Egypt France	Israel United Kingdom
5	Sino-Indian	1962	China	India
6	Kashmir	1965	India	Pakistan
7	Vietnam	1965	Australia North Vietnam South Korea	South Vietnam United States
8	Six-Day	1967	Egypt Israel	Jordan Syria
9	Soviet-Czech	1968	Czechoslovakia	Soviet Union
10	Football	1969	El Salvador	Honduras
11	Indo-Pakistani	1971	India	Pakistan
12	Yom Kippur	1973	Egypt Israel	Syria
13	Cyprus	1974	Cyprus	Turkey
14	Ogaden	1977	Ethiopia	Somalia
15	Cambodian-Vietnamese	1978	Cambodia China	Vietnam
16	Ugandan-Tanzanian	1978	Tanzania	Uganda
17	Afghanistan	1979	Afghanistan	Soviet Union
18	Persian Gulf	1980	Iran	Iraq
19	Angola	1981	Angola Cuba	South Africa
20	Falklands	1982	Argentina	United Kingdom
21	Saharan	1983	Chad	Libya
22	Lebanon	1987	France Israel Lebanon	Syria United States
23	Panama	1989	Panama	United States
24	Persian Gulf	1990	Iraq United Nations: United States and 7 other countries	
25	Yugoslavia	1990	Bosnia-Herzegovina	Croatia Serbia
26	Peruvian-Ecuadorian	1995	Ecuador	Peru
27	Albania	1995	Albania	Yugoslavia (Serbia-Montenegro)
28	Rwanda	1995	Burundi	Rwanda
29	East Timor	1995	Indonesia	New Guinea insurgency
30	Cameroon	1996	Cameroon	Nigeria
31	Northern Iraq	1996	Iraq	Kurdish insurgency
32	Eritrea	1997	Eritrea	Yemen
33	Iraq	1998	Great Britain Iraq	United States
34	Kosovo	1999	Albania NATO	Yugoslavia
35	Democratic Republic of the Congo	1998	Angola Chad Dem. Rep. of Congo	Namibia Uganda Zimbadwe
36	Chechnya	1999	Chechnya	Russia
37	"War on Terrorism"	2001	Afghanistan (Taliban) al-Qaeda organization Great Britain	United States Others
38	Iraq	2003	Great Britain Iraq	United States

[1] "Conflict" implies at least 1,000 battle deaths.
[2] "Belligerent" implies country supplied at least 5% of the combat troops in the conflict.

International Conflicts in the Post World War II World

Area of conflict

10 23 26 20

The Korean War and the Vietnam War dominated the post–World War II period in terms of international military conflict. But numerous smaller conflicts have taken place, with fewer numbers of belligerents and with fewer battle and related casualties. These smaller international conflicts have been mostly territorial conflicts, reflecting the continual readjustment of political boundaries and loyalties brought about by the end of colonial empires, and the dissolution of the Soviet Union. Many of these conflicts were not wars in the more traditional sense, in which two or more countries formally declare war on one

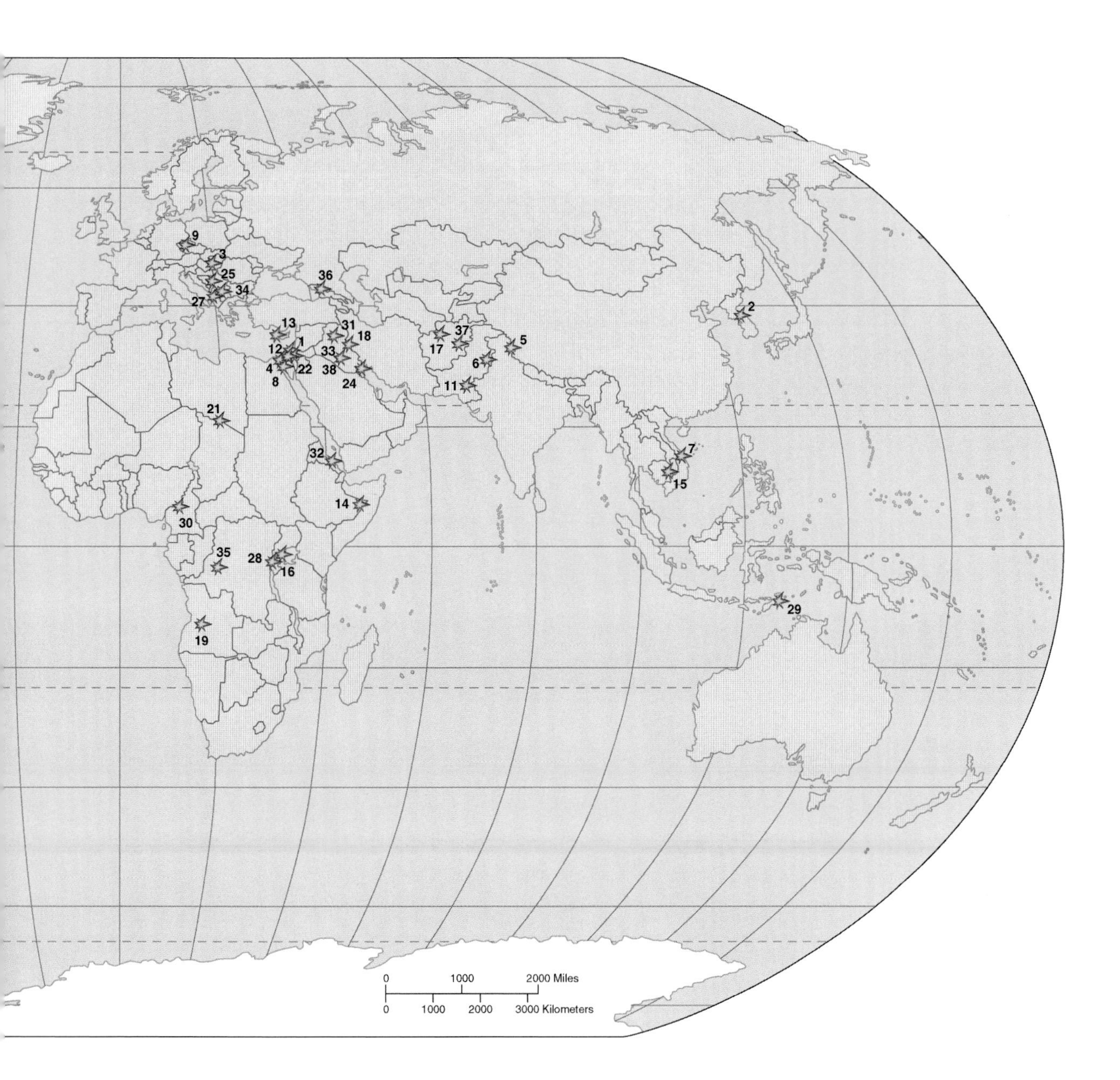

another, severing diplomatic ties and devoting their entire national energies to the war effort. Rather, many of these conflicts were and are undeclared wars, sometimes fought between rival groups within the same country with outside support from other countries. The aftermath of the September 11, 2001, terrorist attacks on the United States indicate the dawn of yet another type of international conflict, namely a "war" fought between traditional nation-states and non-state actors.

Map 28 World Refugees, 2005

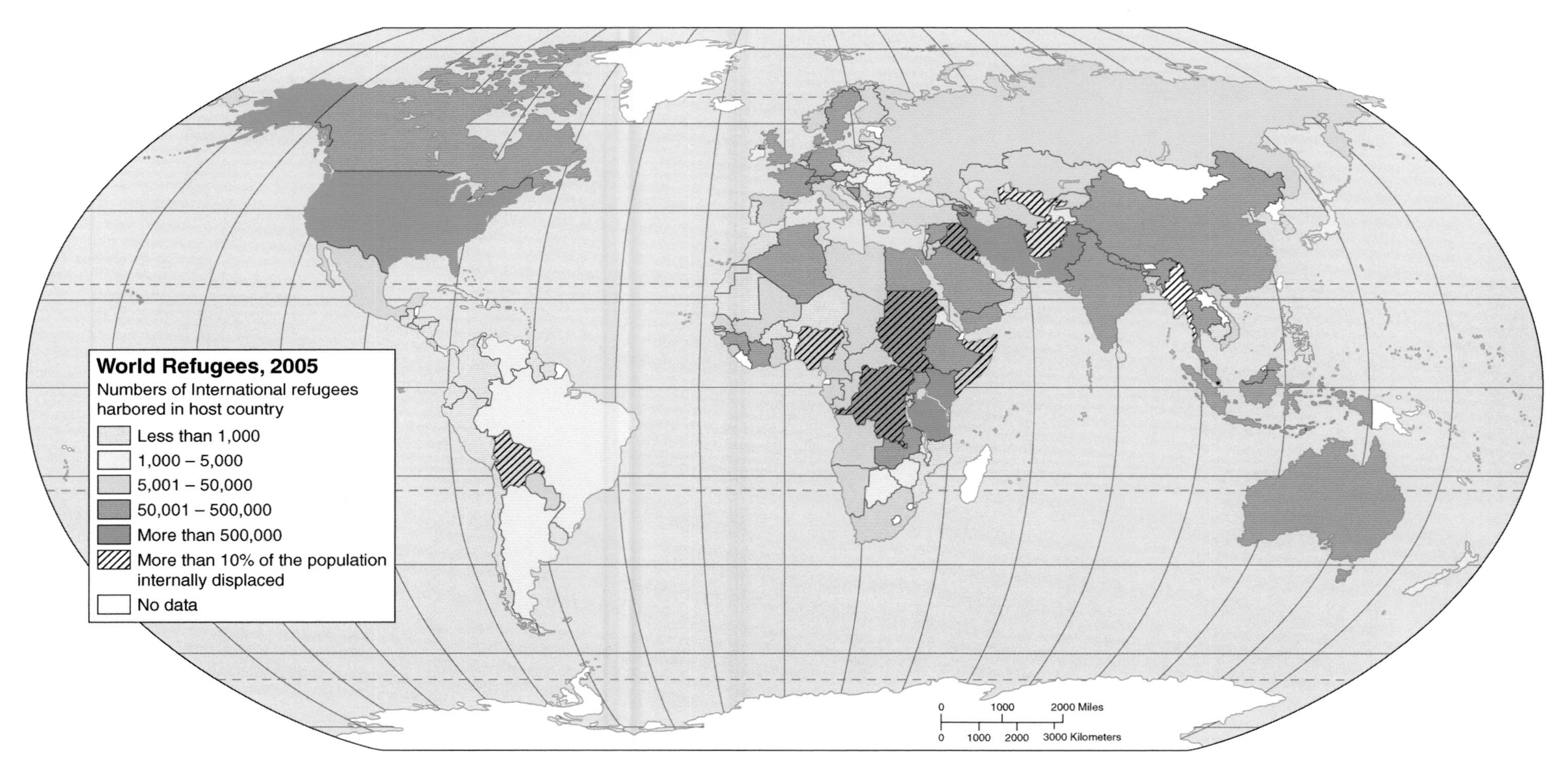

Refugees are persons who have been driven from their homes, normally by armed conflict, and have sought refuge by relocating. The most numerous refugees have traditionally been international refugees, who have crossed the political boundaries of their homelands into other countries. This refugee population is recognized by international agencies, and the countries of refuge are often rewarded financially by those agencies for their willingness to take in externally displaced persons. In recent years, largely because of an increase in civil wars, there have been growing numbers of internally displaced persons—those who leave their homes but stay within their country of origin. There are no rewards for harboring such internal refugee populations.

Map 29 Post-Cold War International Alliances

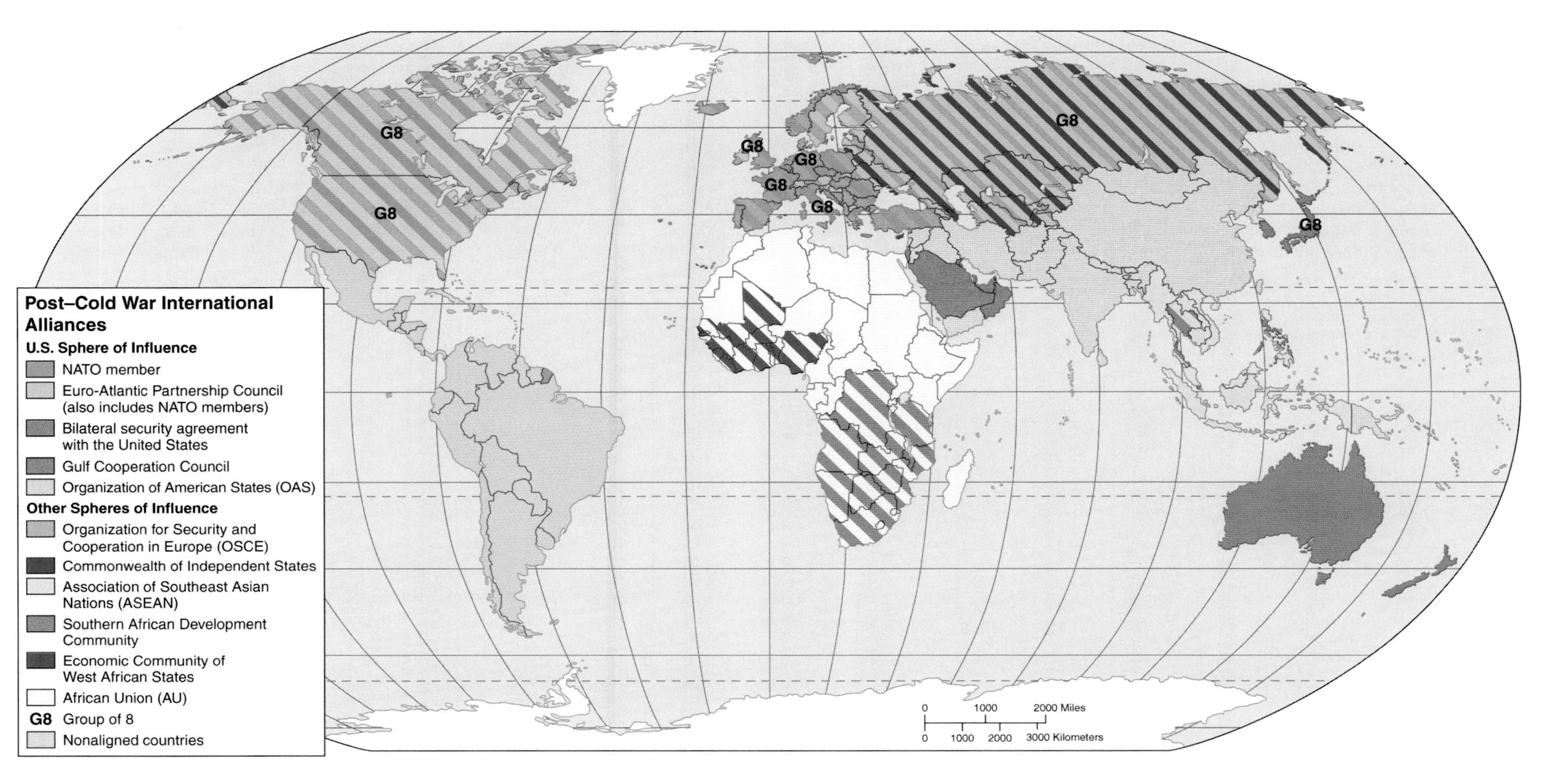

When the Warsaw Pact dissolved in 1992, the North Atlantic Treaty Organization (NATO) was left as the only major military alliance in the world. Some former Warsaw Pact members (Czech Republic, Hungary, and Poland) have joined NATO and others are petitioning for entry. The bipolar division of the world into two major military alliances is over, at least temporarily, leaving the United States alone as the world's dominant political and military power. But other international alliances, such as the Commonwealth of Independent States (including most of the former republics of the Soviet Union), will continue to be important. It may well be that during the first few decades of the twenty-first century economic alliances will begin to overshadow military ones in their relevance for the world's peoples.

Map 30 Flashpoints, 2005

Macedonia: Since the fifteenth-century conquest of much of southeastern Europe by the Turkish Ottoman Empire, conflict in the Balkan region has been precipitated by ethnic and religious enmity between Orthodox Christians and Muslims. An area of particular concern has been the interface between predominantly Orthodox Macedonia and predominantly Muslim Albania, particularly in the border region the two share with the former Yugoslav republic of Kosovo. Tensions between the Macedonian government and its Albanian (Muslim) minority were heightened by the fallout from the conflict in Kosovo in the late 1990s between the Serbs and the Kosovar Albanians, backed by Albania. The resolution of this conflict by a NATO-led force in favor of the Kosovars led to emboldened feelings of Albanian patriotism in the region. Several sporadic incidents occurred along the Macedonian-Albanian border in late 2000, with more protracted and heavier combat between rebels and the Macedonian government forces occurring in the northern part of Macedonia (near the capital of Skopje) throughout 2001. The rebel forces assert they are only seeking to revise the Macedonian constitution and attain better rights for the Albanian minority in Macedonia. The Macedonian government is concerned that the Albanian minority centered in northern and western Macedonia wishes to secede and merge (along with Kosovo) into a Greater Albania, and suspects that Albania itself has encouraged this objective.

Chechnya: The area in southern Russia known as the Caucasus Region is home to a large variety of non-Russian ethnic groups; many are Muslim and resent centuries of Russian domination and Soviet-era totalitarianism. After the Soviet Union disintegrated in 1991, several of these ethnic groups began agitating for more autonomy from Moscow or for outright independence. One of the more vocal groups with a history of opposition to Moscow's rule were the Chechens. The Chechens declared themselves a sovereign nation in 1991 and by 1994 relations between the breakaway government in Chechnya and the Russian government had drastically deteriorated. In December of that year, Russian forces attacked Chechnya, beginning the first of two (1994–96 and 1999–present) full-scale military conflicts that have also crept into the neighboring Russian autonomous area of Dagestan, itself largely Muslim. In the mid- and late 1990s Russia experienced several terrorist attacks in cities throughout the nation, which the Russian government attributed to Islamic extremists supporting Chechen independence. As a result, a second round of the conflict began in August 1999 with a full-scale Russian military assault on Dagestan and Chechnya. This assault is ongoing and continues to face intense resistance, with heavy casualties on both sides.

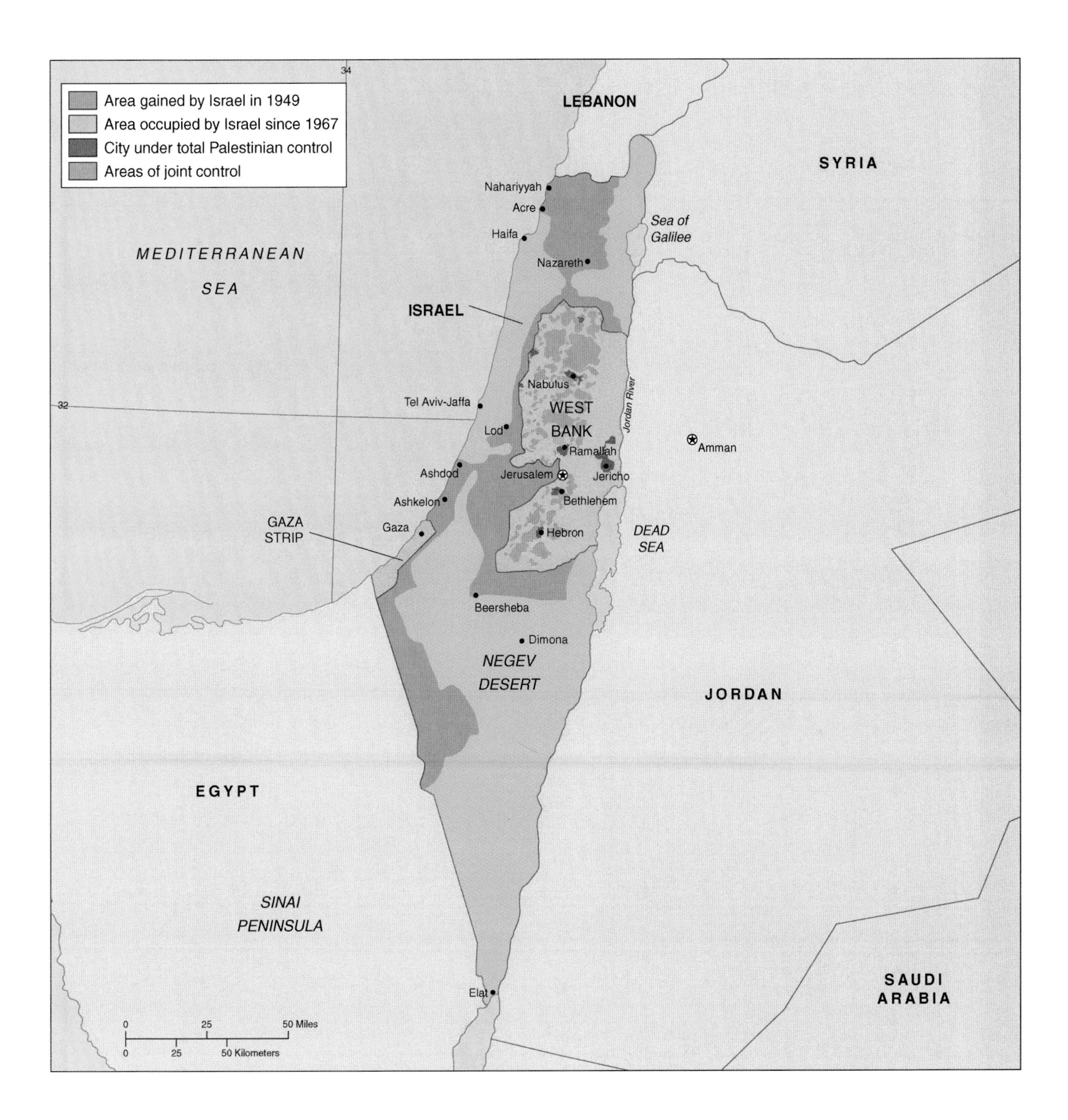

Israel and Its Neighbors: The modern state of Israel was created out of the former British Protectorate of Palestine, inhabited primarily by Muslim Arabs, after World War II. Conflict between Arabs and Israeli Jews has been a constant ever since. Much of the present tension revolves around the West Bank area, not part of the original Israeli state but taken from Jordan, an Arab country, in the Six-Day War of 1967. Many Palestinians had settled this part of Jordan after the creation of Israel and remain as a majority population in the West Bank region today. Israel has established many agricultural settlements within the region since 1967, angering Palestinian Arabs. For Israel, the West Bank is the region of ancient Judea and this region, won in battle, will not be ceded back to Palestinian Arabs without protracted or severe military action. The West Bank, inhabited by nearly 400,000 Israeli settlers and 4 million Palestinians, is also the location of most of the suicide bombings carried out by Islamic militant groups from 2001 to 2003.

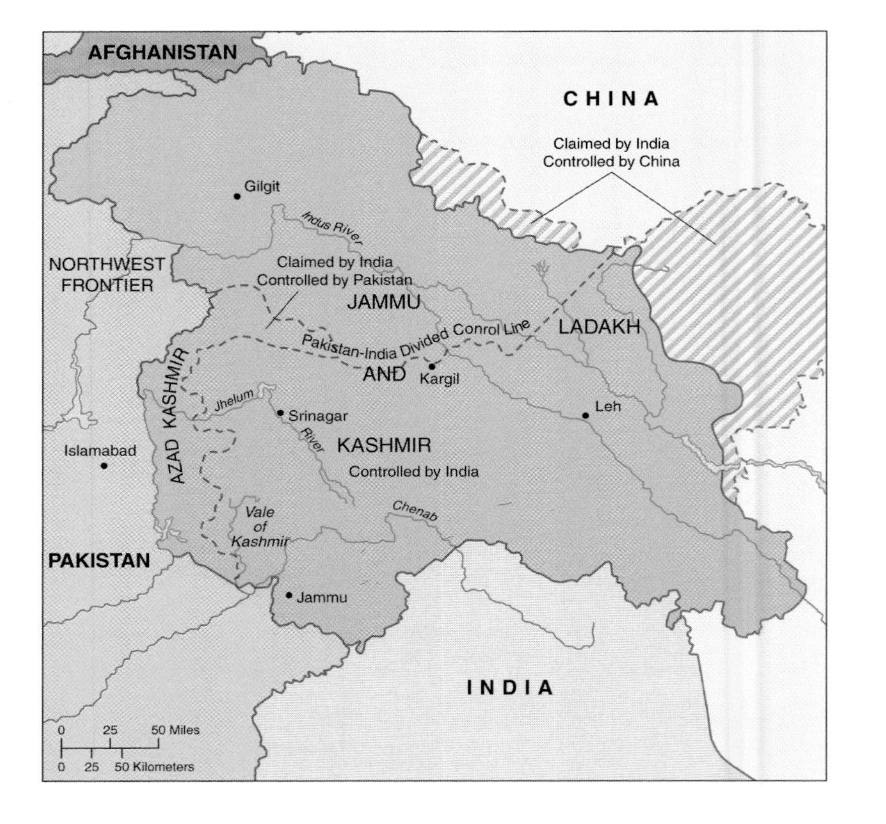

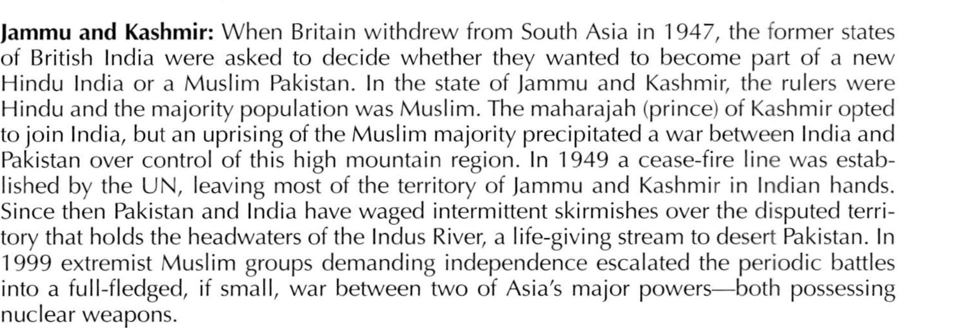

Jammu and Kashmir: When Britain withdrew from South Asia in 1947, the former states of British India were asked to decide whether they wanted to become part of a new Hindu India or a Muslim Pakistan. In the state of Jammu and Kashmir, the rulers were Hindu and the majority population was Muslim. The maharajah (prince) of Kashmir opted to join India, but an uprising of the Muslim majority precipitated a war between India and Pakistan over control of this high mountain region. In 1949 a cease-fire line was established by the UN, leaving most of the territory of Jammu and Kashmir in Indian hands. Since then Pakistan and India have waged intermittent skirmishes over the disputed territory that holds the headwaters of the Indus River, a life-giving stream to desert Pakistan. In 1999 extremist Muslim groups demanding independence escalated the periodic battles into a full-fledged, if small, war between two of Asia's major powers—both possessing nuclear weapons.

Kurdistan: Where Turkey, Iran, and Iraq meet in the high mountain region of the Tauros and Zagros mountains, a nation of 25 million people exists. This nation is "Kurdistan," but the Kurds, the occupants of this area for over 3,000 years, have no state, and receive much less attention than other stateless nations like the Palestinians. Following the 1991 Gulf War between Iraq and a U.S.-led coalition of European and Arabic states, the United Nations demarcated a Kurdish "security zone" in northern Iraq. From 1991 to 2003 the Security Zone was anything but secure as Iraqi militants from the south and Turks from the north infringed on Kurdish territory, and the internal militant extremist groups, such as the Kurdish Workers' Party, staged periodic attacks on rival villages. During the 2003 U.S.-led invasion of Iraq that eliminated the Baathist regime of Saddam Hussein, the Kurds played an important role in securing the northern portions of Iraq for the U.S.-British coalition and fought alongside American troops in expelling elements of the Iraqi army from cities like Mosul and Kirkuk. Rich in oil and history, Kurdistan will probably remain as a nation without a state, shared by Iraq, Turkey, and Iran—none of which is likely to give up substantial portions of territory for the establishment of a Kurdish state.

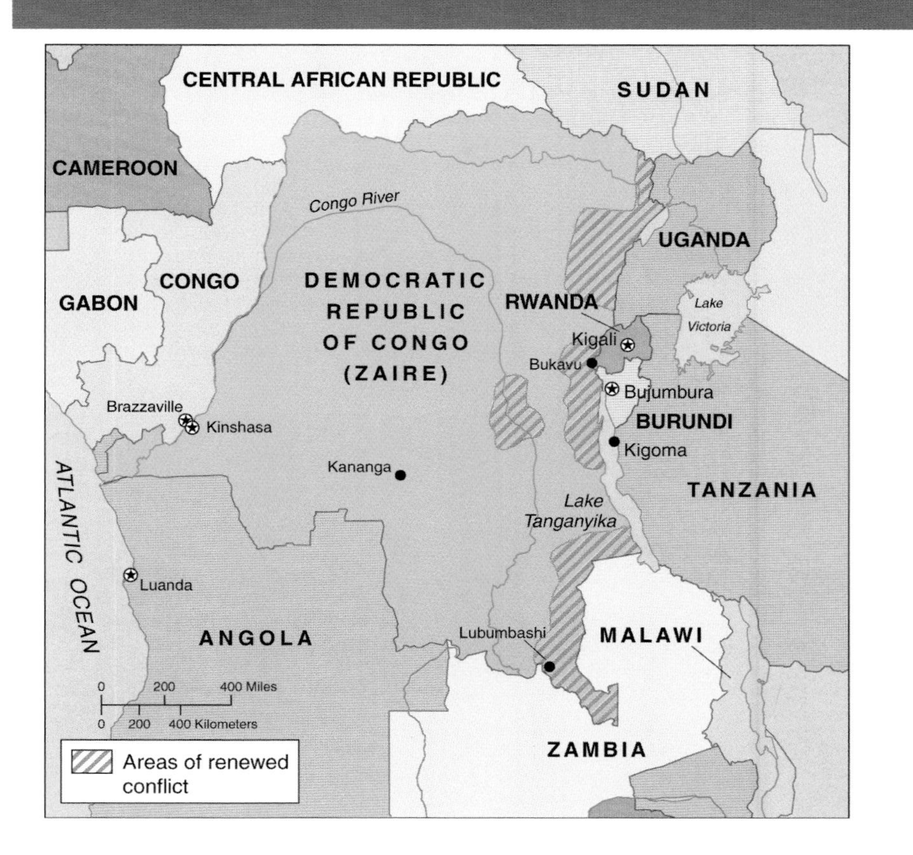

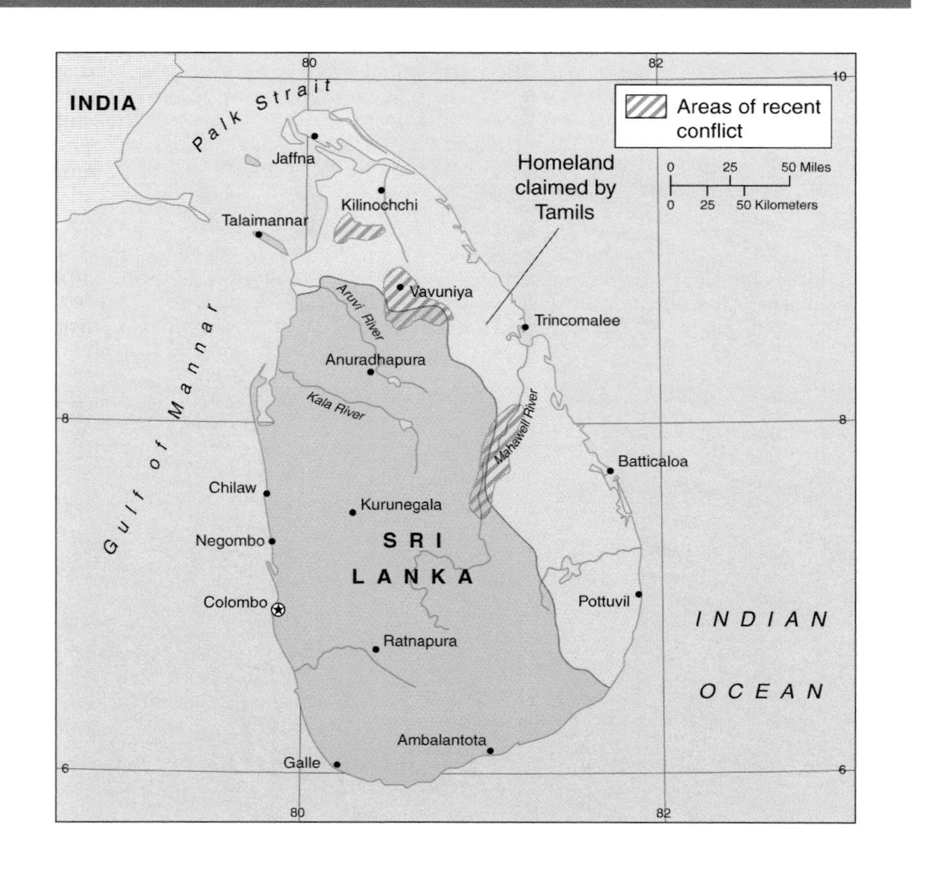

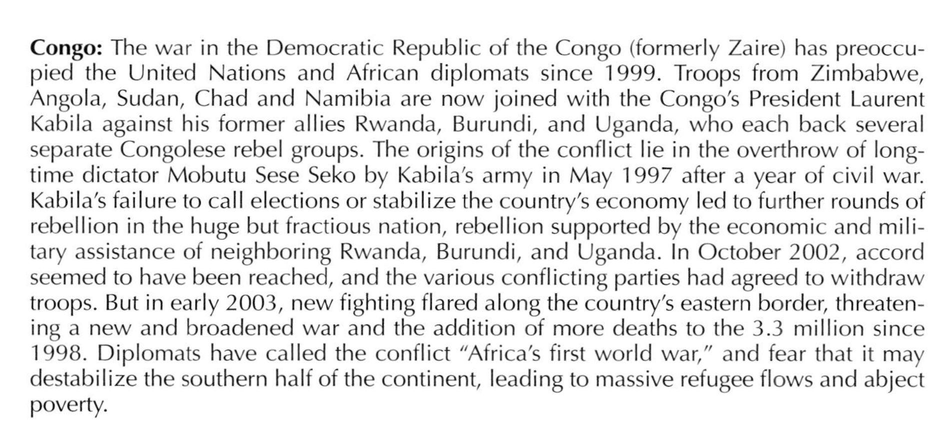
Congo: The war in the Democratic Republic of the Congo (formerly Zaire) has preoccupied the United Nations and African diplomats since 1999. Troops from Zimbabwe, Angola, Sudan, Chad and Namibia are now joined with the Congo's President Laurent Kabila against his former allies Rwanda, Burundi, and Uganda, who each back several separate Congolese rebel groups. The origins of the conflict lie in the overthrow of long-time dictator Mobutu Sese Seko by Kabila's army in May 1997 after a year of civil war. Kabila's failure to call elections or stabilize the country's economy led to further rounds of rebellion in the huge but fractious nation, rebellion supported by the economic and military assistance of neighboring Rwanda, Burundi, and Uganda. In October 2002, accord seemed to have been reached, and the various conflicting parties had agreed to withdraw troops. But in early 2003, new fighting flared along the country's eastern border, threatening a new and broadened war and the addition of more deaths to the 3.3 million since 1998. Diplomats have called the conflict "Africa's first world war," and fear that it may destabilize the southern half of the continent, leading to massive refugee flows and abject poverty.

Sri Lanka: The island state of Sri Lanka, historically known as Ceylon, is potentially one of the most agriculturally productive regions of Asia. Unfortunately for plans related to agricultural development, two quite different peoples have occupied the island country: The Buddhist Sinhalese originally from northern India and long the dominant population in Sri Lanka, and the minority Hindu Tamil, a Dravidian people from south India. Since independence from Britain, Sri Lankan governments have sought to "resettle" the Tamil population in south India, actions that finally precipitated an armed rebellion by Tamils against the Sinhalese-dominated government. The Tamils at present are demanding a complete separation of the state into two parts, with a Tamil homeland in the north and along the east coast. At one time viewed as an island paradise, Sri Lanka is now a troubled country with an uncertain future. While recent developments suggest that the Sri Lanka problem may be solved through peaceful solutions, one of the world's most fertile islands is still a "flashpoint."

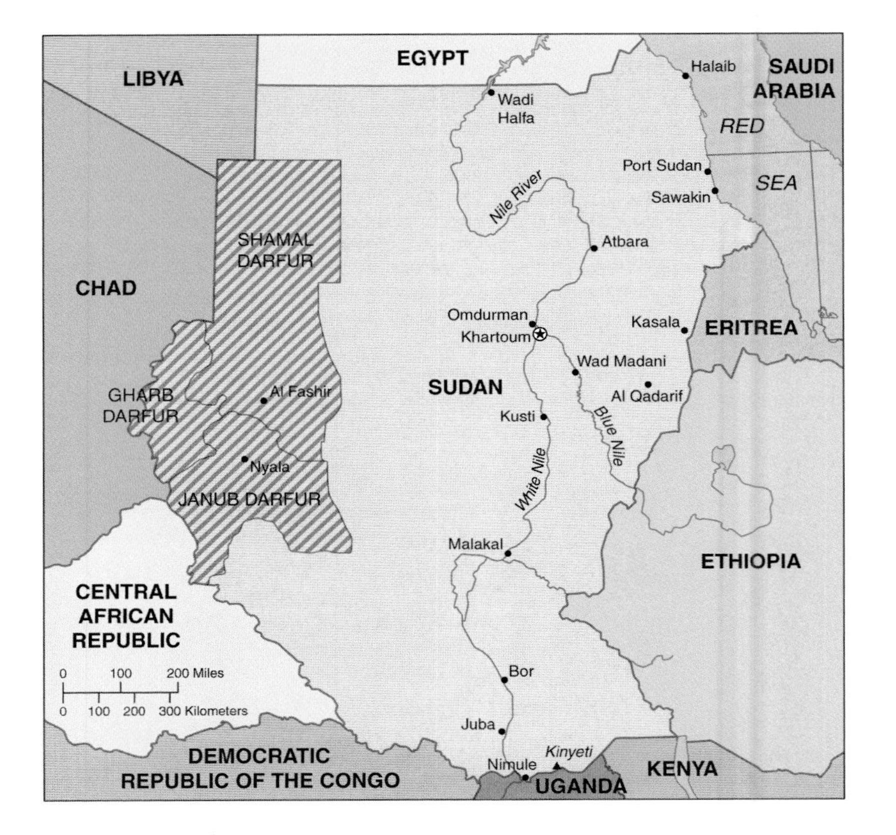

Sudan and Darfur Region: Since Sudan achieved independence from Great Britain in 1956, military regimes favoring Islamic-oriented governments have dominated national politics. These regimes have embroiled the country in a civil war for nearly all of the past half-century. These wars have been rooted in the attmpts of northern economic, political, an social interests dominated by Muslims to control territories occupied by non-Muslim, non-Arab southern Sudanese such as the Dinka tribal groups. Since 1983, the war and war-and famine-related effects have resulted in more than 2 million deaths and over 4 million people displaced. The current ruling regime is a mixture of military elite and an Islamist party that came to power in a 1989 coup. Some northern opposition parties have made common cause with the southern rebels and entered the war as part of an anti-government alliance. Peace talks gained momentum in 2002-03 with the signing of several accords, including a cease-fire agreement. However, conflicts have continued to persist in the 3 Darfur provinces of western Sudan adjacent to the border with Chad and the Central Afrcian Republic where government-backed Muslim militia have attacked and killed tens of thousands of non-Muslim tribal peoples. The Darfur region is relatively water-rich and forested in a country that is chiefly desert and is therefore desired by Muslim pastoral groups from the north for settlement purposes. International organizations have labeled the conflict in Darfur as "genocide" against the non-Muslim populations.

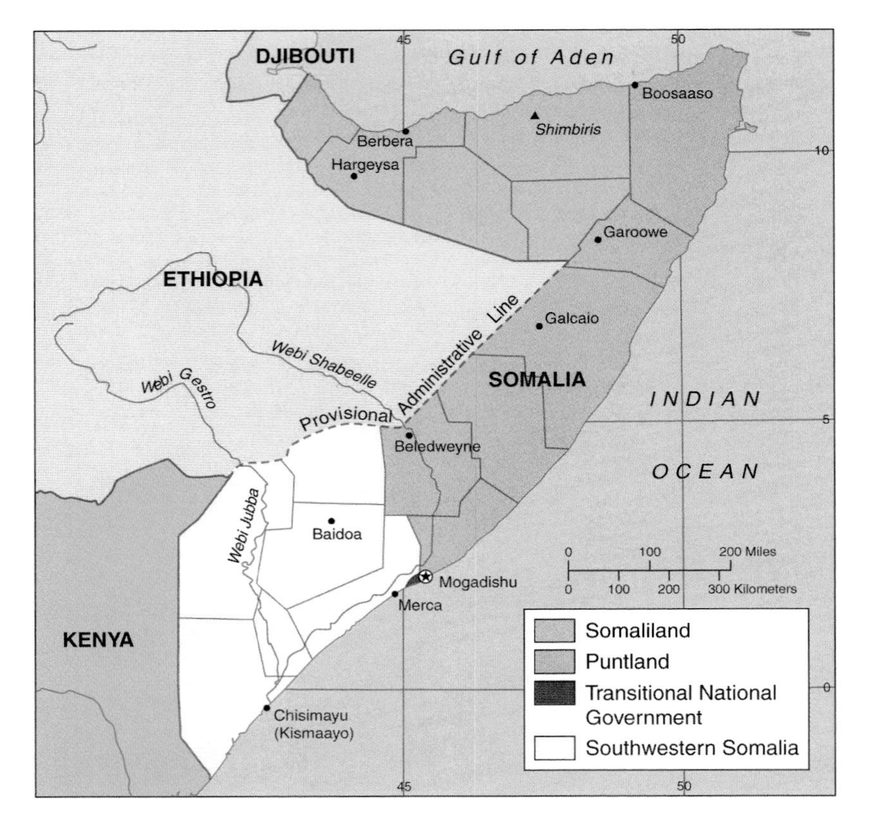

Somalia: With the ouster of the government led by Mohamed Said Barre in January 1991, turmoil, factional fighting, and anarchy have followed in Somalia, with several separate governments arising in different parts of the country. The northern clans declared an independent Republic of Somaliland. which, although not recongnized by any government, has maintained a stable existence, aided by the overwhelming dominance of a ruling clan and economic infrastucture left behind by British, Russian, and American military assistance programs. Puntland, the central portion of Somalia, from the Horn of Africa to the coast of the Indian Ocean and the border with Ethiopia, has been a self-governing autonomous state since 1998. The area of Southwestern Somalia is poorly organized and more conflict-ridden than any other part of the country, with much of that conflict centered on attempts to control the nomial Somalia capital of Mogadishu and ongoing famine. UN humanitarian efforts were unable to either quell guerilla activity or alleviate famine and the UN withdrew in 1995. In 2004 a new government, the Transitional Federal Government (TFG), was created for the entire country. The president and parliament of the new government have not yet moved to Mogadishu and discussions regarding the government's establishment are ongoing in Kenya. Numerous warlords and factions are still fighting for control of the capital city as well as for other southern regions.

Afghanistan: In the aftermath of the September 11, 2001, terrorist attacks on the World Trade Center and the Pentagon, the United States (backed to varying degrees by its allies) has declared a massive and global "war on terrorism" and any states that may provide "safe harbor" to terrorists. To date, the most prominent target of this U.S. declaration of war has been the Taliban regime of Islamic extremists who controll about 95 percent of the territory of Afghanistan. International observers believe that the Taliban regime has welcomed and provided a base for the al-Qaeda terrorist network dominated by Saudi expatriate and millionaire Osama bin Laden since the late 1990s. As a result of this intelligence, the U.S. and Britain pursued a daily bombardment of key al-Qaeda and Taliban installations inside Afghanistan for several months, with key logistical support in the form of air bases and supply depots provided by the government of Pakistan (in exchange for financial considerations and political support of the non-elected Pakistani government). U.S. and allied ground troops, aided by members of the Northern Alliance of Afghan rebels opposed to the Taliban regime, expelled the Taliban government in 2002. While now under home rule following free elections, Afghanistan still has within its borders a NATO military presence, dominated by U.S. troops. There are still pockets of resistance by Taliban and al-Qaeda forces in the forests and mountains along the border with Pakistan. NATO forces are still searching for the supposed perpetrator of the World Trade Center attack, Osama Bin Laden, in that area.

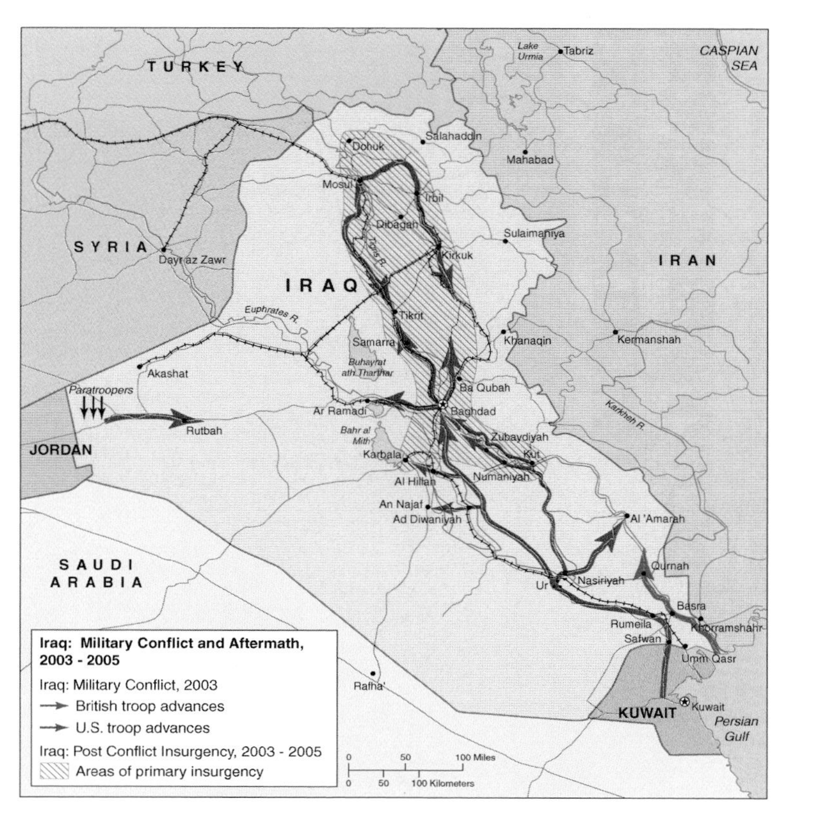

Iraq: Military Conflict, 2003: Following the failure of the United States and allies Great Britain and Spain to secure approval from the United Nations to begin a UN-sponsored military conflict to "disarm" Iraq, the United States and its allies launched an independent military attack on Iraq in April 2003. The military campaign began with massive air and sea bombardments on government and military targets, then ground troops moved from Kuwait to secure oil fields and major urban areas. By early May 2003, virtually all of the country was under the control of the U.S.-led coalition of forces. The Iraqi military, for the most part, melted away into the general civilian population and there were no major battles for territory. A period of insurgency began with predominantly Sunni Muslim insurgents acting against the American and British forces who, by mid-2004, represented over 98% of the coalition forces in Iraq. Sunni insurgents also directed attacks toward the Shi'ite and Kurdish poulations who were loosely allied in the elections of January 2005 (which were largely boycotted by Sunni Muslims). Although an elected Iraqi government was in place early in 2005, insurgencies continued with significant casualities among both the Iraqi citizenry and U.S. military forces. U.S. military and political officials estimate that an American-British troop presence will be required in Iraq for the foreseeable future.

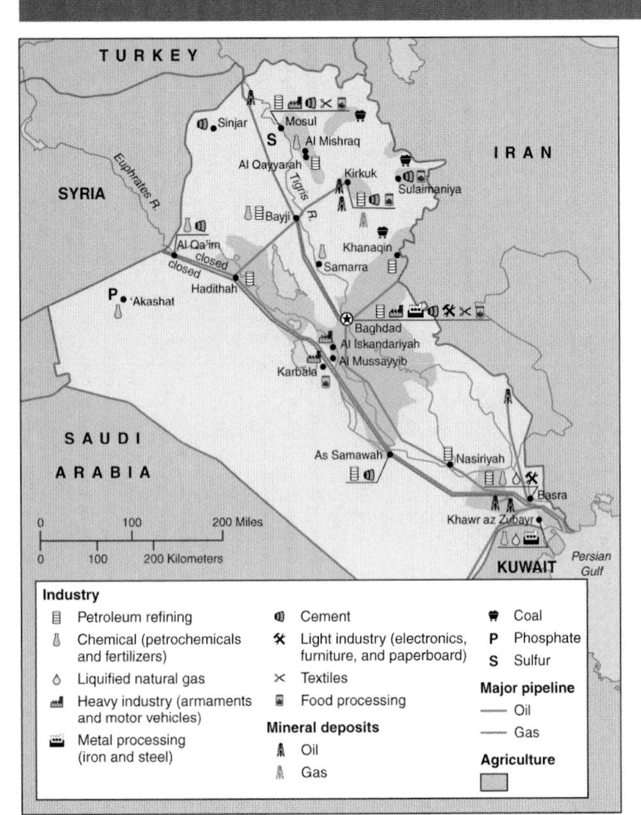

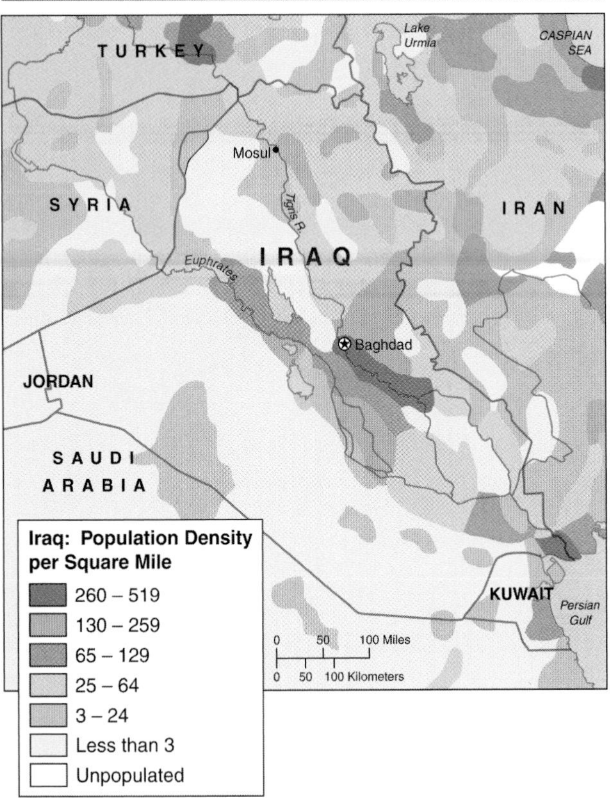

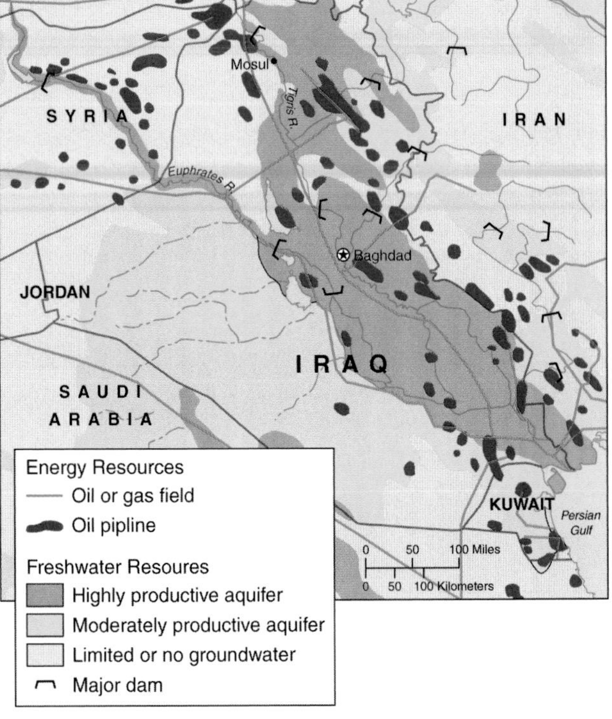

Iraq: Prior to the 1990–91 invasion of Kuwait by Iraq and the subsequent United Nations coalition's military expulsion of Iraq from its neighbor, Iraq was one of the most prosperous countries in the Middle East and the only one with full capacity to feed itself, even without the vast oil revenues generated by the country's immense reserves. Despite the inefficiencies of the Baathist dictatorship of Saddam Hussein, the country has a solid agricultural base and a burgeoning industry. The combination of military adventurism and conflict, in the form of a lengthy war with Iran and the ill-advised invasion of Kuwait, limited further economic development, however. Development was also problematic given the country's internal tensions between Arabic Sunni Muslims and Arabic Shiite Muslims, and between Arabs and Kurds and a few other minority populations in the northern parts of the country.

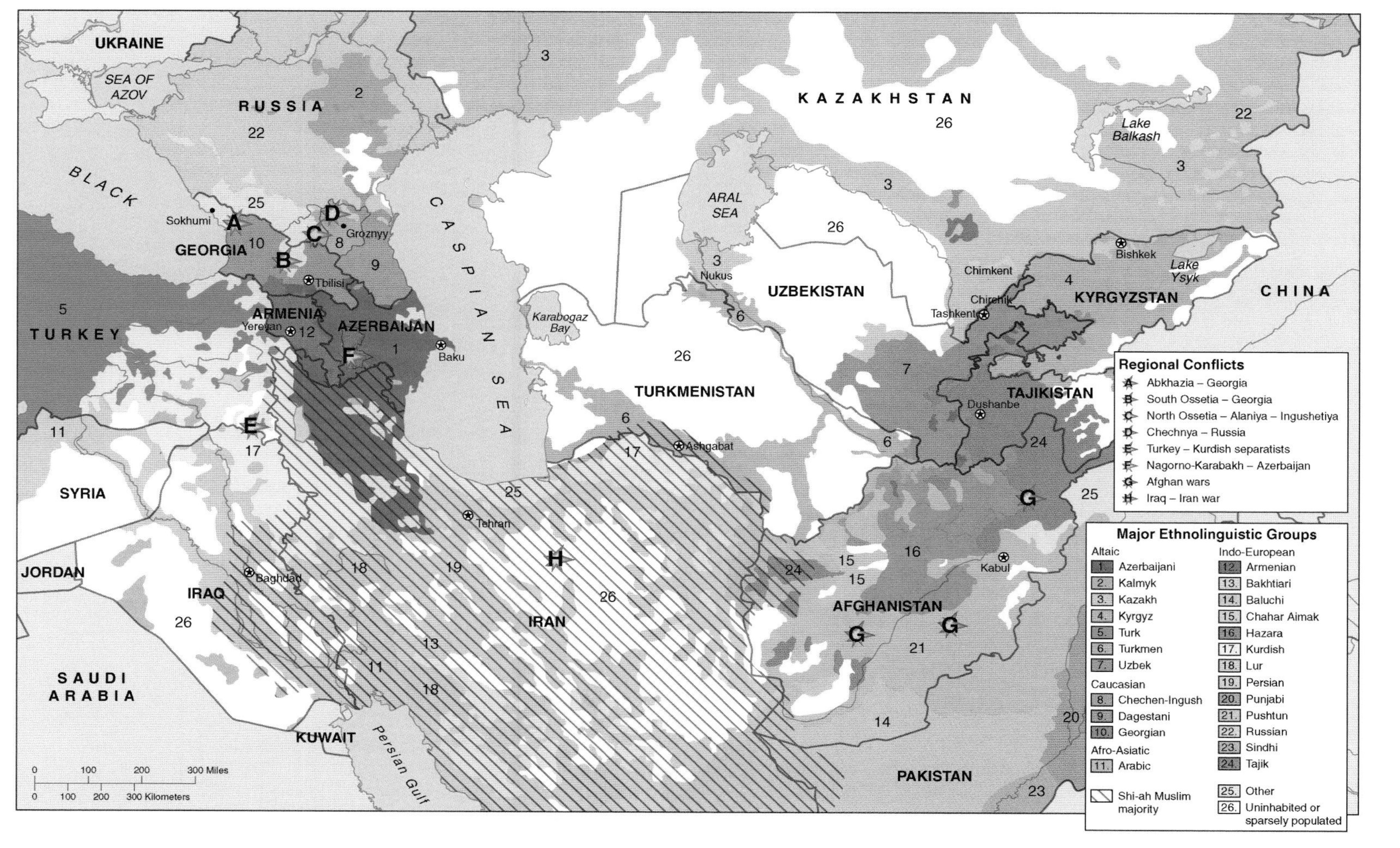

South-Central Eurasia: The area of south-central Eurasia represents what is perhaps the world's most volatile area in terms of potential military conflict. A crazy-quilt of ethnic and linguistic groups, this region contains politically-defined "states," but nothing approaching "nation-states" as they are understood elsewhere. Even reasonably well-consolidated states, such as Iran, are hampered by the mixture of languages and ethnic populations within their borders. For some countries, such as Afghanistan, the ethnolinguistic mix is so historically fixed as to render any attempts at modern state building nearly hopeless. This vast area—although nearly universally Muslim—is also split between the two primary Islamic sects, Sunni and Shia, with several smaller sectarian divisions as well. The division of one of the world's global religions into two principal factions occurred in the seventh century, originally over the question of the source of authority in the religious hierarchy. But it has since come to be a theological and metaphysical separation, and Sunni and Shia Muslims now bear somewhat the same relationship to one another as did Catholics and Protestants in seventeenth- and eighteenth-century Europe—a mutual antipathy that periodically flares into civil unrest and conflict that goes far beyond the bounds of religious debate. The area is rich in natural resources, but the mixture of ethnicity, language, and religion has and will continue to produce the human conflicts that inhibit human development.

Map 31 International Terrorism Incidents, 1998–2003

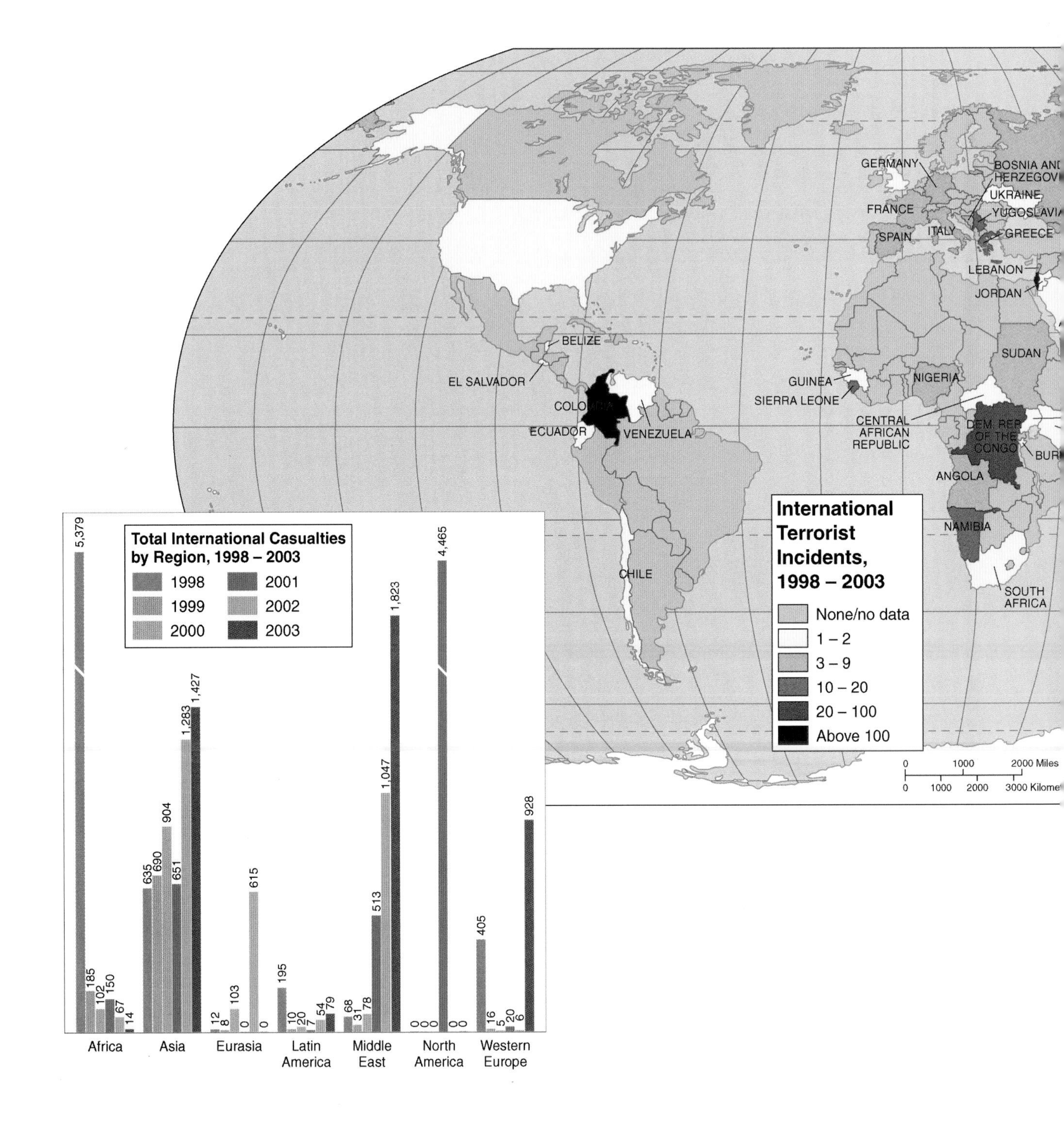

Americans have made a mantra of the saying "the world has changed" as a consequence of the terrorist attacks on the World Trade Center and the Pentagon on September 11, 2001. As the map above and the accompanying graphs of terrorist activities before and after 9/11/01 indicate, however, the world did not change, although the focus of a major terrorist attack shifted from Africa, Asia, and the Middle East to North America. Many other areas of the world have lived with terrorism and terrorist activity for years. In 2000 and 2001, despite the enormous losses in the United States in the 9/11 attacks, more lives were lost in Asia and Africa as a result of terrorism than were lost in North America. The world did not change, but Americans' perception of that world and their place in it has certainly

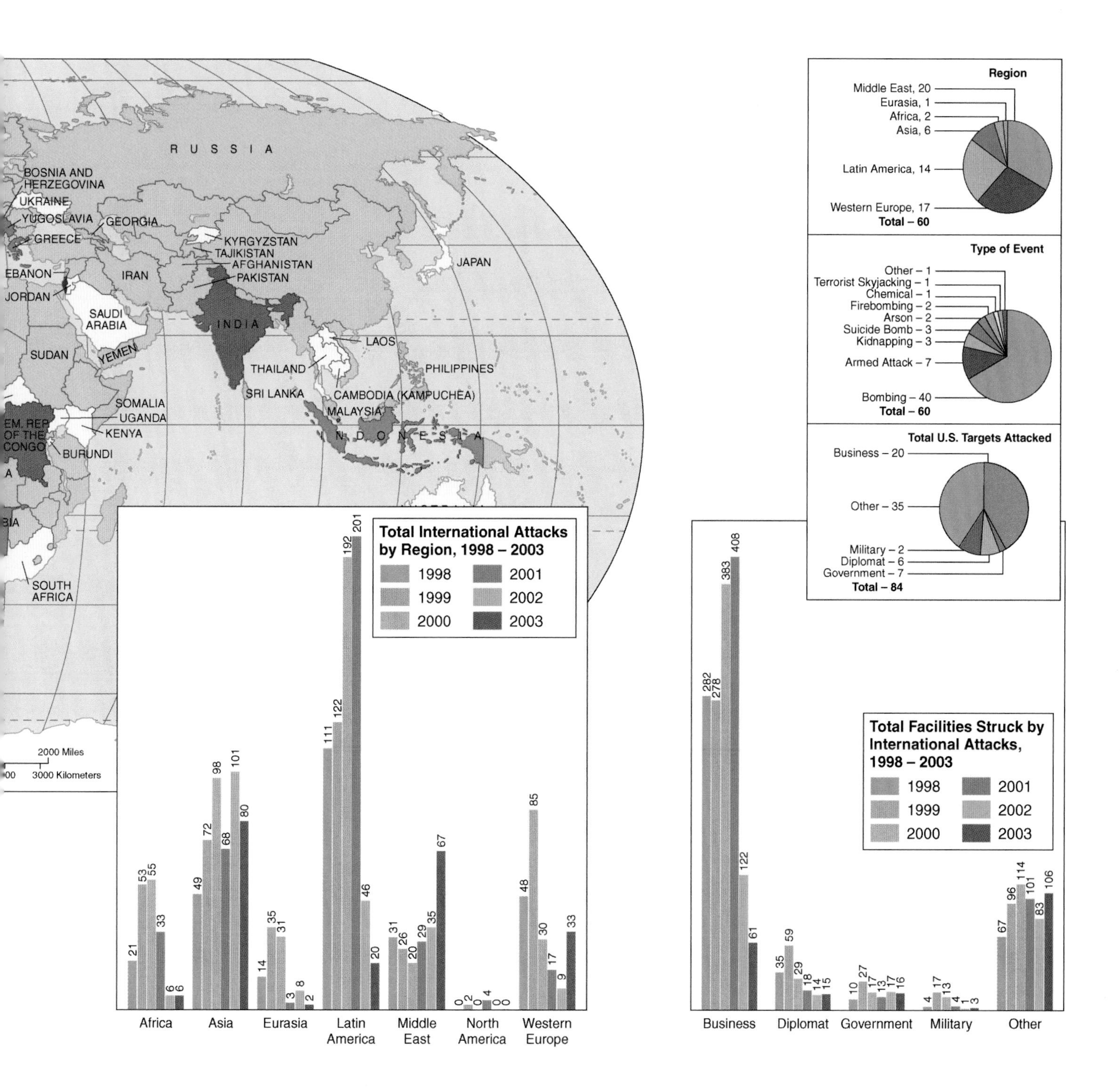

changed. Events subsequent to 9/11 indicate a shift back to more "normal" patterns in North America in 2002, and 2003 showed no terrorist activities—in sharp contrast to many other world regions, including Europe. Two qualifiers need to be added regarding the data presented: the term "casualties" refers to dead and wounded (the number of the dead represent a relatively small percentage of total casualities by region) the term "terrorist attacks" does not include isurgency events in areas of military conflict and, therefore, do not include activities carried out against U.S. forces or others in places like Afghanistan or Iraq where military action continued throughout the time frame of the data.

Map 32 The Political Geography of a Global Religion: The Islamic World

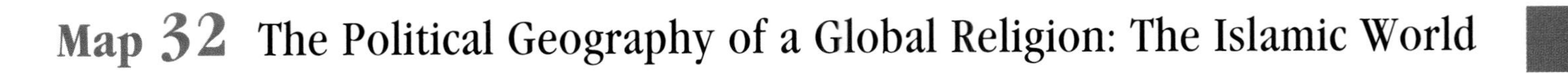

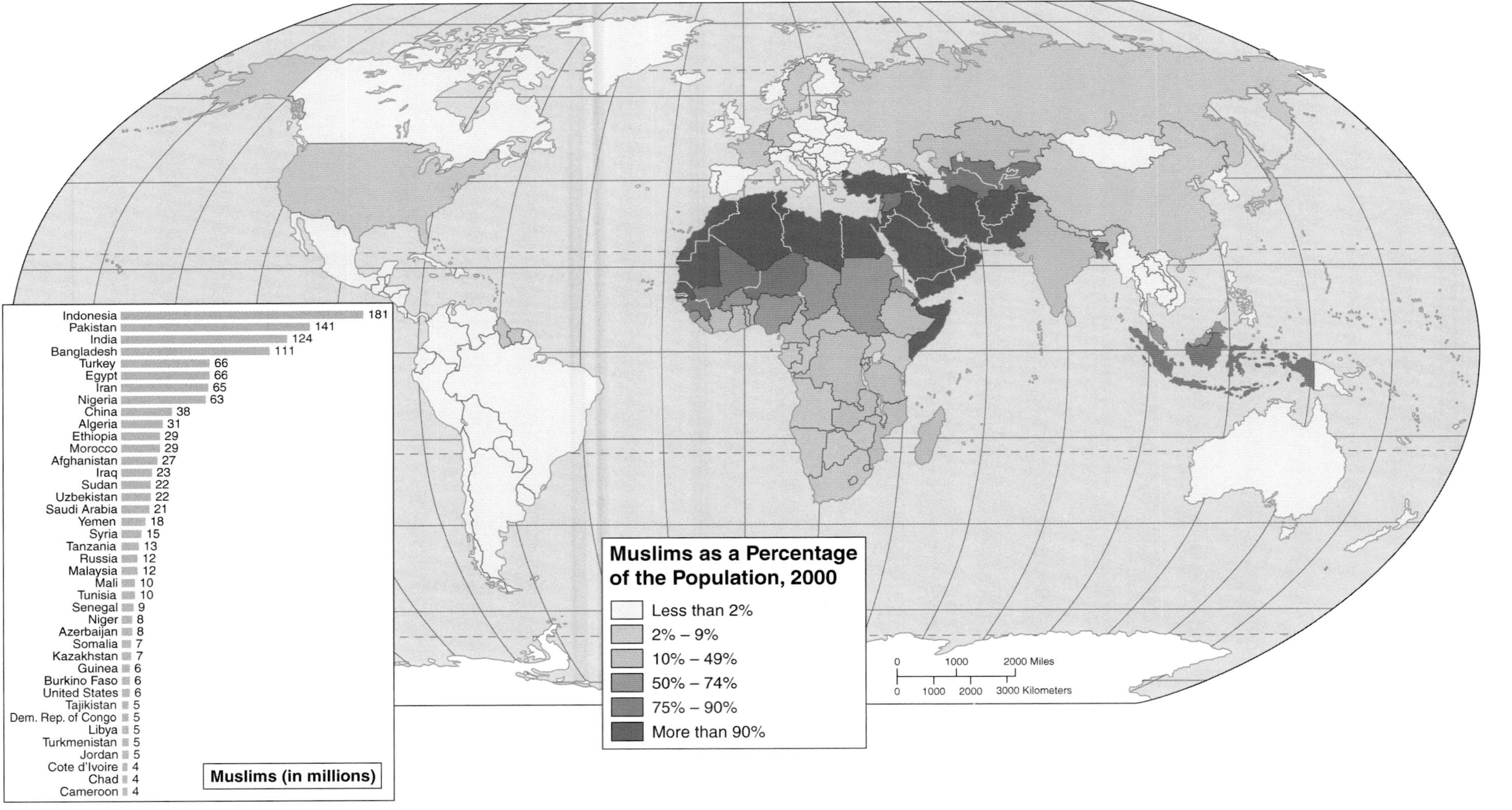

Islam, as a religion, does not promote conflict. The term *jihad*, often mistranslated to mean "holy war," in fact refers to the struggle to find God and to promote the faith. In spite of the beneficent nature of Islamic teachings, the tensions between Muslims and adherents of other faiths often flare into warfare. A comparison of this map with the map of international conflict will show a disproportionate number of wars in that portion of the world where Muslims are either majority or significant minority populations. The reasons for this are based more in the nature of government, cultures, and social structure, than in the tenets of the faith of Islam. Nevertheless, the spatial correlations cannot be ignored. Similarly, terrorist incidents falling considerably short of open armed warfare, are spatially consistent with the distribution of Islam and even more consistent with the presence of Islamic fundamentalism or "Islamism," which tends to be less tolerant and more aggressive than the mainstream of the religion. Terrorism is also consistent with those areas where the legacy of colonialism or the persistent presence of non-Islamic cultures intrude into the Islamic world.

Map 33 Nations With Nuclear Weapons

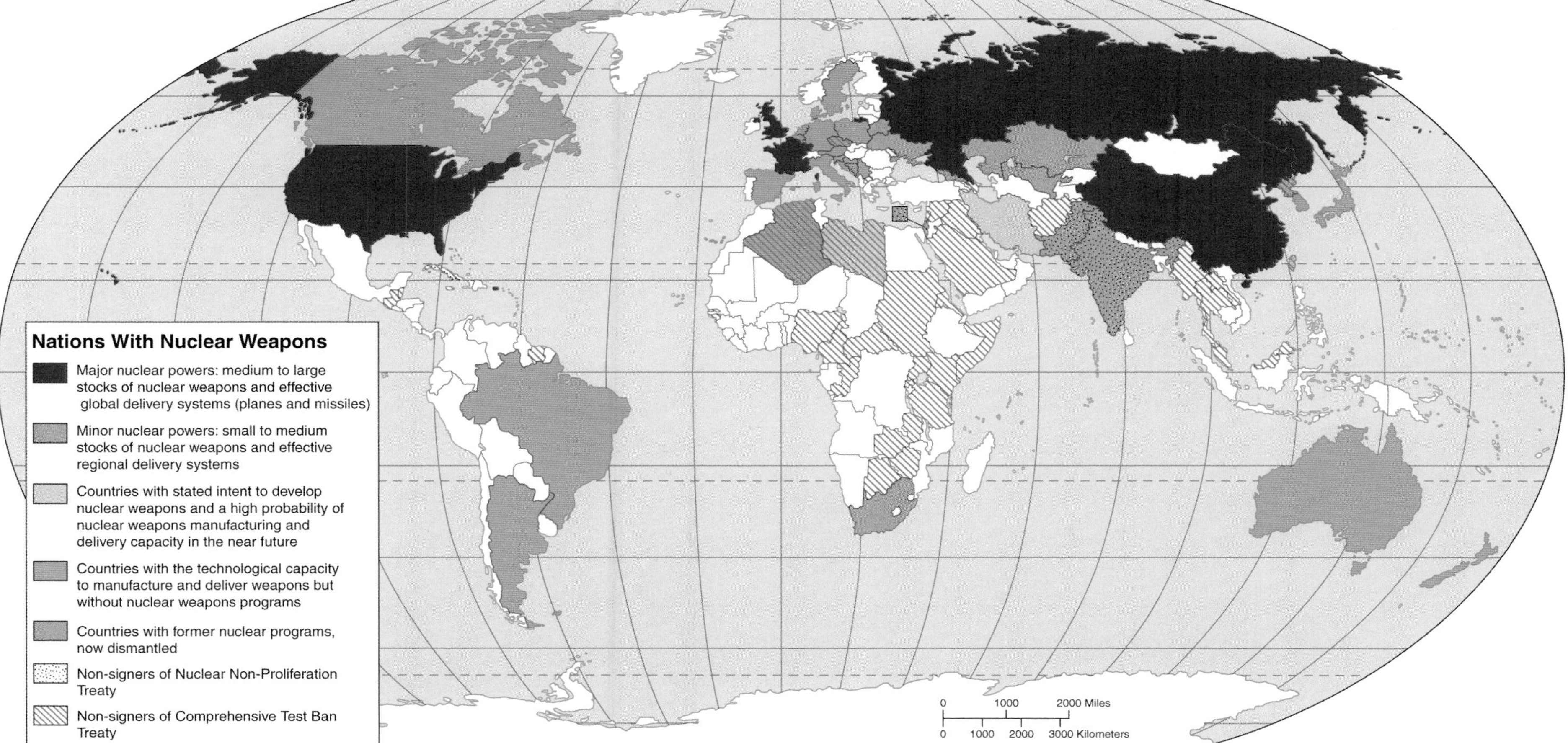

Since 1980, the number of countries possessing the capacity to manufacture and deliver nuclear weapons has grown dramatically, increasing the chances of accidental or intentional nuclear exchanges. In addition to the traditional nuclear powers of the United States, Russia, China, the United Kingdom, and France, must now be added Israel, India, and Pakistan as countries that, without possessing the large stocks of weapons of the major powers, nor the extensive delivery systems of the United States and Russia, still have effective regional (and possibly global) delivery systems and medium stocks of warheads. Countries such as Kazakhstan, Ukraine, Georgia, and Belarus that were created out of what had been the Soviet Union did have some nuclear capacity in the 1991–1995 period but have since had all nuclear weapons removed from their territories. However, North Korea has recently announced the re-establishment of its suspended nuclear weapons programs and may possess a small stock of nuclear warheads, along with the capacity to deliver those weapons regionally. Both Iran and Libya have nuclear ambitions, as did Iraq until the overthrow of the Baathist regime of Saddam Hussein by a U.S.-led military coalition in 2003. The proliferation of nuclear states threatens global security, and the objective of the Nuclear Non-Proliferation Treaty was to reduce the chances for expanding nuclear arsenals worldwide. This treaty has been partially successful in that a number of countries in the developed world certainly have the capacity to manufacture and deliver nuclear weapons but have chosen not to do so. These countries include Canada, European countries other than the United Kingdom and France, South Korea, Japan, Australia, and New Zealand, and Brazil and Argentina in South America. On the other side of the coin, the intent of North Korea to emerge as a nuclear power may force countries such as South Korea and Japan to re-think their positions as non-nuclear countries.

Map 34 Size of Armed Forces

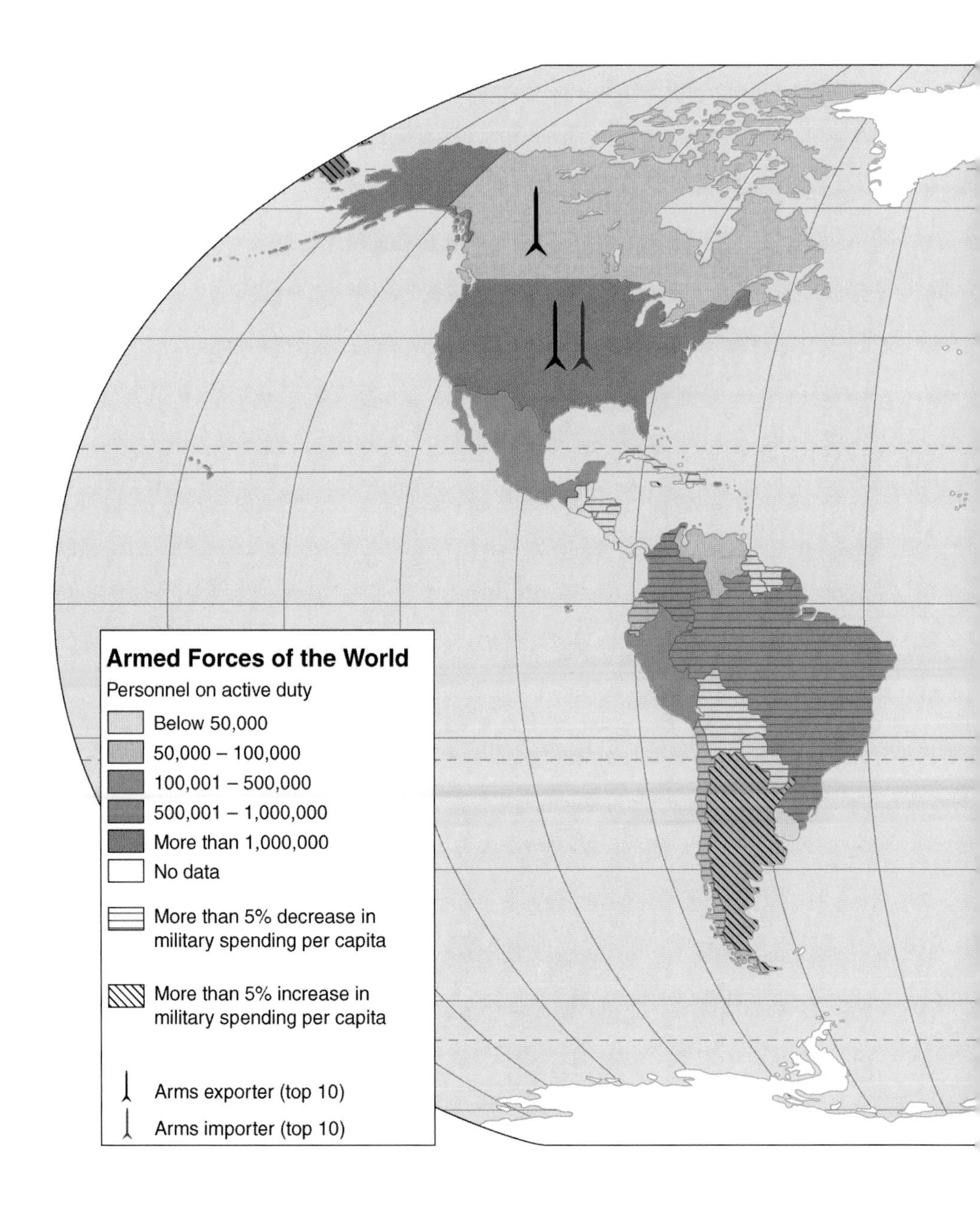

While the size of a country's armed forces is still an indicator of national power on the international scene, it is no longer as important as it once was. The increasing high technology of military hardware allows smaller numbers of military personnel to be more effective. There are some countries, such as China, with massive numbers of military personnel but with relatively limited military power because of a lack of modern weaponry. Additionally, the use of rapid transportation allows personnel to be deployed about the globe or any region of it quickly; this also increases effectiveness of highly trained and well-armed smaller military units. Nevertheless, the world is still a long way from the pre-

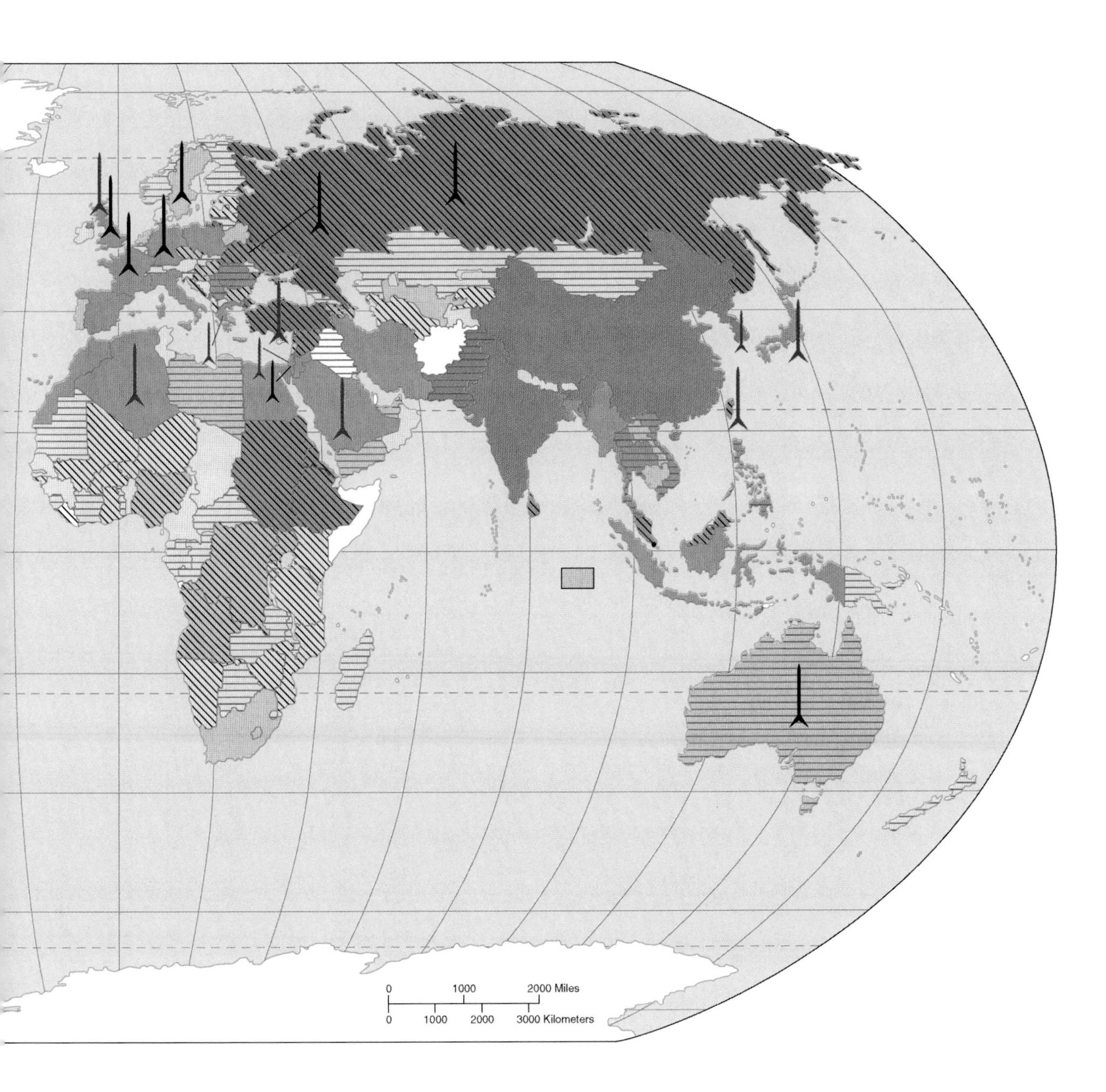

dicted "push-button warfare" that many experts have long anticipated. Indeed, the pattern of the last few years has been for most military conflicts to involve ground troops engaged in fairly traditional patterns of operation. Even in the Persian Gulf conflict, with its highly-publicized "smart bombs," the bulk of the military operation that ended the conflict was carried out by infantry and armor operating on the ground and supported by traditional air cover using conventional weaponry. Thus, while the size of a country's armed forces may not be as important as it once was, it is still a major factor in measuring the ability of nations to engage successfully in armed conflict.

Map 35 Military Expenditures as a Percentage of Gross National Product

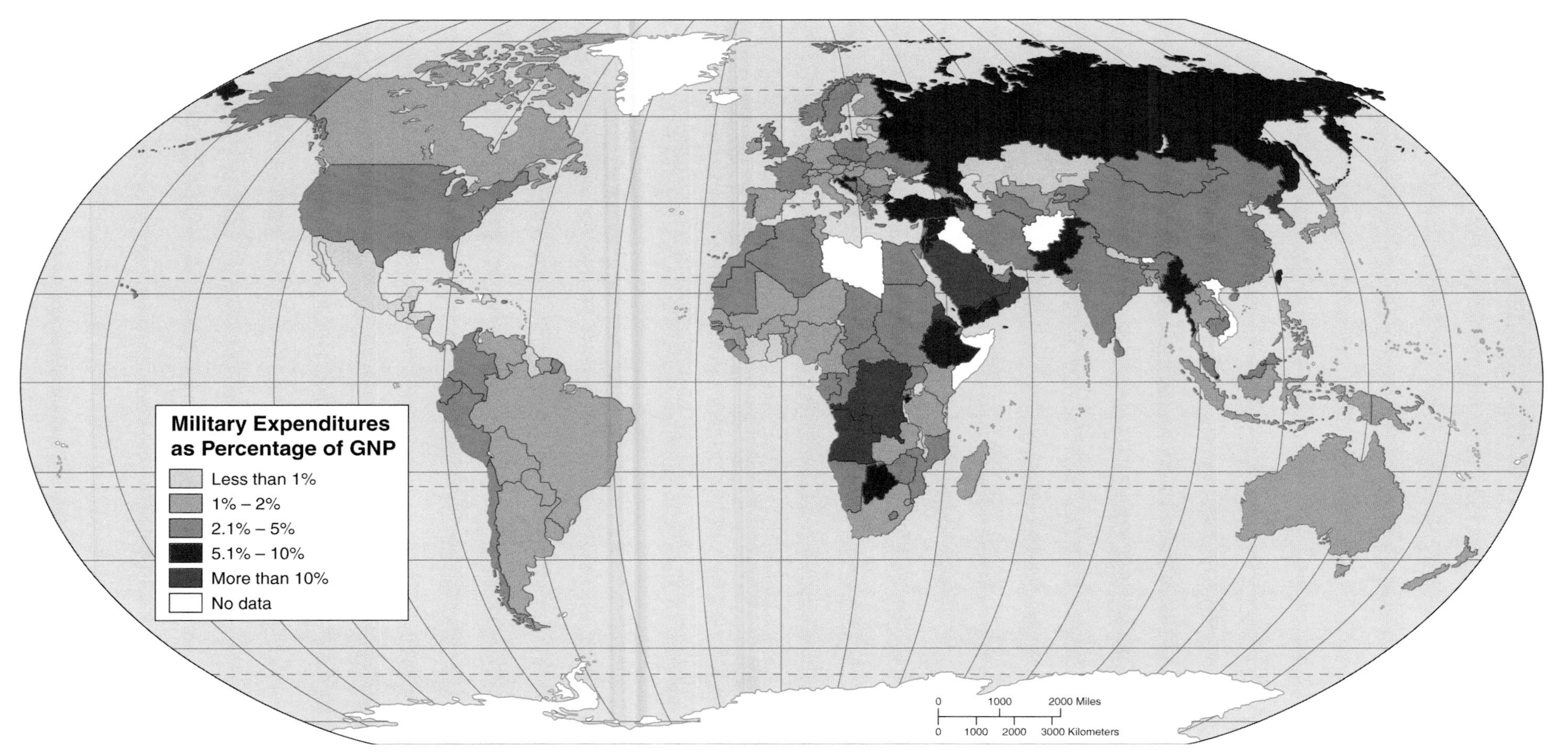

Many countries devote a significant proportion of their total central governmental expenditures to defense: weapons, personnel, and research and development of military hardware. A glance at the map reveals that there are a number of regions in which defense expenditures are particularly high, reflecting the degree of past and present political tension between countries. The clearest example is the Middle East. The steady increase in military expenditures by developing countries is one of the most alarming (and least well known) worldwide defense issues. Where the end of the cold war has meant a substantial reduction of military expenditures for the countries in North America and Europe and for Russia, in many of the world's developing countries military expenditures have risen between 15 percent and 20 percent per year for the past few years, averaging out to 7.5 percent per year for the past quarter century. Even though many developing countries still spend less than 5 percent of their gross national product on defense, these funds could be put to different uses in such human development areas as housing, land reform, health care, and education.

Part III

The Global Economy

Map 36 Membership in the World Trade Organization

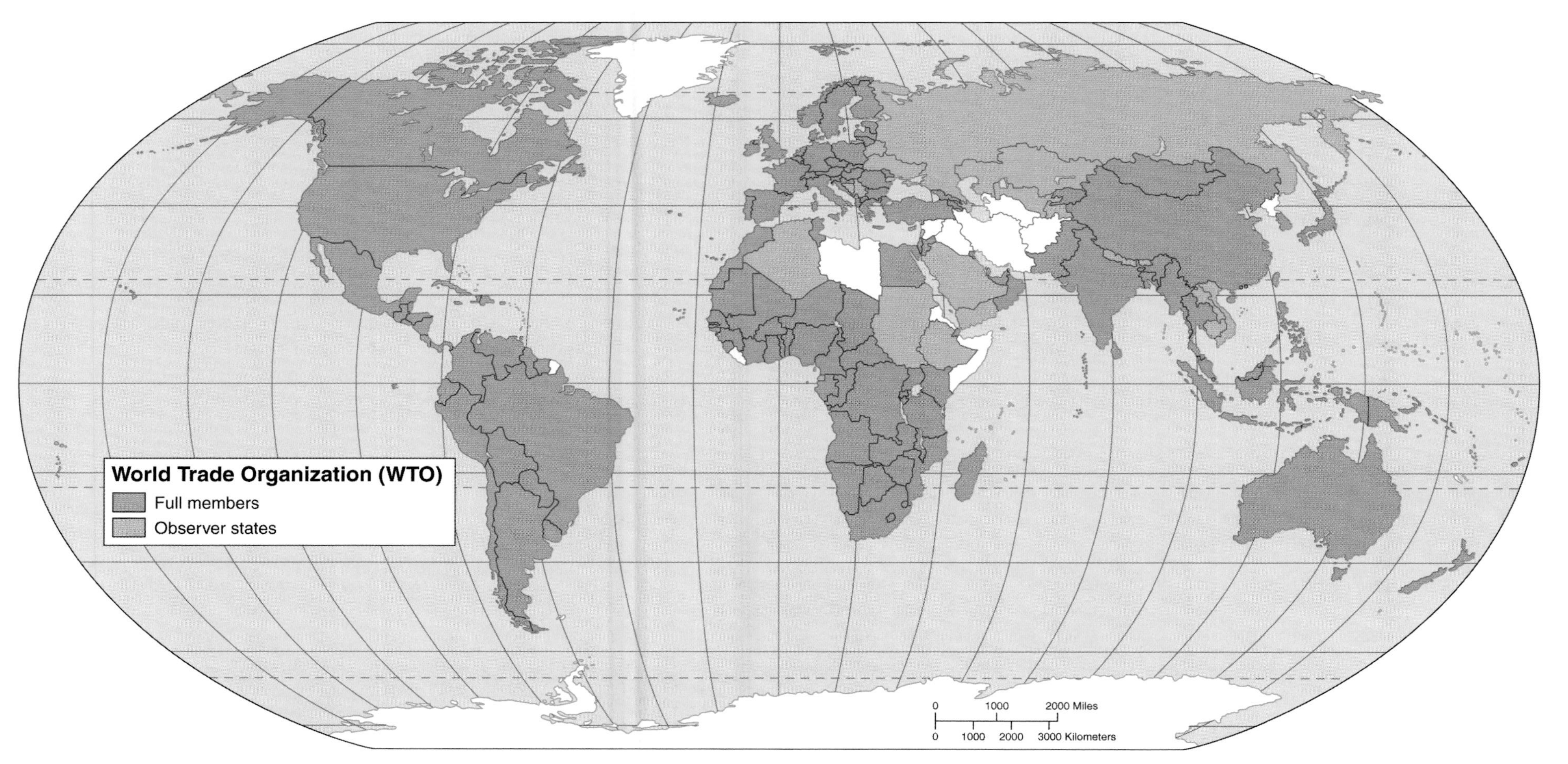

After World War II, the General Agreement on Tariffs and Trade (GATT) sponsored several rounds of negotiations, especially related to lower tariffs but also considering issues such as dumping and other nontariff questions. The last round of negotiations under GATT took place in Uruguay in 1986–1994 and set the stage for the World Trade Organization (WTO), which was formally established in 1995. Today the WTO has 146 members with more than 30 "observer" governments. With the exception of the Holy See (Vatican), observer governments are expected to begin negotiations for full membership within five years of becoming observers. The objective of the WTO is to help international trade flow smoothly and fairly and to assure more stable and secure supplies of goods to consumers. To this end, it administers trade agreements, acts as a forum for trade negotiations, settles trade disputes, reviews national trade policies, assists developing countries through technical assistance and training programs, and cooperates with other international organizations. Increased globalization of the world's economy makes the administrative role of the WTO of increasing importance in the twenty-first century. The headquarters of the WTO is in Geneva, Switzerland.

Map 37 Regional Trade Organizations

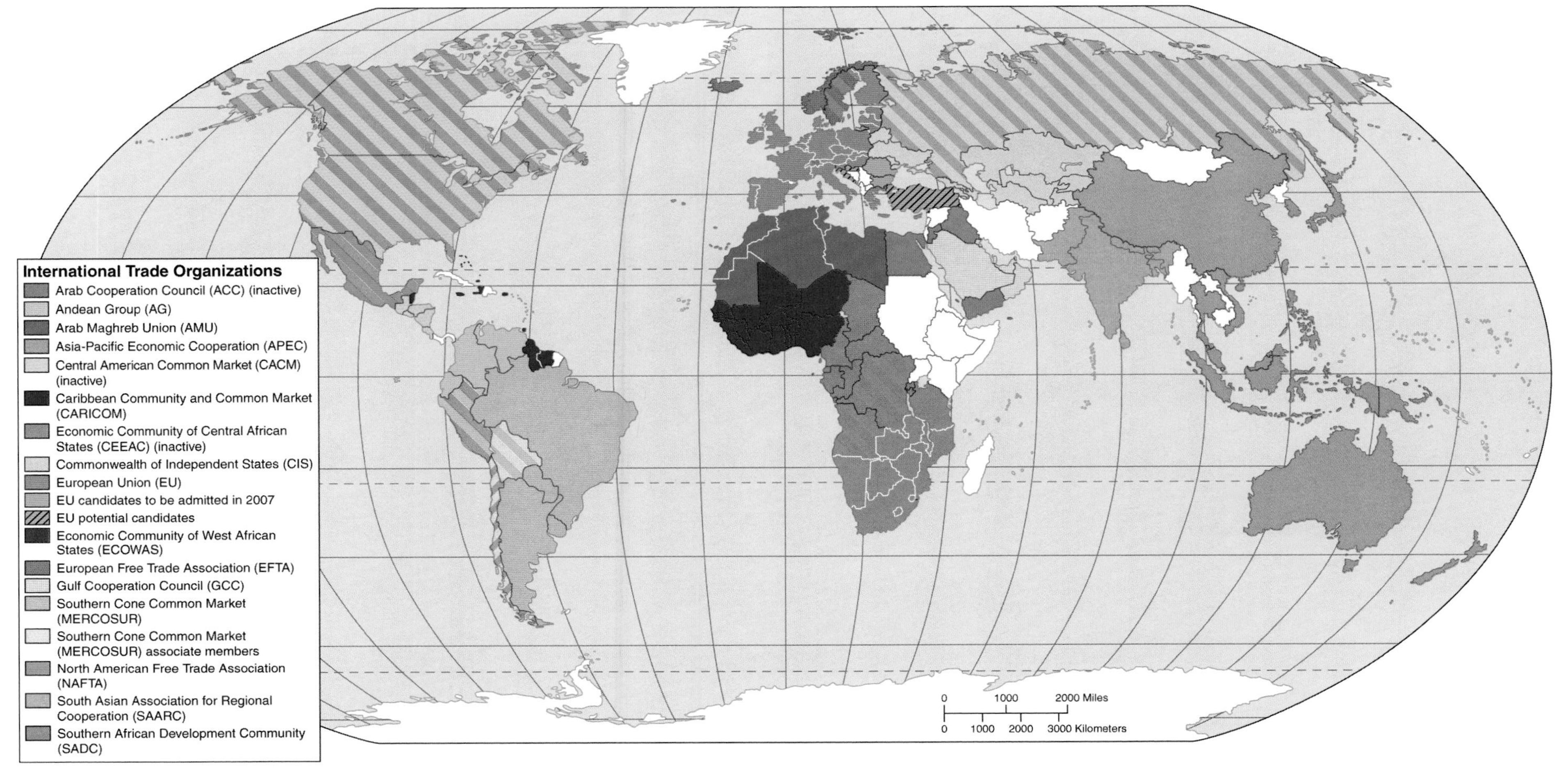

One of the most pervasive influences in the global economy over the last half century has been the rapid rise of international trade organizations. Pioneered by the European Economic Community, founded in part to assist in rebuilding the European economy after World War II, these organizations have become major players in global movements of goods, services, and labor. Some have integrated to form financial and political unions, such as the European Community, which has grown into the European Union. Others, like the Commonwealth of Independent States, are attempts to hold on to the remnants of better economic times. Still others, like the Asia-Pacific Economic Cooperation, are infant organizations that incorporate vastly different regions, states, and even economic systems, and are attempts to anticipate the direction of future economic growth. The role of international trade organizations is likely to grow greater in the twenty-first century. Witness the emergence of the North American Free Trade Agreement (NAFTA) that has restructured many segments of the Canadian, American, and Mexican economies (particularly in Mexico) and the probability of the adoption of the Central American Free Trade Agreement (CAFTA) that would link more closely the economies of the United States, El Salvador, Nicaragua, Guatemala, Honduras, Costa Rica, and, possibly, The Dominican Republic.

Map 38 Gross National Income Per Capita

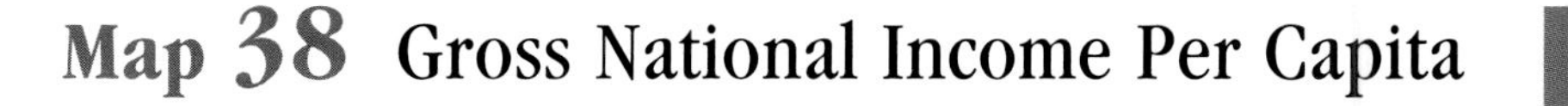

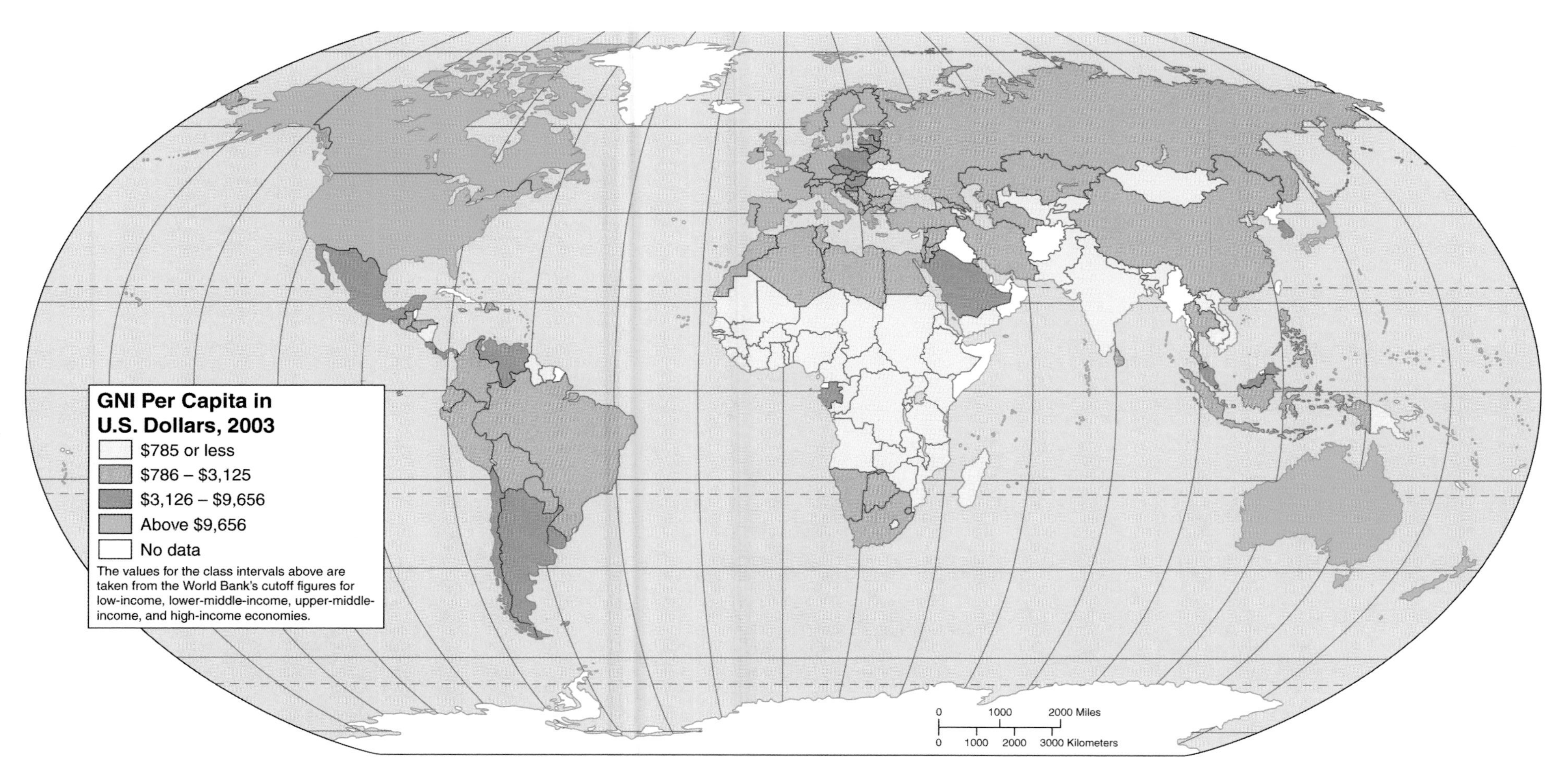

Gross National Income in either absolute or per capita form should be used cautiously as a yardstick of economic strength because it does not measure the distribution of wealth among a population. There are countries (most notably, the oil-rich countries of the Middle East) where per capita GNI is high but where the bulk of the wealth is concentrated in the hands of a few individuals, leaving the remainder in poverty. Even within countries in which wealth is more evenly distributed (such as those in North America or Western Europe), there is a tendency for dollars or pounds sterling or francs or marks to concentrate in the bank accounts of a relatively small percentage of the population. Yet the maldistribution of wealth tends to be greatest in the less developed countries, where the per capita GNI is far lower than in North America and Western Europe, and poverty is widespread. In fact, a map of GNI per capita offers a reasonably good picture of comparative economic well-being. It should be noted that a low per capita GNI does not automatically condemn a country to low levels of basic human needs and services. There are a few countries, such as Costa Rica and Sri Lanka, that have relatively low per capita GNI figures but rank comparatively high in other measures of human well-being, such as average life expectancy, access to medical care, and literacy.

Map 39 Relative Wealth of Nations: Purchasing Power Parity

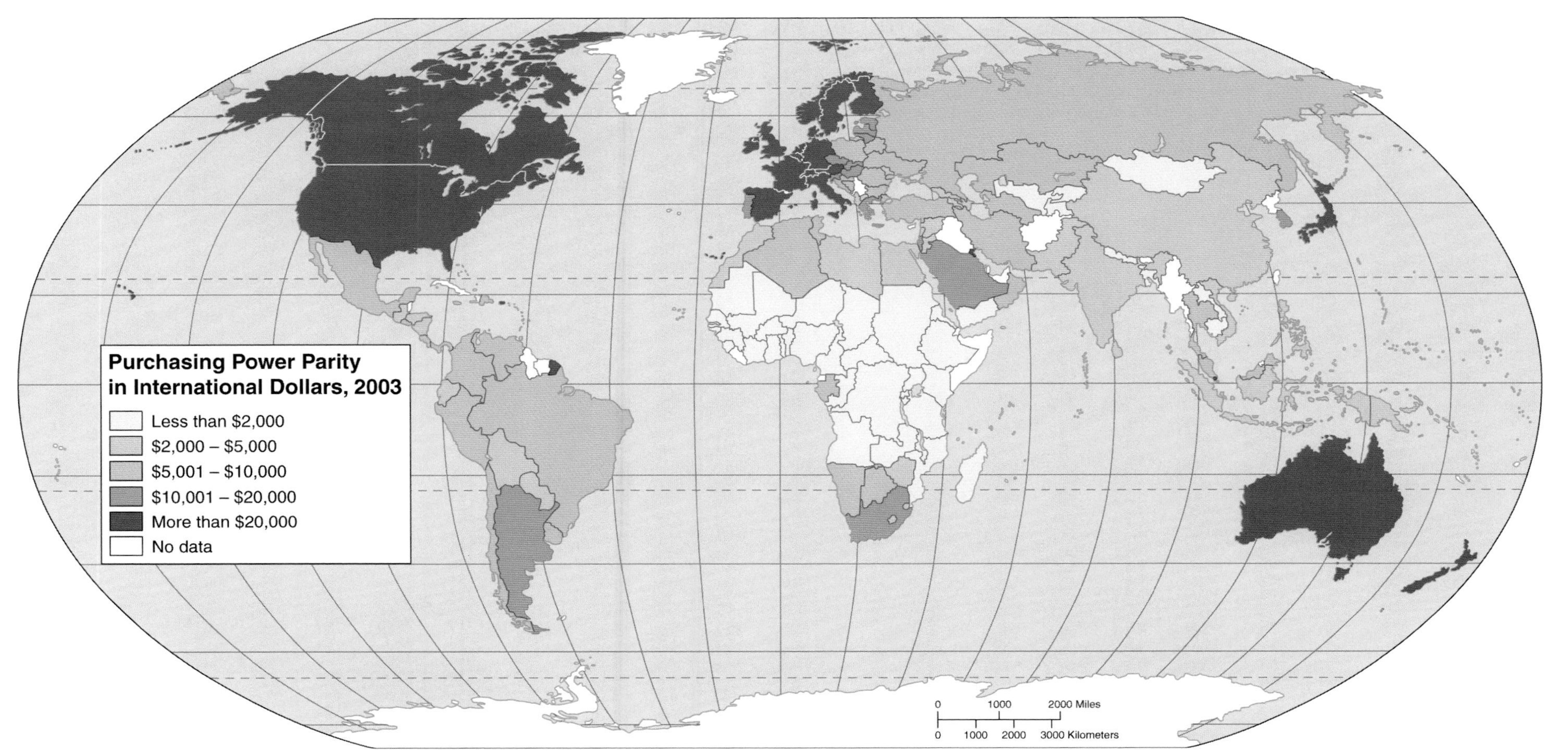

Of all the economic measures that separate the "haves" from the "have-nots," perhaps per capita Purchasing Power Parity (PPP) is the most meaningful. While per capita figures can mask significant uneven distributions within a country, they are generally useful for demonstrating important differences between countries. Per capita GNP and GDP (Gross Domestic Product) figures, and even per capita income, have the limitation of seldom reflecting the true purchasing power of a country's currency at home. In order to get around this limitation, international economists seeking to compare national currencies developed the PPP measure, which shows the level of goods and services that holders of a country's money can acquire locally. By converting all currencies to the "international dollar," the World Bank and other organizations using PPP can now show more truly comparative values, since the new currency value shows the number of units of a country's currency required to buy the same quantity of goods and services in the local market as one U.S. dollar would buy in an average country. The use of PPP currency values can alter the perceptions about a country's true comparative position in the world economy. More than per capita income figures, PPP provides a valid measurement of the ability of a country's population to provide for itself the things that people in the developed world take for granted: adequate food, shelter, clothing, education, and access to medical care. A glance at the map shows a clear-cut demarcation between temperate and tropical zones, with most of the countries with a PPP above $5,000 in the midlatitude zones and most of those with lower PPPs in the tropical and equatorial regions. Where exceptions to this pattern occur, they usually stem from a tremendous maldistribution of wealth among a country's population.

Map 40 International Flows of Capital

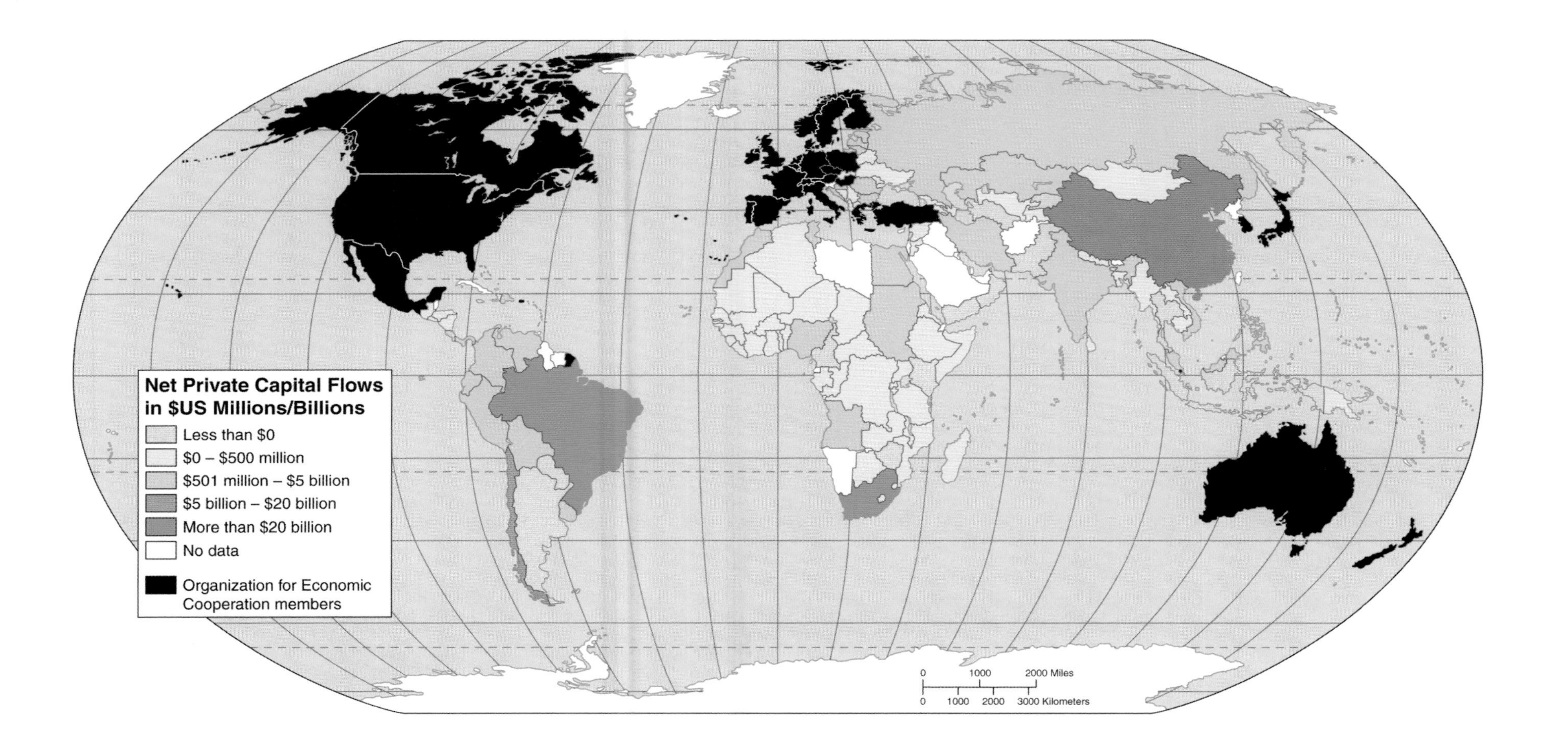

International capital flows include private debt and nondebt flows from one country to another, shown on the map as flows into a country. Nearly all of the capital comes from those countries that are members of the Organization for Economic Cooperation and Development (OECD), shown in black on the map. Capital flows include commercial bank lending, bonds, other private credits, foreign direct investment, and portfolio investment. Most of these flows are indicators of the increasing influence developed countries exert over the developing economies. Foreign direct investment or FDI, for example, is a measure of the net inflow of investment monies used to acquire long-term management interest in businesses located somewhere other than in the economy of the investor. Usually this means the acquisition of at least 10 percent of the stock of a company by a foreign investor and is, then, a measure of what might be termed "economic colonialism": control of a region's economy by foreign investors that could, in the world of the future, be as significant as colonial political control was in the past. International capital flows have increased greatly in the last decade as the result of the increasing liberalization of developing countries, the strong economic growth exhibited by many developing countries, and the falling costs and increased efficiency of communication and transportation services.

Map 41 Economic Output per Sector

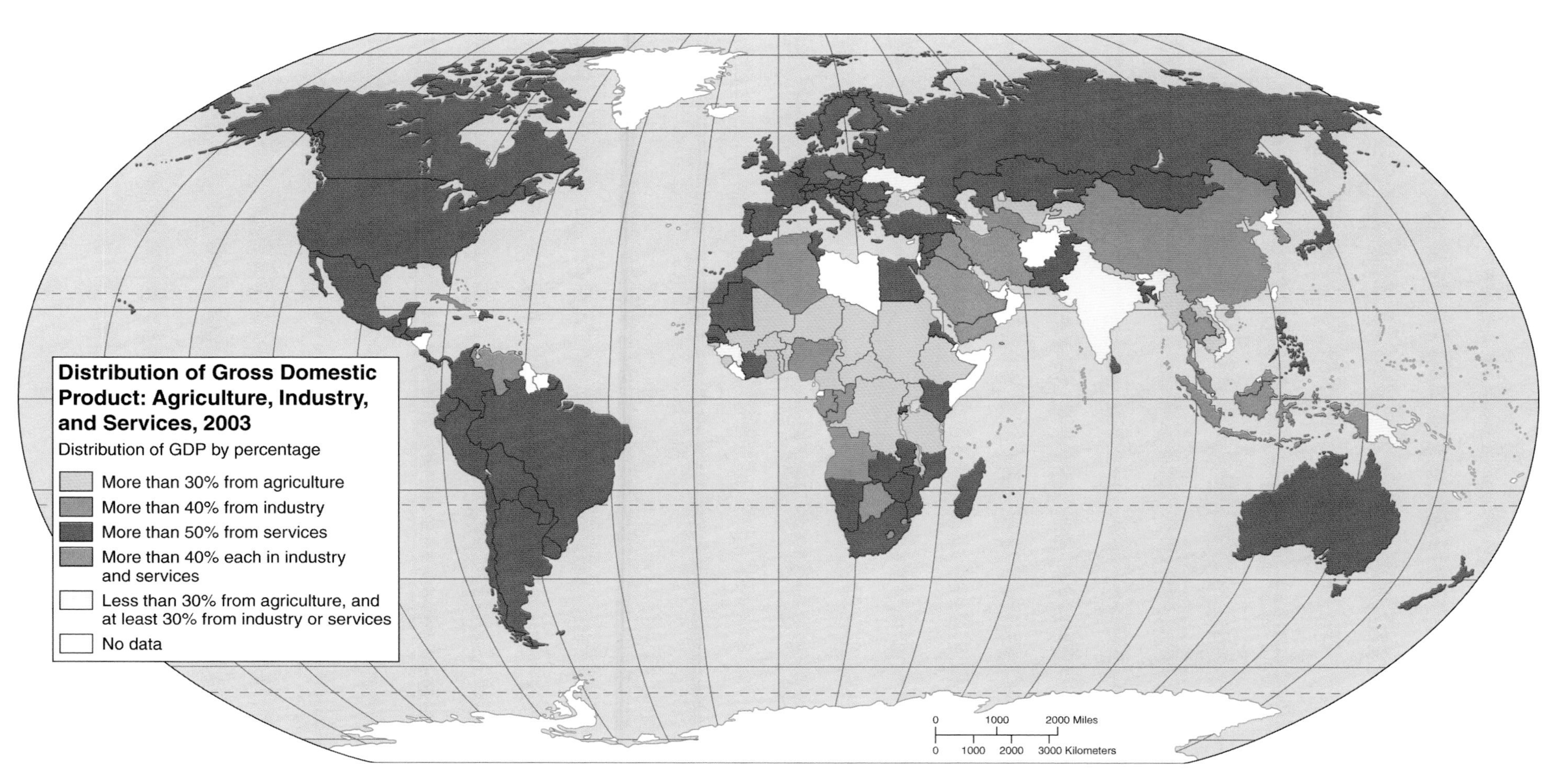

The percentage of the gross domestic product (the final output of goods and services produced by the domestic economy, including net exports of goods and nonfactor—nonlabor, noncapital—services) that is devoted to agricultural, industrial, and service activities is considered a good measure of the level of economic development. In general, countries with more than 40 percent of their GDP derived from agriculture are still in a "colonial dependency" economy—that is, raising agricultural goods primarily for the export market and dependent upon that market (usually the richer countries). Similarly, countries with more than 40 percent of GDP devoted to both agriculture and services often emphasize resource extractive (primarily mining and forestry) activities. These also tend to be "colonial dependency" countries, providing raw materials for foreign markets. Countries with more than 40 percent of their GDP obtained from industry are normally well along the path to economic development. Countries with more than half of their GDP based on service activities fall into two ends of the development spectrum. On the one hand are countries heavily dependent upon both extractive activities and tourism and other low-level service functions. On the other hand are countries that can properly be termed "postindustrial": they have already passed through the industrial stage of their economic development and now rely less on the manufacture of products than on finance, research, communications, education, and other service-oriented activities.

Map 42 Employment by Economic Activity

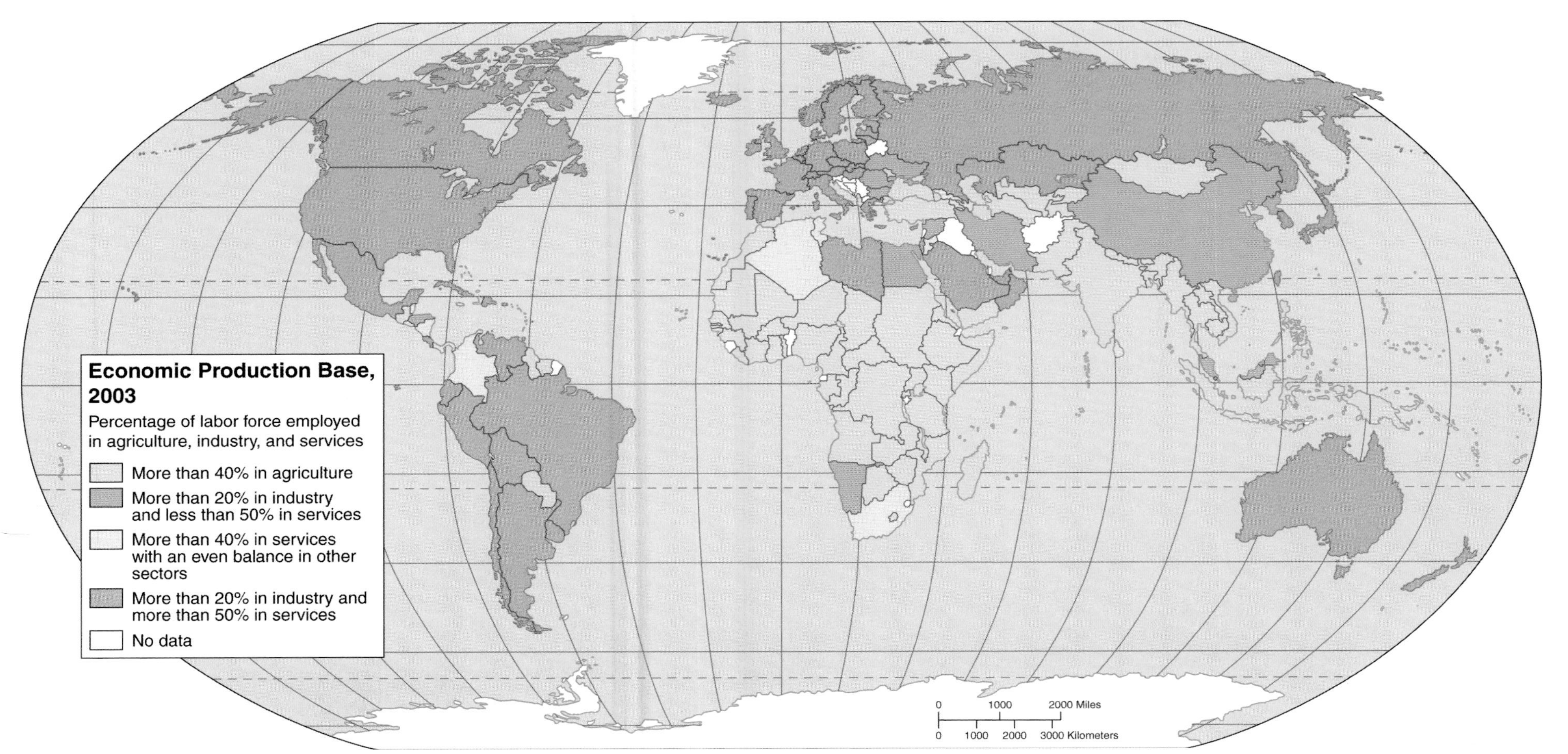

The employment structure of a country's population is one of the best indicators of the country's position on the scale of economic development. At one end of the scale are those countries with more than 40 percent of their labor force employed in agriculture. These are almost invariably the least developed, with high population growth rates, poor human services, significant environmental problems, and so on. In the middle of the scale are two types of countries: those with more than 20 percent of their labor force employed in industry and those with a fairly even balance among agricultural, industrial, and service employment but with at least 40 percent of their labor force employed in service activities. Generally, these countries have undergone the industrial revolution fairly recently and are still developing an industrial base while building up their service activities. This category also includes countries with a disproportionate share of their economies in service activities primarily related to resource extraction. On the other end of the scale from the agricultural economies are countries with more than 20 percent of their labor force employed in industry and more than 50 percent in service activities. These countries are, for the most part, those with a highly automated industrial base and a highly mechanized agricultural system (the "postindustrial," developed countries). They also include, particularly in Middle and South America and Africa, industrializing countries that are also heavily engaged in resource extraction as a service activity.

Map 43 Central Government Expenditures Per Capita

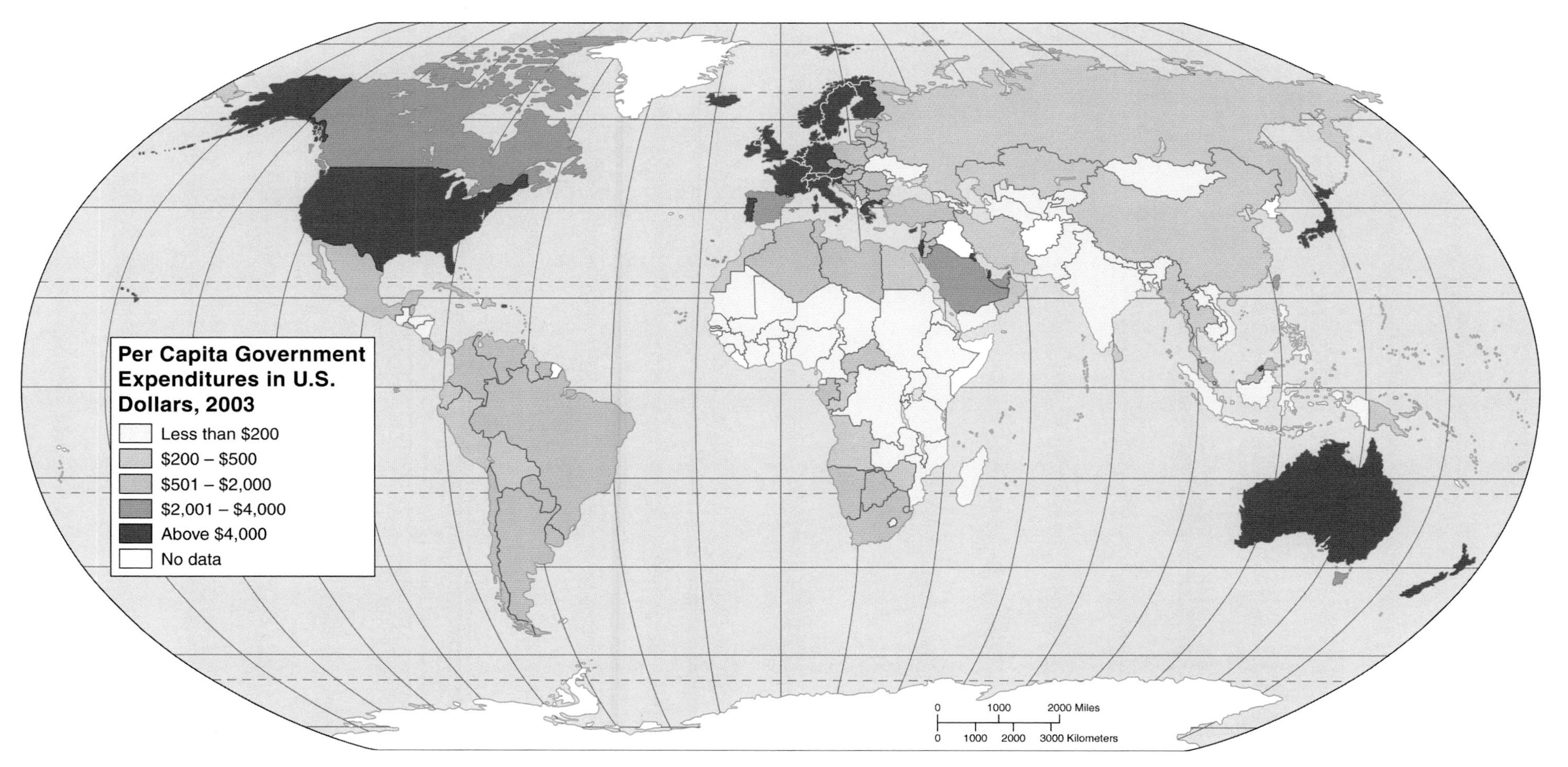

The amount of money that the central government of a country spends upon a variety of essential governmental functions is a measure of relative economic development, particularly when it is viewed on a per-person basis. These functions include such governmental responsibilities as agriculture, communications, culture, defense, education, fishing and hunting, health, housing, recreation, religion, social security, transportation, and welfare. Generally, the higher the level of economic development, the greater the per capita expenditures on these services. However, the data do mask some internal variations. For example, countries that spend 20 percent or more of their central government expenditures on defense will often show up in the more developed category when, in fact, all that the figures really show is that a disproportionate amount of the money available to the government is devoted to purchasing armaments and maintaining a large standing military force. Thus, the fact that Libya spends more than the average for Africa does not suggest that the average Libyan is much better off than the average Tanzanian. Nevertheless, this map—particularly when compared with Map 62, Energy Consumption Per Capita—does provide a reasonable approximation of economic development levels.

Map 44 The Indebtedness of States

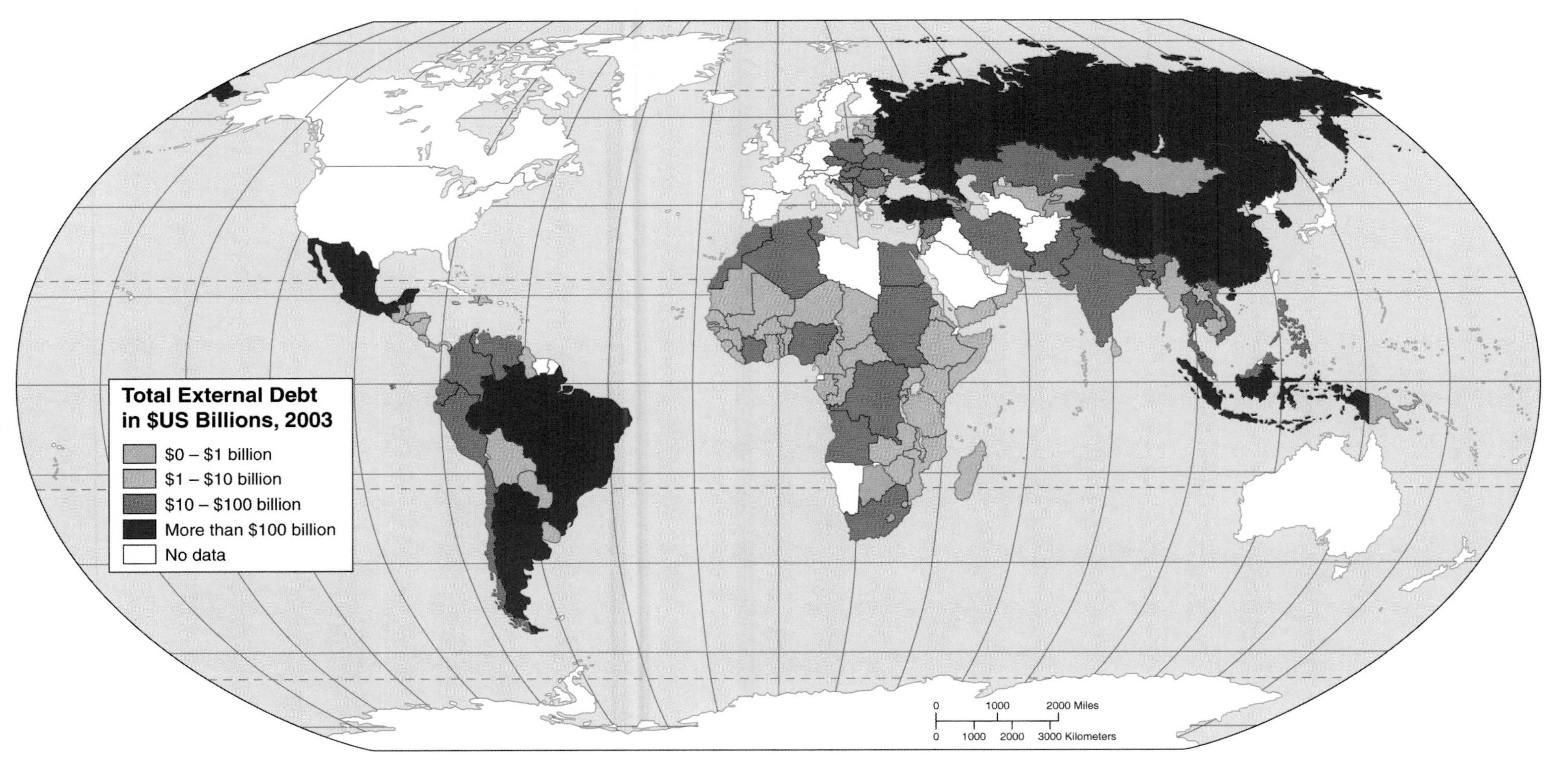

Many governments spend more on a wide variety of services and activities than they collect in taxes and other revenues. In order to finance this deficit spending, governments borrow money—often from banks or other investors outside their country. Repayment of these debts, or even meeting interest payments on them, often means expending a country's export income—in other words, exchanging a country's wealth in production or, more often, resources, for debt service. Where the debt is external, as it is in most developing countries, governments become more open to outside influence in political as well as economic terms. Even internal debt service or repayment of monies owed to investors within a country gives financial establishments a measure of influence over government decisions. The amounts of debt shown on the map indicate the total external indebtedness of states.

Map 45 Exports of Primary Products

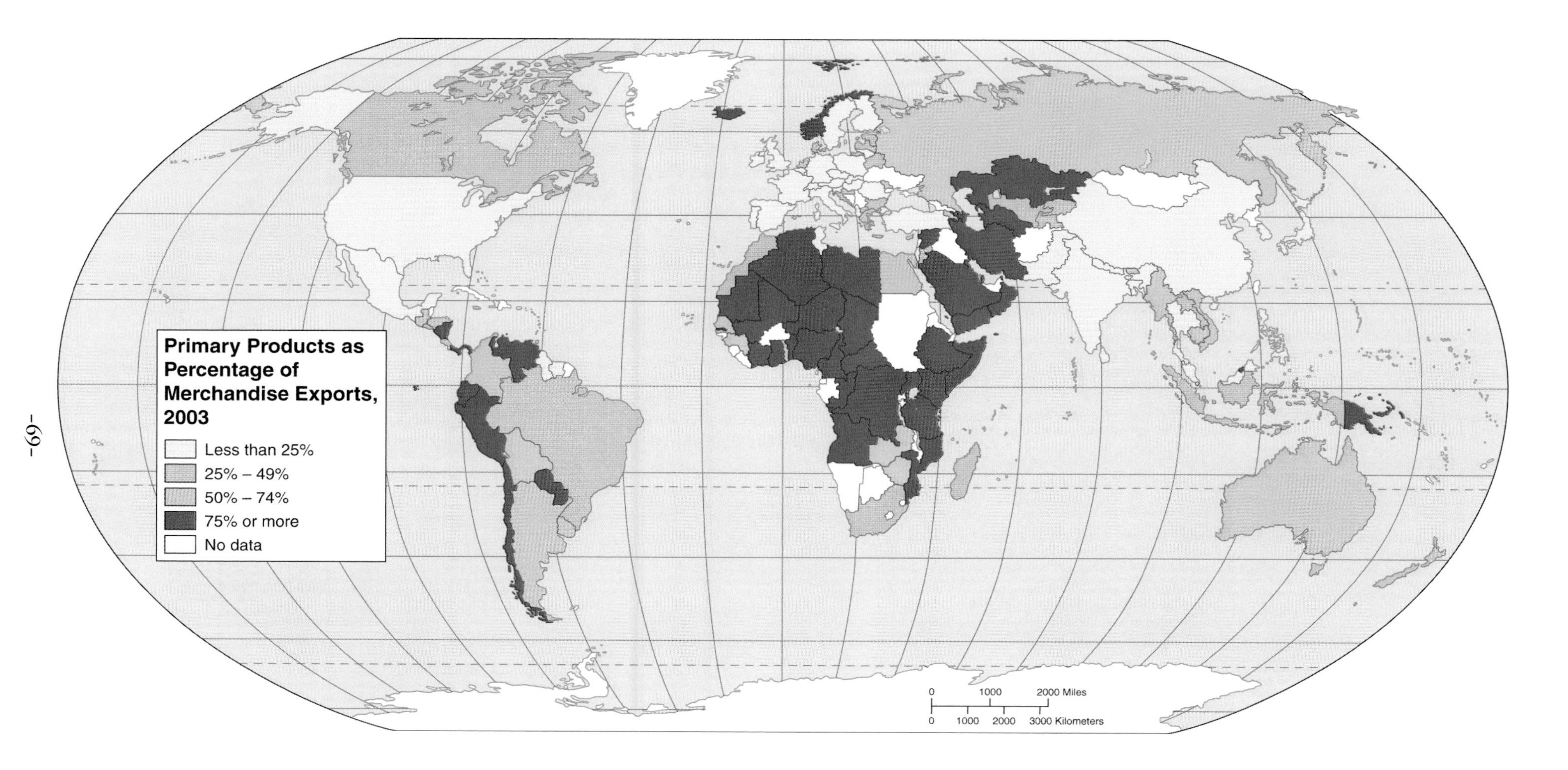

Primary products are those that require additional processing before they enter the consumer market: metallic ores that must be converted into metals and then into metal products such as automobiles or refrigerators; forest products such as timber that must be converted to lumber before they become suitable for construction purposes; and agricultural products that require further processing before being ready for human consumption. It is an axiom in international economics that the more a country relies on primary products for its export commodities, the more vulnerable its economy is to market fluctuations. Those countries with only primary products to export are hampered in their economic growth. A country dependent on only one or two products for export revenues is unprotected from economic shifts, particularly a changing market demand for its products. Imagine what would happen to the thriving economic status of the oil-exporting states of the Persian Gulf, for example, if an alternate source of cheap energy were found. A glance at this map, together with Map 57, shows that those countries with the lowest levels of economic development tend to be concentrated on primary products and, therefore, have economies that are especially vulnerable to economic instability.

Map 46 Dependence on Trade

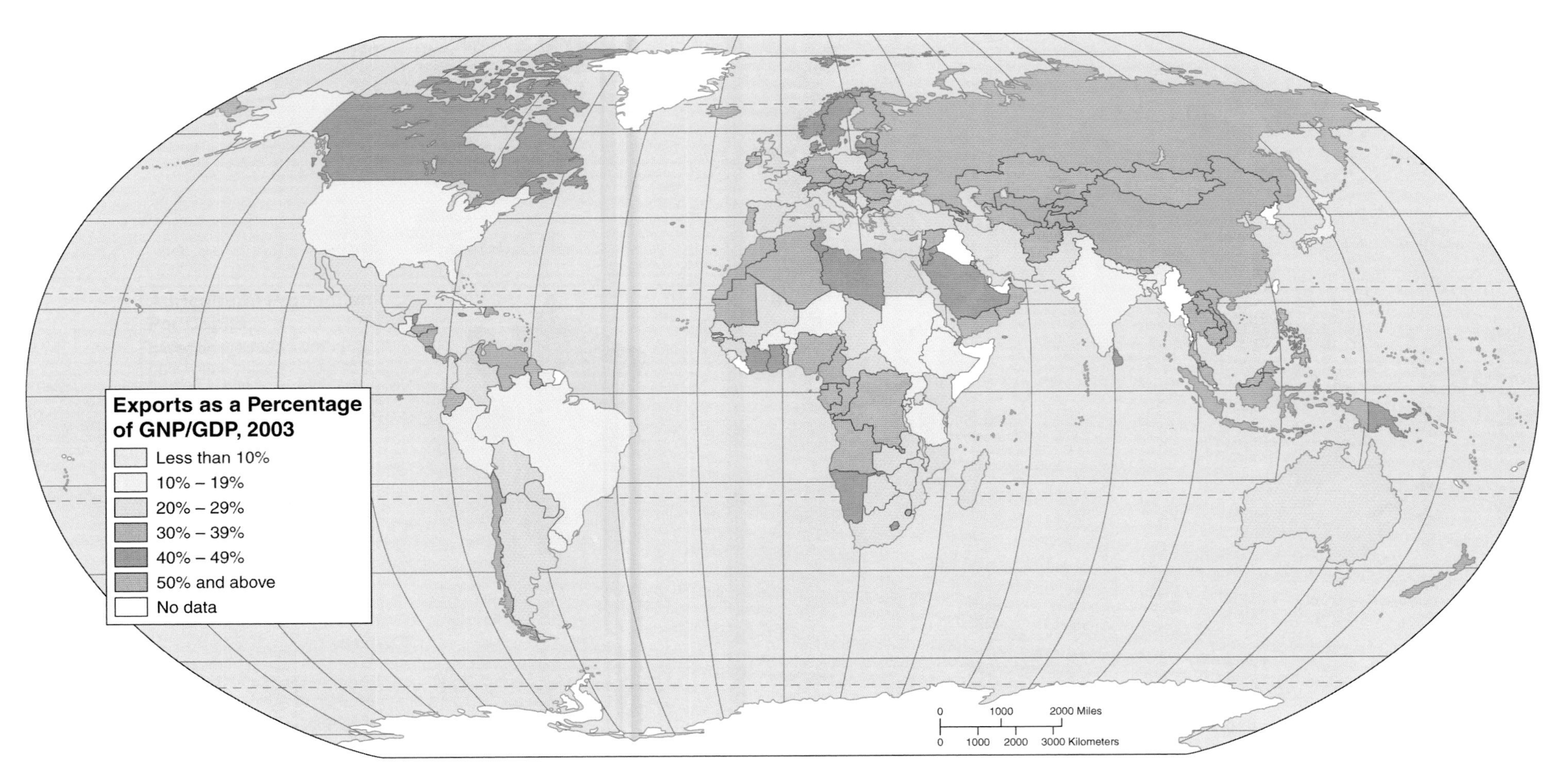

As the global economy becomes more and more a reality, the economic strength of virtually all countries is increasingly dependent upon trade. For many developing nations, with relatively abundant resources and limited industrial capacity, exports provide the primary base upon which their economies rest. Even countries like the United States, Japan, and Germany, with huge and diverse economies, depend on exports to generate a significant percentage of their employment and wealth. Without imports, many products that consumers want would be unavailable or more expensive; without exports, many jobs would be eliminated.

Part IV

Population and Human Development

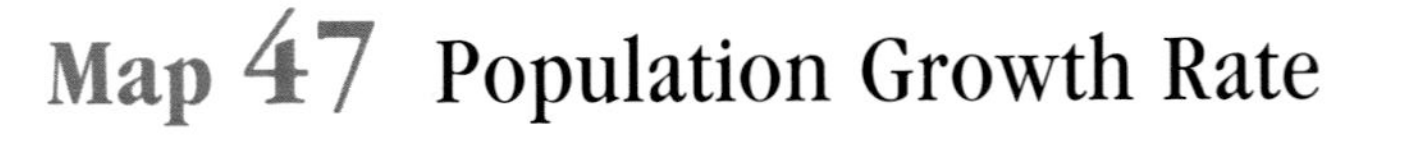

Map 47 Population Growth Rate

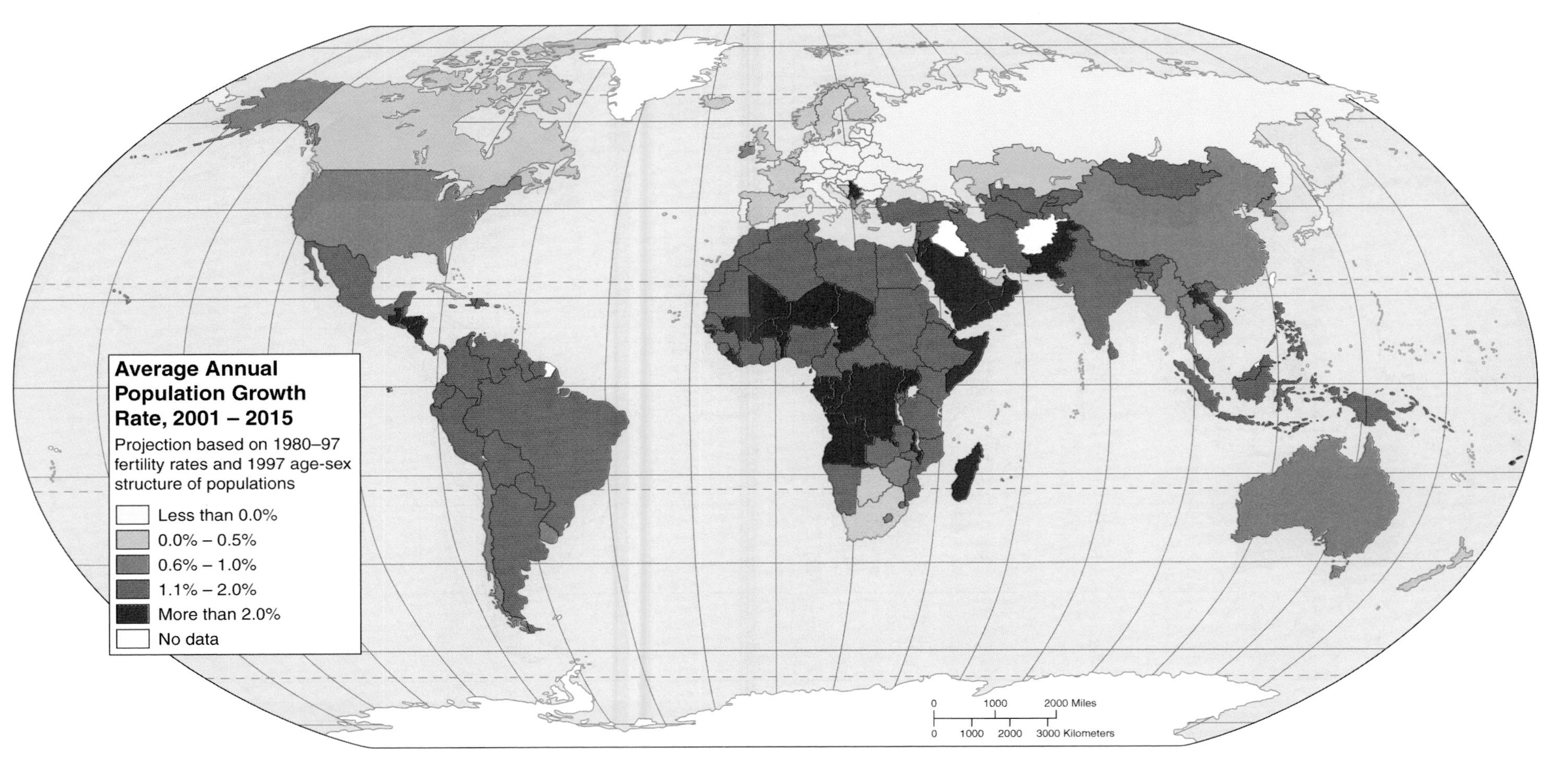

Of all the statistical measurements of human population, that of the rate of population growth is the most important. The growth rate of a population is a combination of natural change (births and deaths), in-migration, and out-migration; it is obtained by adding the number of births to the number of immigrants during a year and subtracting from that total the sum of deaths and emigrants for the same year. For a specific country, this figure will determine many things about the country's future ability to feed, house, educate, and provide medical services to its citizens. Some of the countries with the largest populations (such as India) also have high growth rates. Since these countries tend to be in developing regions, the combination of high population and high growth rates poses special problems for political stability and continuing economic development; the combination also carries heightened risks for environmental degradation. Many people believe that the rapidly expanding world population is a potential crisis that may cause environmental and human disaster by the middle of the twenty-first century.

Map 48 Infant Mortality Rate

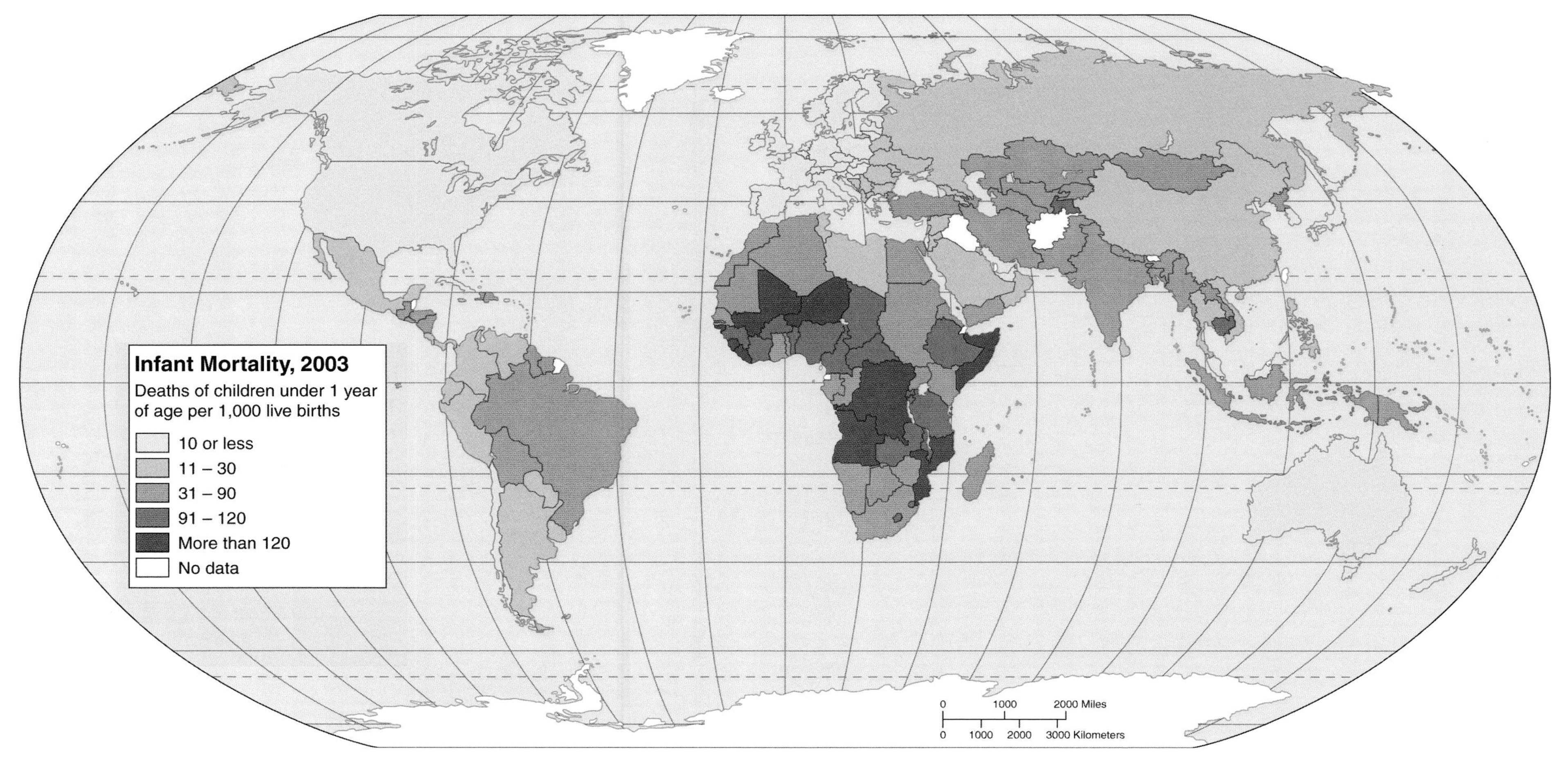

Infant mortality rates are calculated by dividing the number of children born in a given year who die before their first birthday by the total number of children born that year and then multiplying by 1,000; this shows how many infants have died for every 1,000 births. Infant mortality rates are prime indicators of economic development. In highly developed economies, with advanced medical technologies, sufficient diets, and adequate public sanitation, infant mortality rates tend to be quite low. By contrast, in less developed countries, with the disadvantages of poor diet, limited access to medical technology, and the other problems of poverty, infant mortality rates tend to be high.

Although worldwide infant mortality has decreased significantly during the last 2 decades, many regions of the world still experience infant mortality above the 10 percent level (100 deaths per 1,000 live births). Such infant mortality rates not only represent human tragedy at its most basic level, but also are powerful inhibiting factors for the future of human development. Comparing infant mortality rates in the midlatitudes and the tropics shows that children in most African countries are more than 10 times as likely to die within a year of birth as children in European countries.

Map 49 Average Life Expectancy at Birth

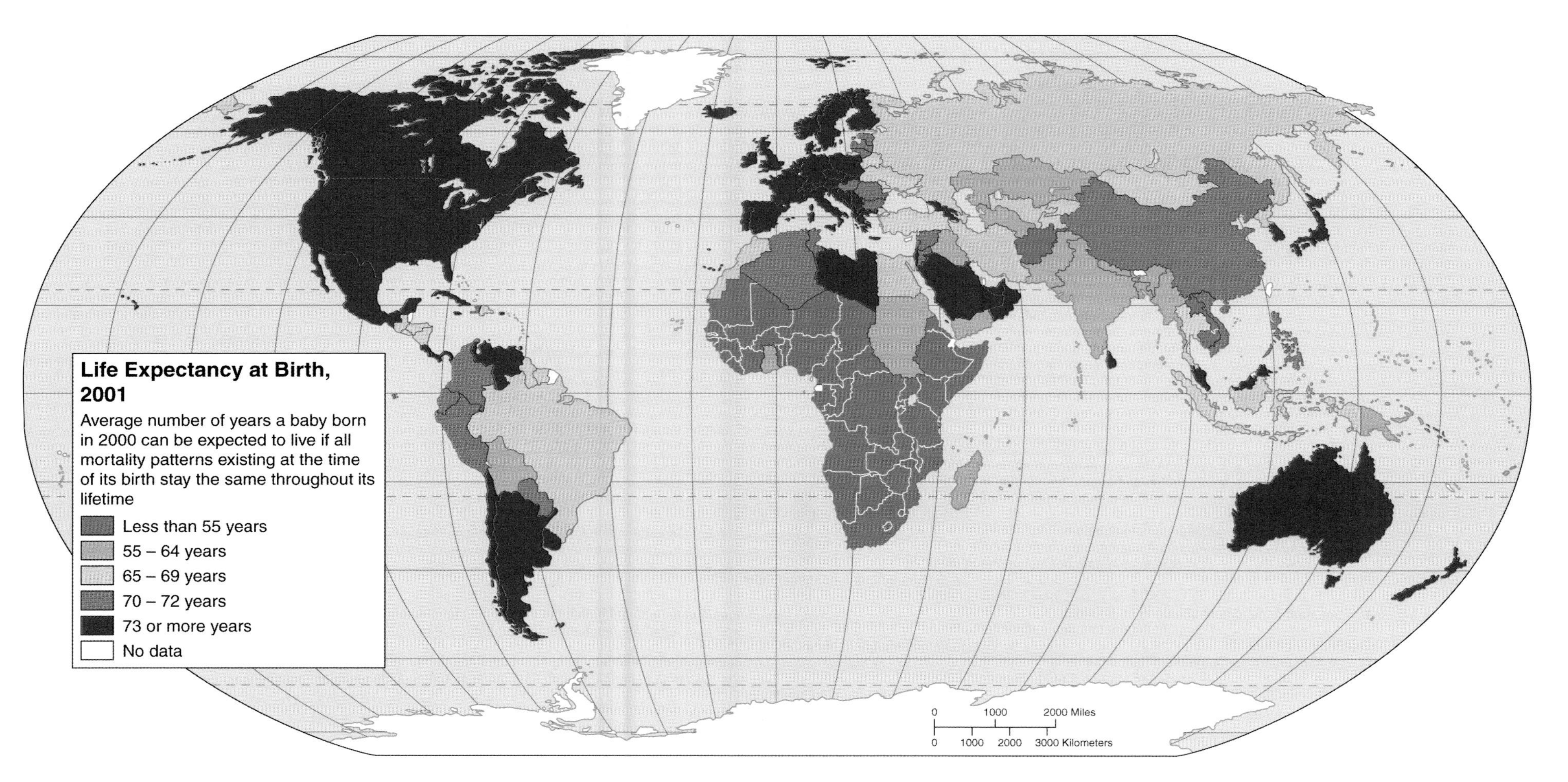

Average life expectancy at birth is a measure of the average longevity of the population of a country. Like all average measures, it is distorted by extremes. For example, a country with a high mortality rate among children will have a low average life expectancy. Thus, an average life expectancy of 45 years does not mean that everyone can be expected to die at the age of 45. More normally, what the figure means is that a substantial number of children die between birth and 5 years of age, thus reducing the average life expectancy for the entire population. In spite of the dangers inherent in misinterpreting the data, average life expectancy (along with infant mortality and several other measures) is a valid way of judging the relative health of a population. It reflects the nature of the health care system, public sanitation and disease control, nutrition, and a number of other key human need indicators. As such, it is a measure of well-being that is significant in indicating economic development and predicting political stability.

Map 50 Population by Age Group

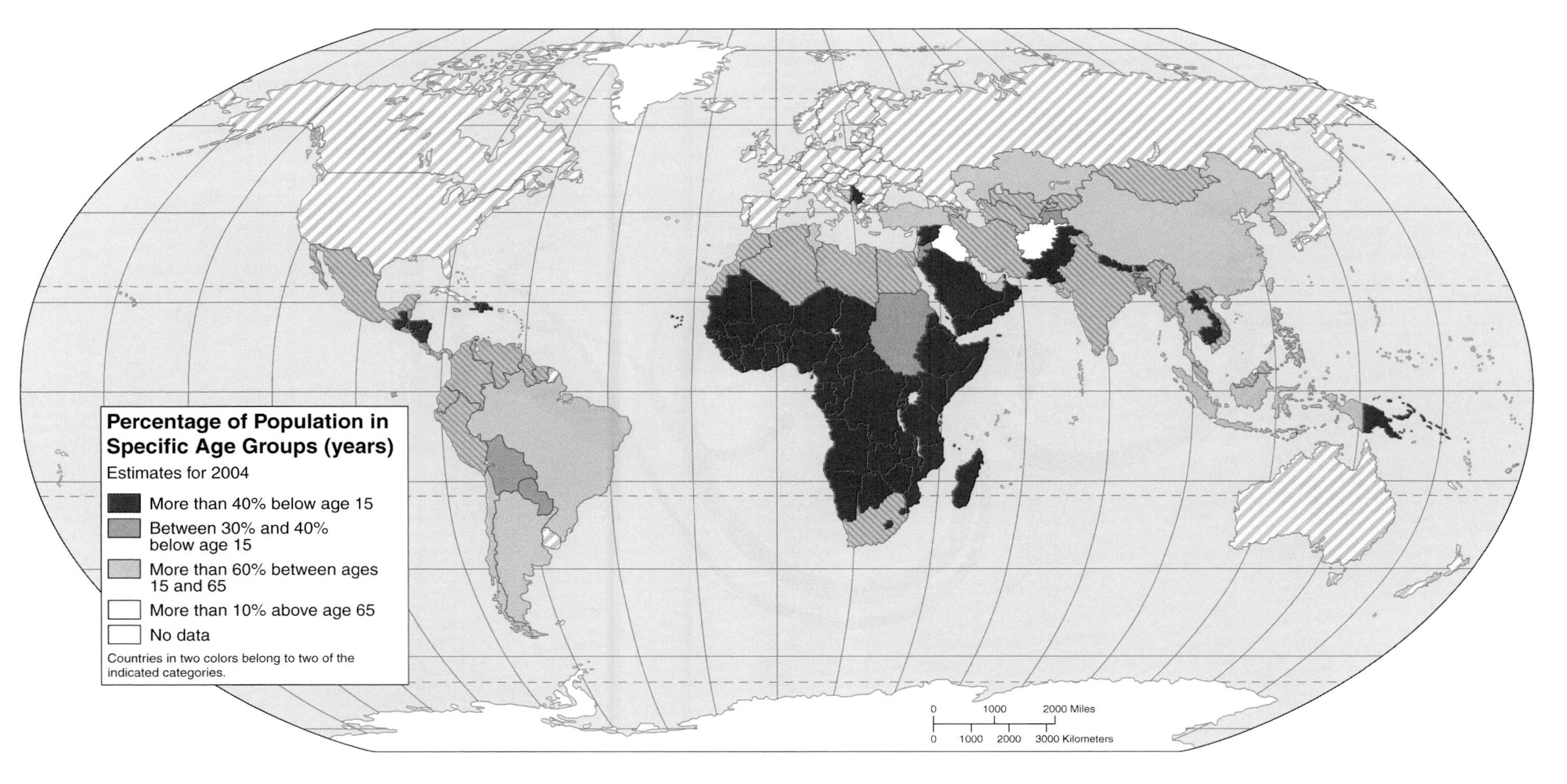

Of all the measurements that illustrate the dynamics of a population, age distribution may be the most significant, particularly when viewed in combination with average growth rates. The particular relevance of age distribution is that it tells us what to expect from a population in terms of growth over the next generation. If, for example, approximately 40–50 percent of a population is below the age of 15, that suggests that in the next generation about one-quarter of the total population will be women of childbearing age. When age distribution is combined with fertility rates (the average number of children born per woman in a population), an especially valid measurement of future growth potential may be derived. A simple example: Nigeria, with a 2002 population of 130 million, has 43.6 percent of its population below the age of 15 and a fertility rate of 5.5; the United States, with a 2002 population of 280 million, has 21 percent of its population below the age of 15 and a fertility rate of 2.07. During the period in which those women presently under the age of 15 are in their childbearing years, Nigeria can be expected to add a total of approximately 155 million persons to its total population. Over the same period, the United States can be expected to add only 61 million.

Map 51 Urban Population

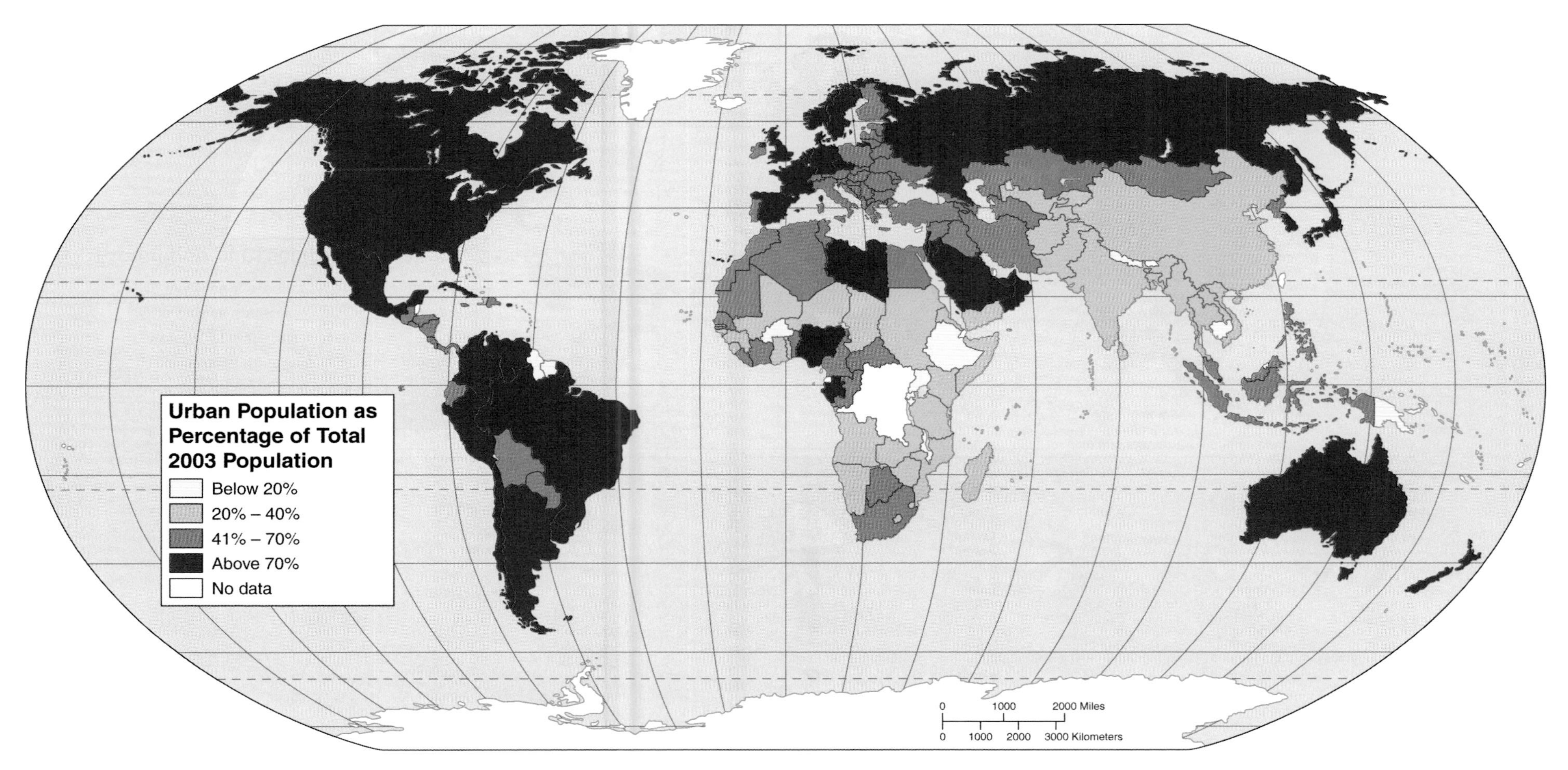

The proportion of a country's population that resides in urban areas was formerly considered a measure of relative economic development, with countries possessing a large urban population ranking high on the development scale and countries with a more rural population ranking low. Given the rapid rate of urbanization in developing countries, however, this traditional measure is no longer so valuable. What relative urbanization rates now tell us is something about levels of economic development in a negative sense. Latin American, African, and Asian countries with more than 40 percent of their populations living in urban areas generally suffer from a variety of problems: rural overpopulation and flight from the land, urban poverty and despair, high unemployment, and poor public services. The rate of urbanization in less developed nations is such that many cities in these nations will outstrip those in North America and Europe by the end of this century. It has been estimated, for example, that Mexico City—now the world's second largest metropolis—has over 18 million inhabitants. Urbanization was once viewed as an indicator of economic health and political maturity. For many countries it is instead a harbinger of potential economic and environmental disaster.

Map 52 Illiteracy Rates

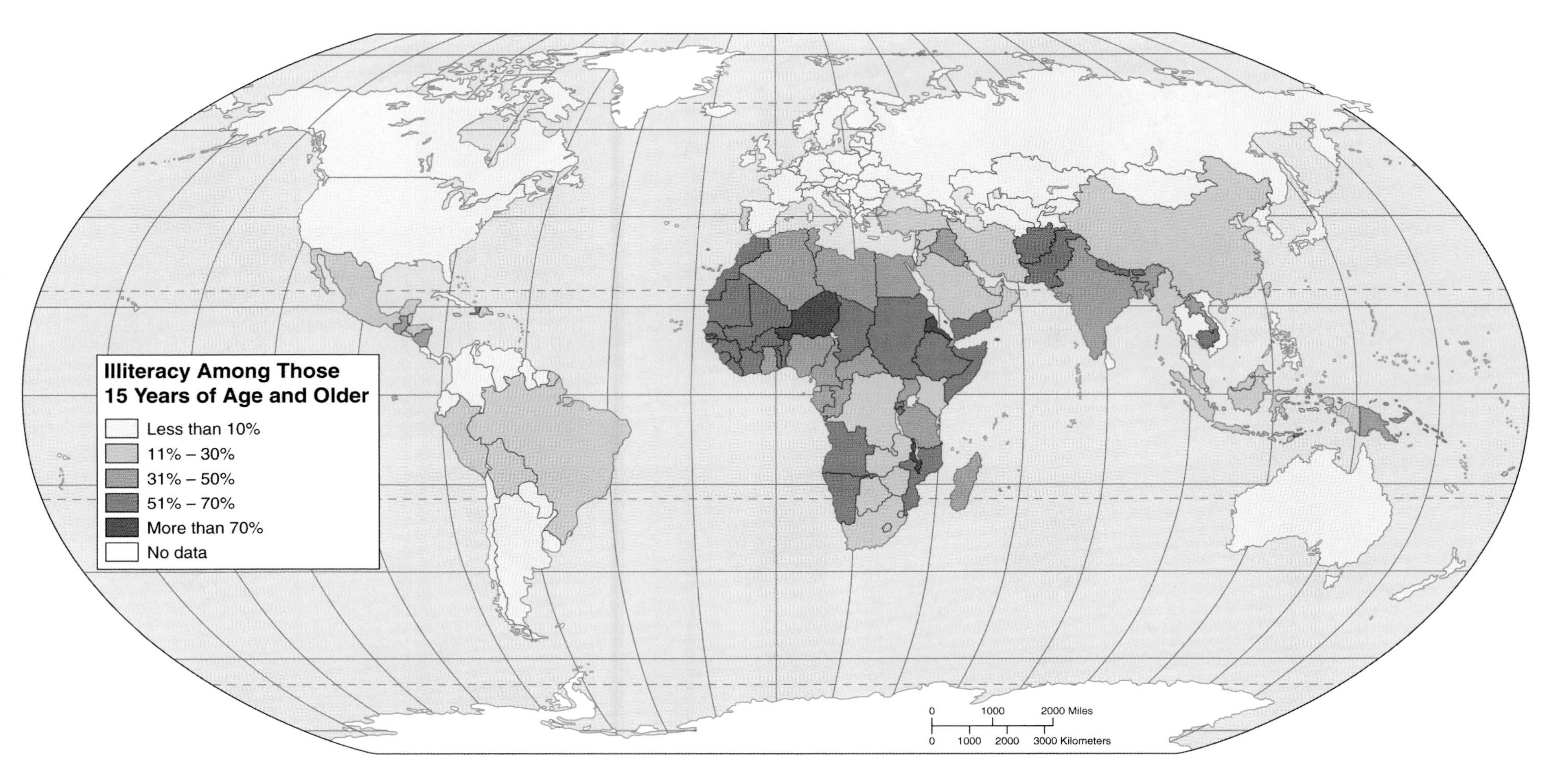

Illiteracy rates are based on the percentages of people age 15 or above (classed as adults in most countries) who are not able to write and read, with understanding, a brief, simple statement about everyday life written in their home- or official language. As might be expected, illiteracy rates tend to be higher in the lesser-developed states, where educational systems are a low government priority. Rates of literacy or illiteracy also tend to be gender-differentiated, with women in many countries experiencing educational neglect or discrimination that makes it more likely they will be illiterate. In many developing countries, between five and ten times as many women will be illiterate as men, and the illiteracy rate for women may even exceed 90 percent. Both male and female illiteracy severely compromise economic development.

Map 53 Unemployment in the Labor Force

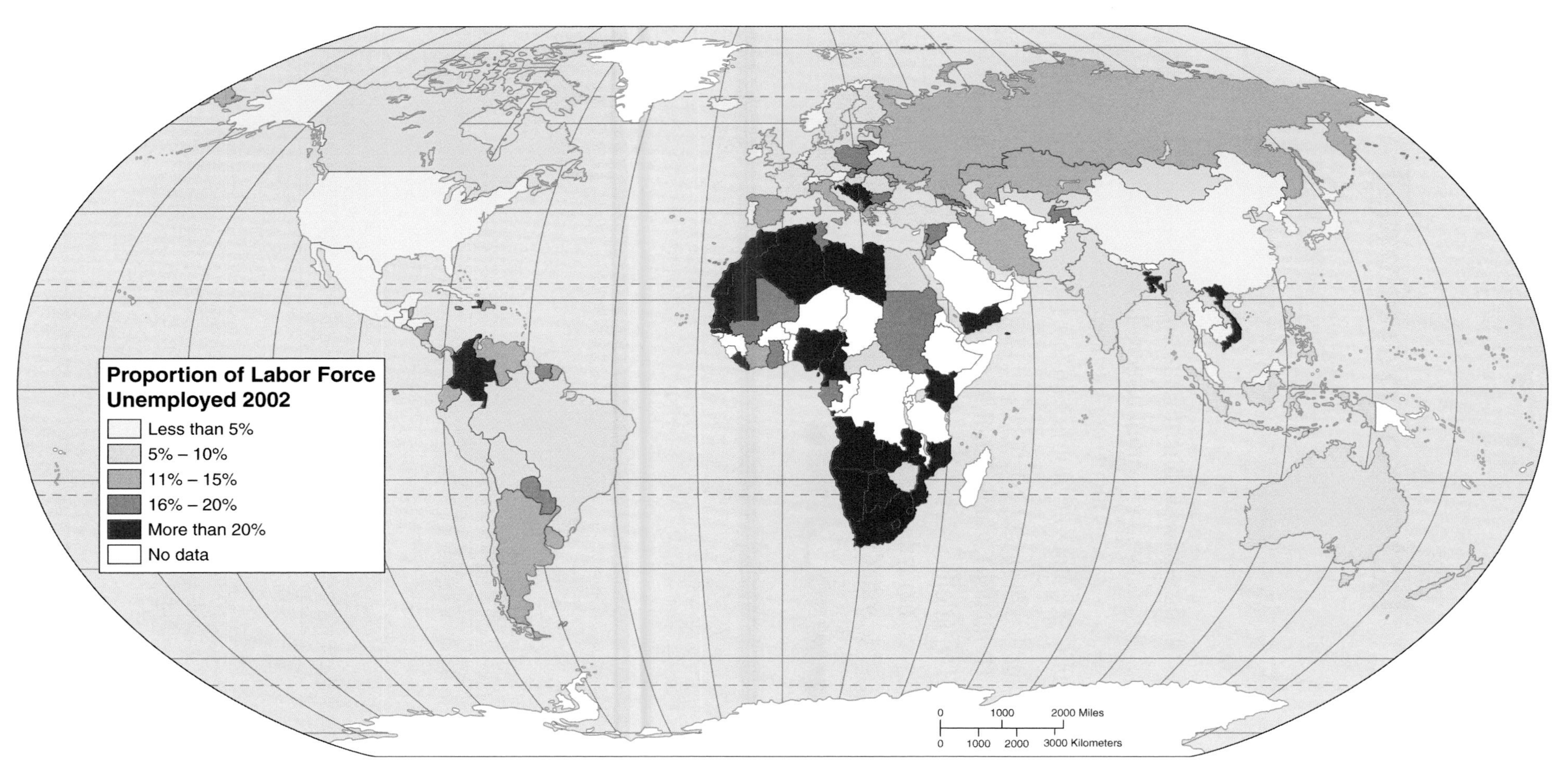

The percentage of a country's labor force that is classified as "unemployed" include those without work but who are available for work and seeking employment. Countries may define the labor force in different ways, however. In many developing countries, for example, "employability" based on age may be more extensive than in more highly developed economies with stringent child labor laws. Generally, countries with higher percentages of their labor forces employed will be countries with higher levels of economic development. Where unemployment tends to be high, the out-migration of labor also tends to be high as workers unable to find employment at home cross international boundaries in search of work. Again, there tends to be a difference based on levels of economic development with the more developed countries experiencing inflows of labor while the reverse is true in the less developed world.

Map 54 The Gender Gap: Inequalities in Education and Employment

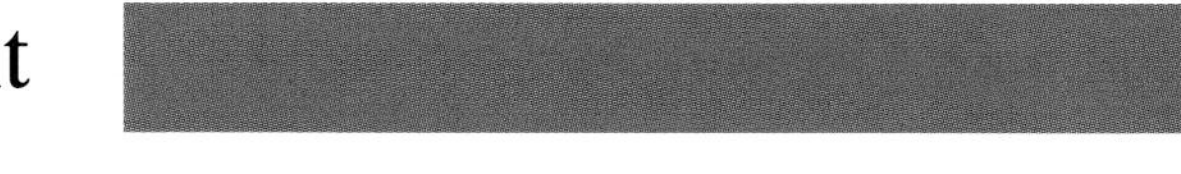

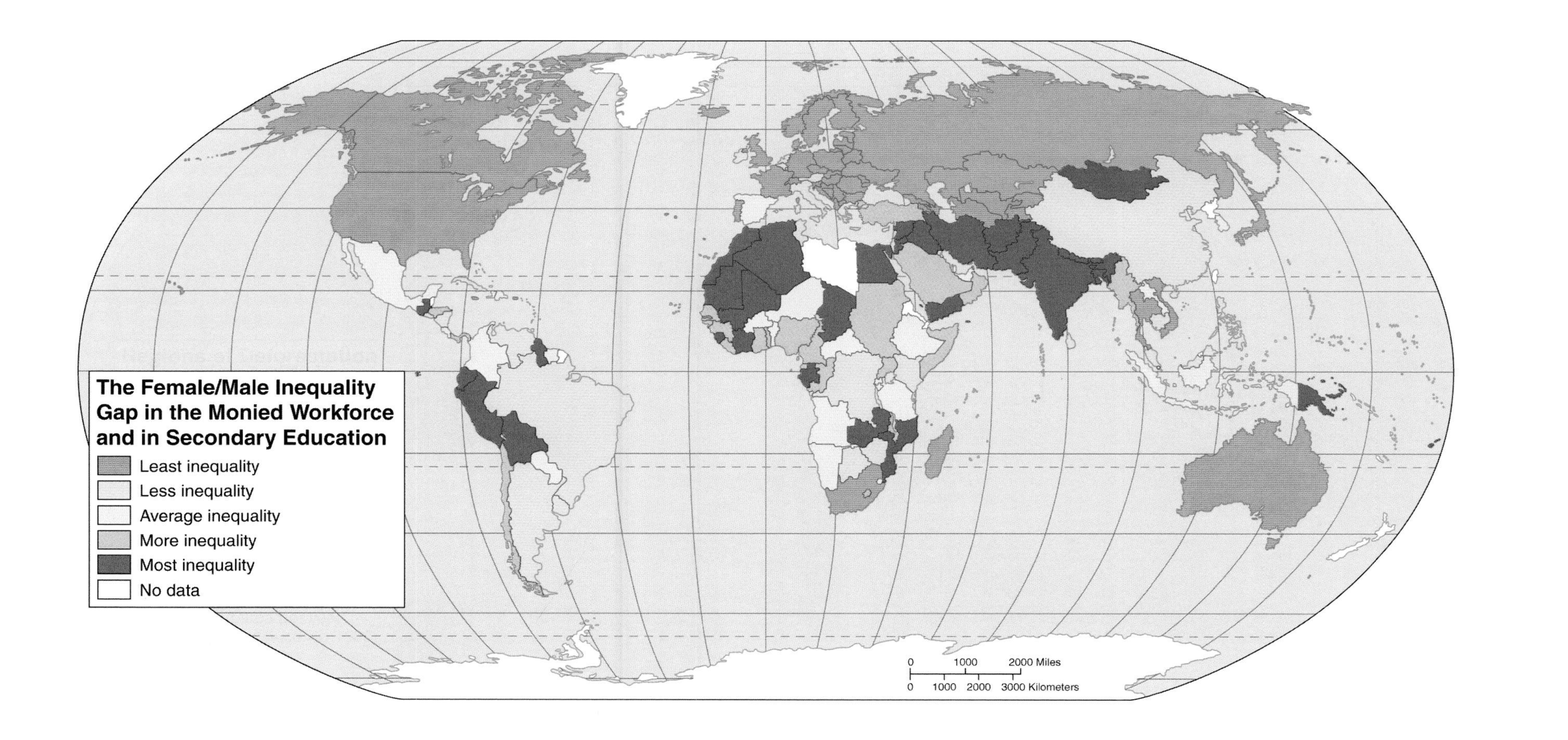

Although women in developed countries, particularly in North America and Europe, have made significant advances in socioeconomic status in recent years, in most of the world females suffer from significant inequality when compared with their male counterparts. Women have received the right to vote in most of the world's countries, but in over 90 percent of these countries that right has only been granted in the last 50 years. In most regions, literacy rates for women still fall far short of those for men; in Africa and Asia, for example, only about half as many women are literate as are men. Women marry considerably younger than men and attend school for shorter periods of time. Inequalities in education and employment are perhaps the most telling indicators of the unequal status of women in most of the world. Lack of secondary education in comparison with men prevents women from entering the workforce with equally high-paying jobs. Even where women are employed in positions similar to those held by men, they still tend to receive less compensation. The gap between rich and poor involves not only a clear geographic differentiation, but a clear gender differentiation as well.

Map 55 Human Rights: Political Rights and Civil Liberties

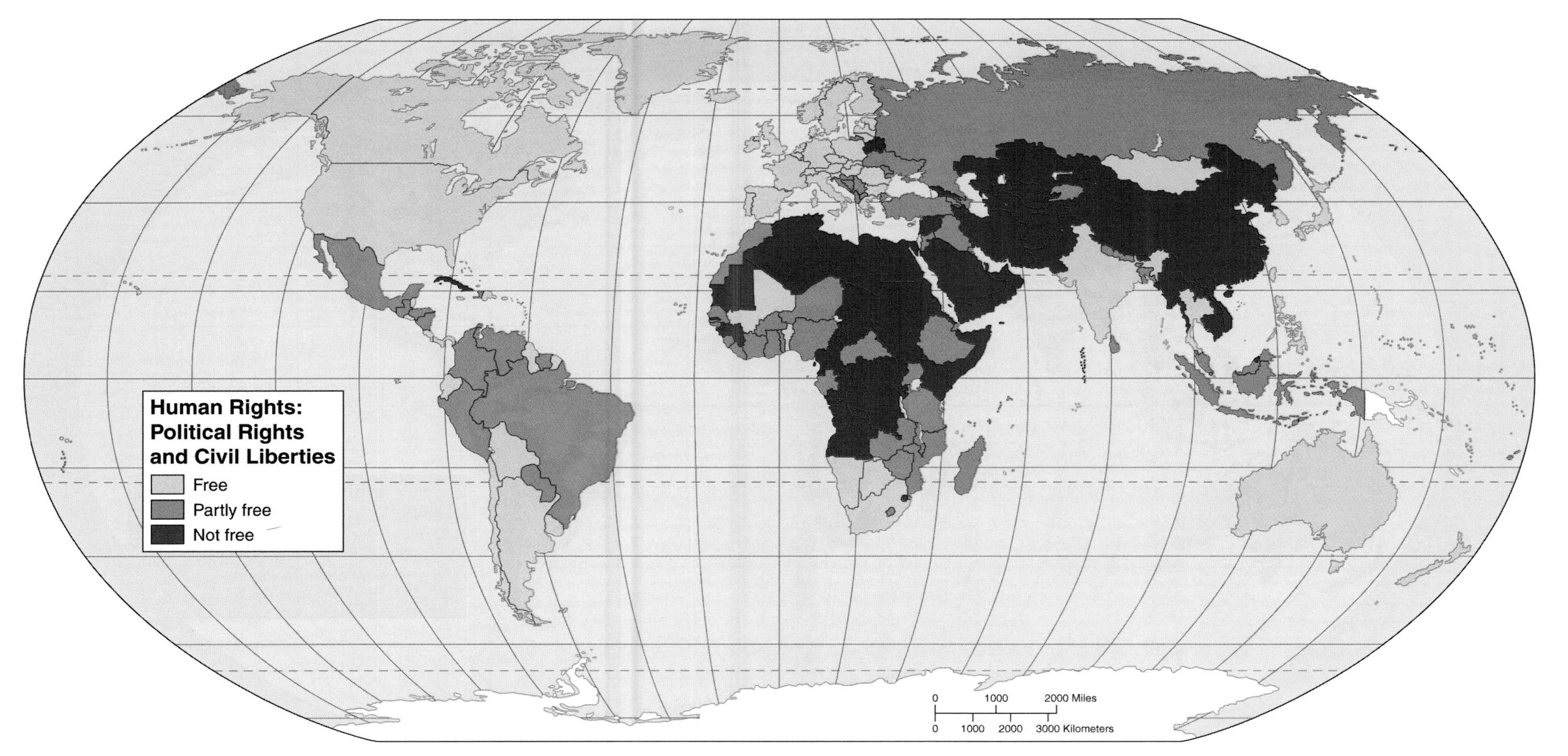

Increasing contact among the world's peoples has brought to everyone more awareness of the variations in human rights from place to place. Democratic forms of government have spread to more and more countries, and public policy insists on attention to human rights. "Human rights" comprise both political rights and civil liberties. Political rights are related to democracy, which, at a minimum involves rights of the people to choose their own authoritative leaders and to endow those chosen leaders with real power. Elections must be open and fair, and ballots must be counted honestly. Candidates may represent political parties, or families, clans, or personalities. Civil liberties are freedoms, including freedom of expression and belief, freedom of assembly and organization, rule of law (including an independent judiciary), and personal autonomy and economic rights (including privacy). Every year, Freedom House ranks all countries of the world in these two sets of rights, using a checklist of questions about real experience, not just legislation. The compilation of these scores produces the three categories shown on this map. Obviously, this process is extremely generalized, but the two sets of rights do tend to go together: countries with strong political rights also have civil liberties.

Map 56 Capital Punishment

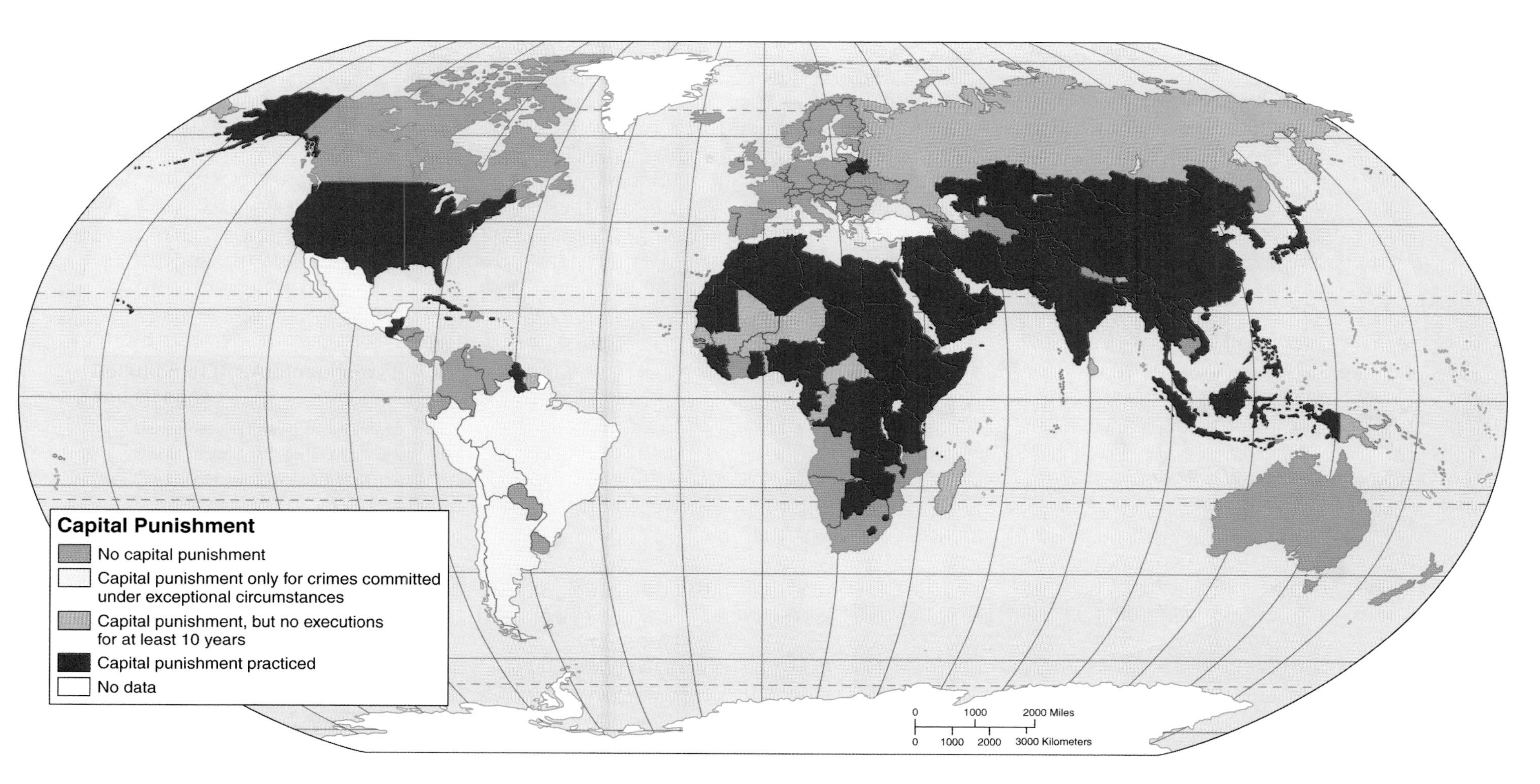

The most basic human right is life itself. More than half the countries of the world have abolished capital punishment by law or in practice. In some of these countries, capital punishment remains on the books, but no one has been executed in so long that in practice, capital punishment can be considered abolished. A few countries retain capital punishment only for crimes committed in extraordinary circumstances, such as military law. About three countries per year have abolished capital punishment in the last decade. Some states in the United States retain capital punishment; in others it has been abolished.

Map 57 Deployment of Peace Corps Volunteers

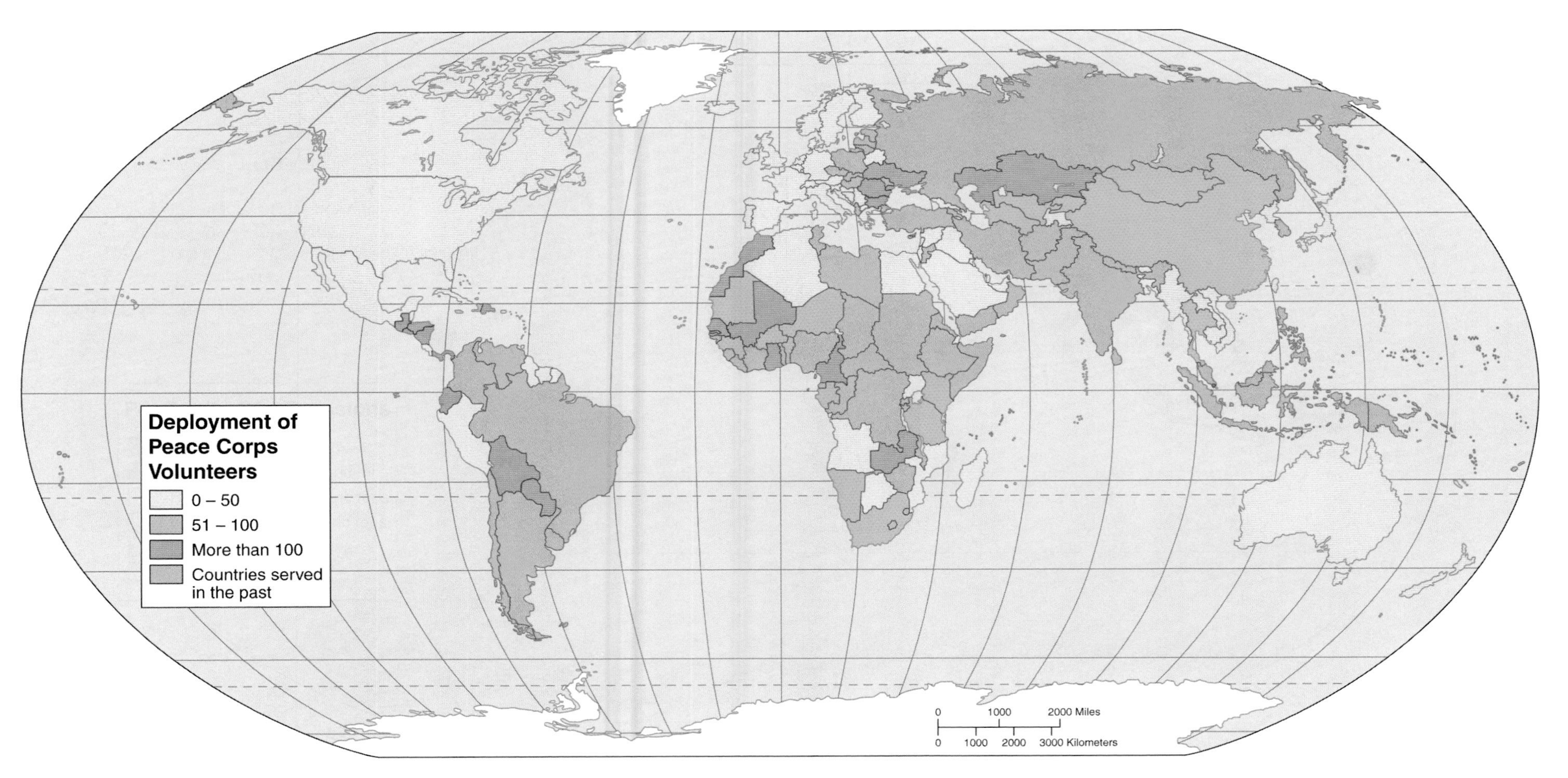

Among the many programs that enable volunteers to serve in less developed countries, the largest is the Peace Corps. John F. Kennedy proposed the Peace Corps when he was running for president in 1960; in 1961 Congress authorized the program, and the first 51 volunteers were selected, trained, and sent to assignments in six countries. The purpose of the Peace Corps is, according to the legislation, to "promote world peace and friendship" through three goals. First is to help people in interested countries to meet their needs for trained workers. Second is to promote better understanding of Americans among peoples in foreign countries. Third is to promote better understanding of other peoples among Americans. Since 1961, 168,000 volunteers have served in 136 countries. Today more than 6,000 volunteers are serving in 70 countries. Among their projects are providing clean water to communities, teaching in schools, helping people start small businesses, and stopping the spread of AIDS.

Map 58 The Index of Human Development

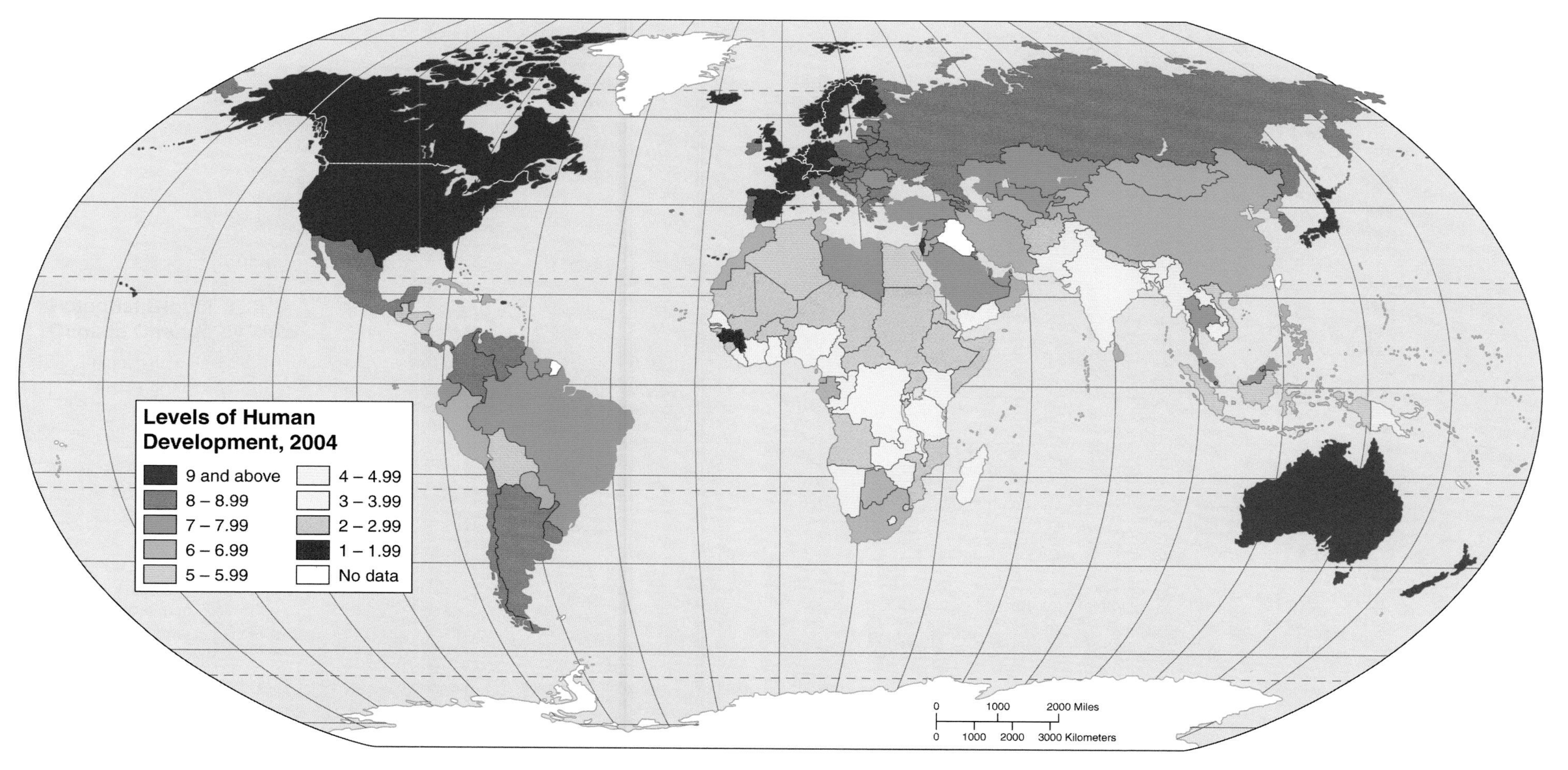

The development index upon which this map is based takes into account a wide variety of demographic, health, and educational data, including population growth, per capita gross domestic income, longevity, literacy, and years of schooling. The map reveals significant improvement in the quality of life in Middle and South America, although it is questionable whether the gains made in those regions can be maintained in the face of the dramatic population increases expected over the next 30 years. More clearly than anything else, the map illustrates the near-desperate situation in Africa and South Asia. In those regions, the unparalleled growth in population threatens to overwhelm all efforts to improve the quality of life. In Africa, for example, the population is increasing by 20 million persons per year. With nearly 45 percent of the continent's population aged 15 years or younger, this growth rate will accelerate as the women reach childbearing age. Africa, along with South Asia, faces the very difficult challenge of providing basic access to health care, education, and jobs for a rapidly increasing population. The map also illustrates the striking difference in quality of life between those who inhabit the world's equatorial and tropical regions and those fortunate enough to live in the temperate zones, where the quality of life is significantly higher.

Part V

Food, Energy, and Materials

Map 59 Production of Staples: Cereals, Roots, and Tubers

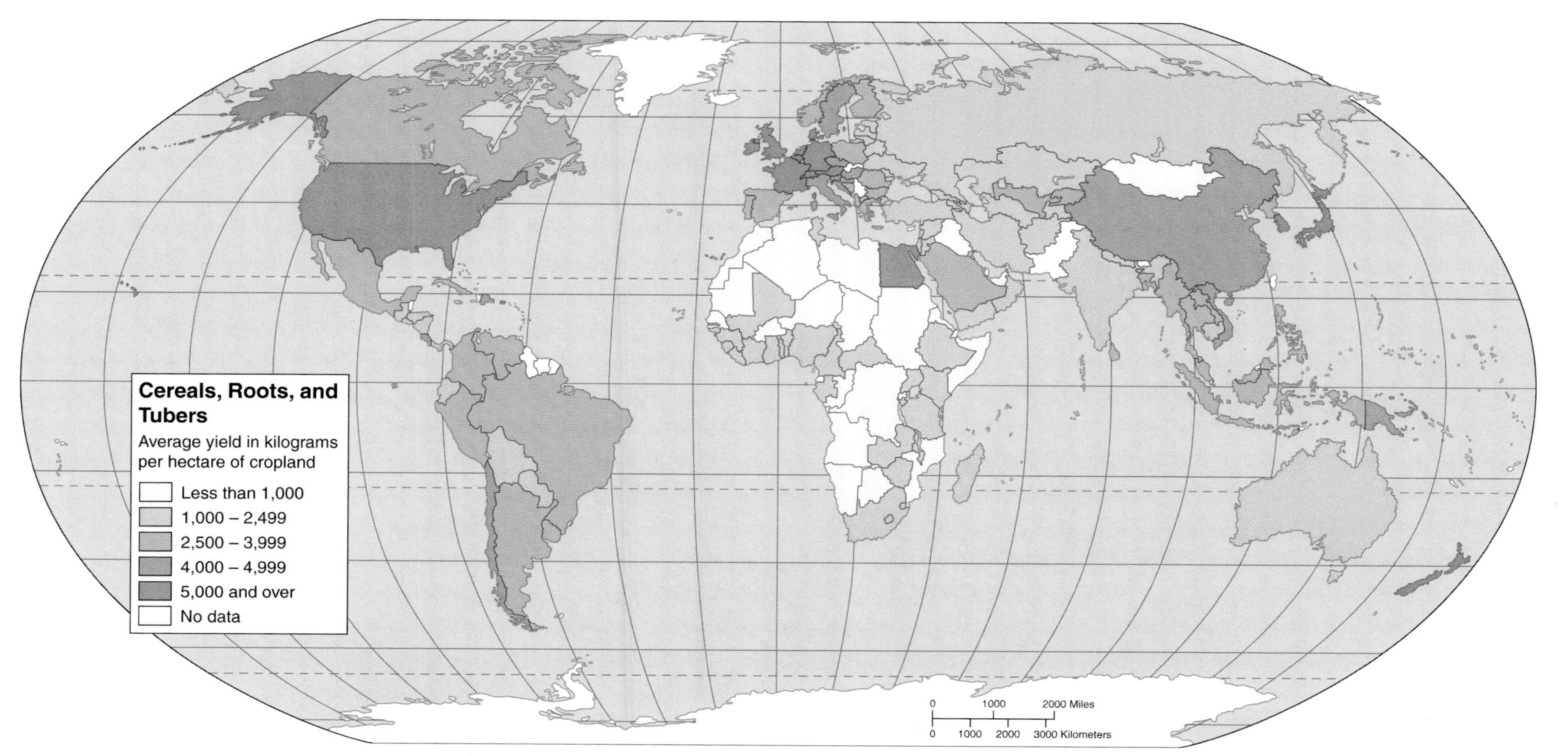

For most of the world's population, food crops (as opposed to livestock foods) provide the bulk of dietary intake. Good agricultural land is simply too scarce to be used for the inefficient process of raising food to feed animals, which, in turn, feed people. Global production of the staple (most important) food crops has increased over the last 10 years—but so has global population. In Africa, for example, despite a 30 percent increase in staple crop production since 1981, per capita food output has dropped more than 5 percent because of population growth that is faster than the growth in agricultural output. The map illustrates considerable regional differences in outputs of food staples per areal unit of cropland. Globally, 1 hectare (2.47 acres) of cropland in 1990 yielded, on average, about 2.1 metric tons (2,100 kilograms or about 4,600 pounds) of cereals. Yet in Africa, 1 hectare yielded only 1.2 metric tons of cereals. In Europe, on the other hand, 1 hectare yielded 4.4 tons of cereals. Such great differences are explainable primarily in terms of agricultural inputs: different farming methods, varying levels of fertilizers, agricultural chemicals, irrigation, and machinery. These conditions are not likely to change and the map may be viewed as an indicator not just of present agricultural output but of potential food production as well.

Map 60 Agricultural Production Per Capita

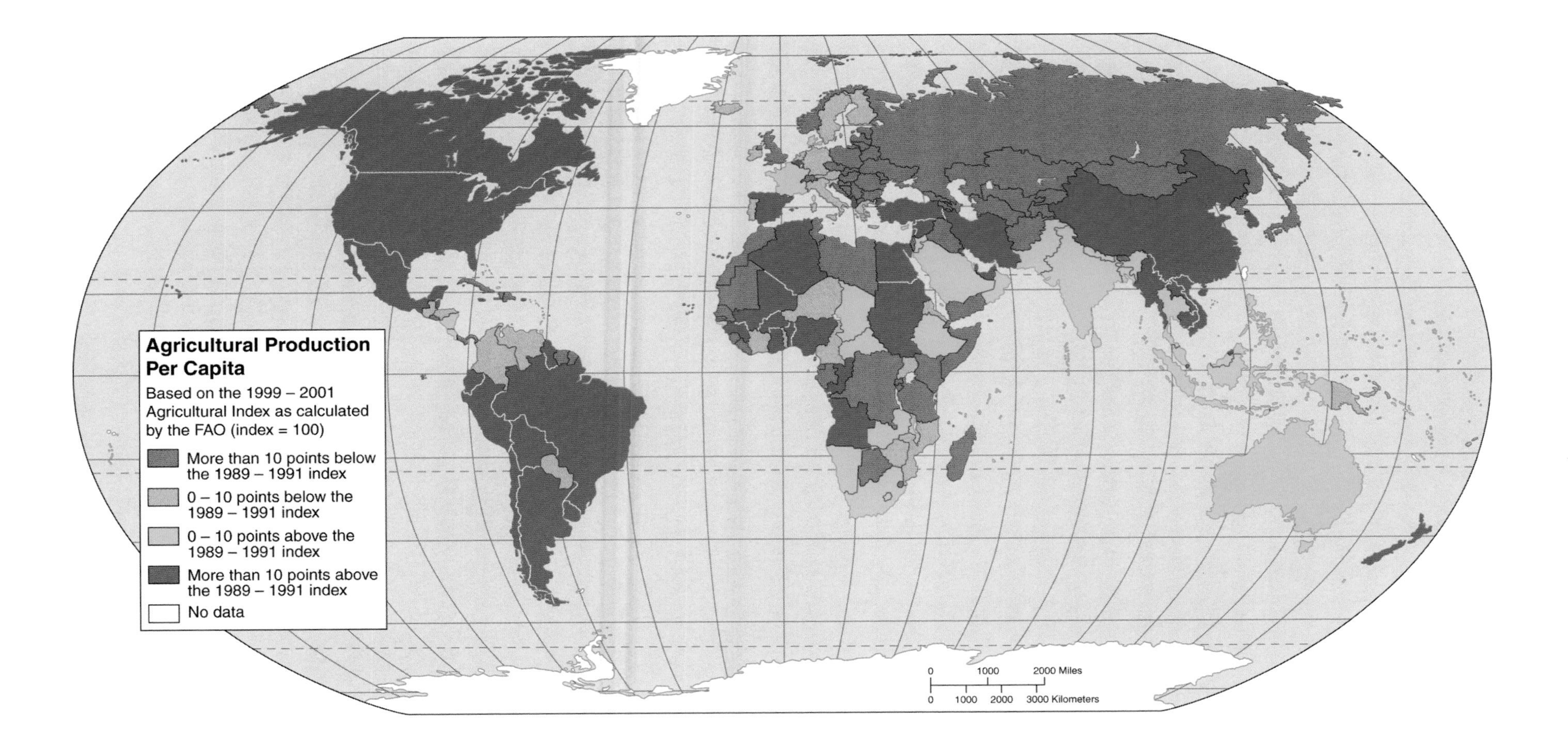

Agricultural production includes the value of all crops and livestock products originating within a country for the base year of 2002. The index value portrays the disposable output (after deductions for livestock feed and seed for planting) of a country's agriculture in comparison with the base period 1989–1991. Thus, the production values show not only the relative ability of countries to produce food but also show whether or not that ability has increased or decreased over a 10-year period. In general, global food production has kept up with or very slightly exceeded population growth. However, there are significant regional variations in the trend of food production keeping up with or surpassing population growth. For example, agricultural production in Africa and in Middle America has fallen, while production in South America, Asia, and Europe has risen. In the case of Africa, the drop in production reflects a population growing more rapidly than agricultural productivity. Where rapid increases in food production per capita exist (as in certain countries in South America, Asia, and Europe), most often the reason is the development of new agricultural technologies that have allowed food production to grow faster than population. In much of Asia, for example, the so-called Green Revolution of new, highly productive strains of wheat and rice made positive index values possible. Also in Asia, the cessation of major warfare allowed some countries (Cambodia, Laos, and Vietnam) to show substantial increases over the 1989–1991 index. In some cases, a drop in production per capita reflects government decisions to limit production in order to maintain higher prices for agricultural products. The United States and Japan fall into this category.

Map 61 Average Daily Per Capita Supply of Calories (Kilocalories)

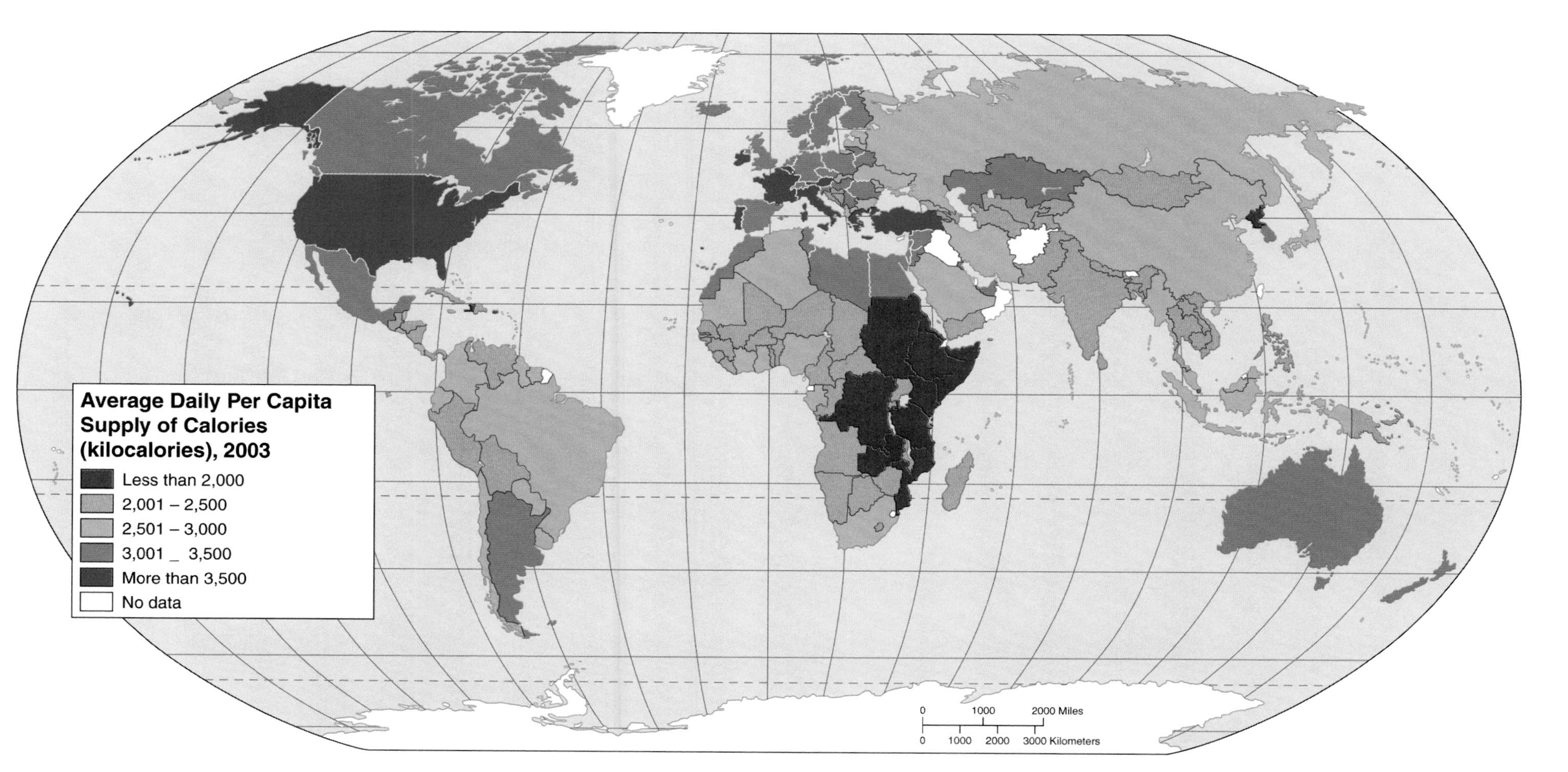

The data shown on this map, which indicate the presence or absence of critical food shortages, do not necessarily indicate the presence of starvation or famine. But they certainly do indicate potential problem areas for the next decade. The measurements are in calories from *all* food sources: domestic production, international trade, drawdown on stocks or food reserves, and direct foreign contributions or aid. The quantity of calories available is that amount, estimated by the UN's Food and Agriculture Organization (FAO), that reaches consumers. The calories actually consumed may be lower than the figures shown, depending on how much is lost in a variety of ways: in home storage (to pests such as rats and mice), in preparation and cooking, through consumption by pets and domestic animals, and as discarded foods, for example. The estimate of need is not a global uniform value but is calculated for each country on the basis of the age and sex distribution of the population and the estimated level of activity of the population. Compare this map with Map 59 for a good measure of potential problem areas for food shortages within the next decade.

Map 62 Energy Production Per Capita

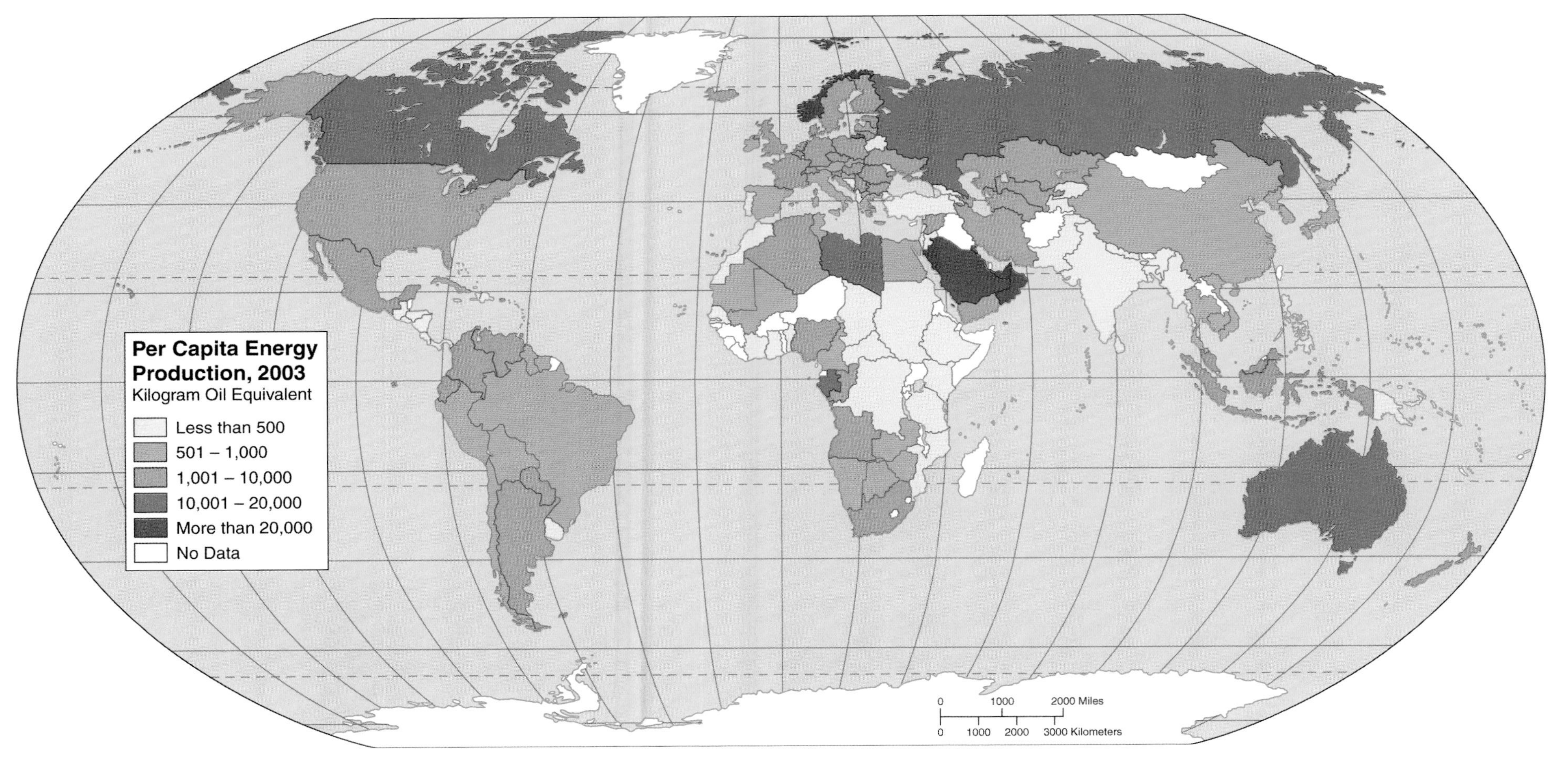

Energy production per capita is a measure of the availability of mechanical energy to assist people in their work. This map shows the amount of all kinds of energy—solid fuel (primarily coal), liquid fuel (primarily petroleum), natural gas, geothermal, wind, solar, hydroelectric, nuclear, waste recycling, and indigenous heat pumps—produced per person in each country. With some exceptions, wealthier countries produce more energy per capita than poor ones. Countries such as Japan and many European states rank among the world's wealthiest, but are energy-poor and produce relatively little of their own energy. They have the ability, however, to pay for imports. On the other hand, countries such as those of the Persian Gulf or the oil-producing states of Central and South America may rank relatively low on the scale of economic development but rank high as producers of energy. In many poor countries, especially in Central and South America, Africa, South Asia, and East Asia, large proportions of energy come from traditional fuels such as firewood and animal dung. Indeed for many in the developing world, the real energy crisis is a shortage of wood for cooking and heating.

Map 63 Energy Consumption Per Capita

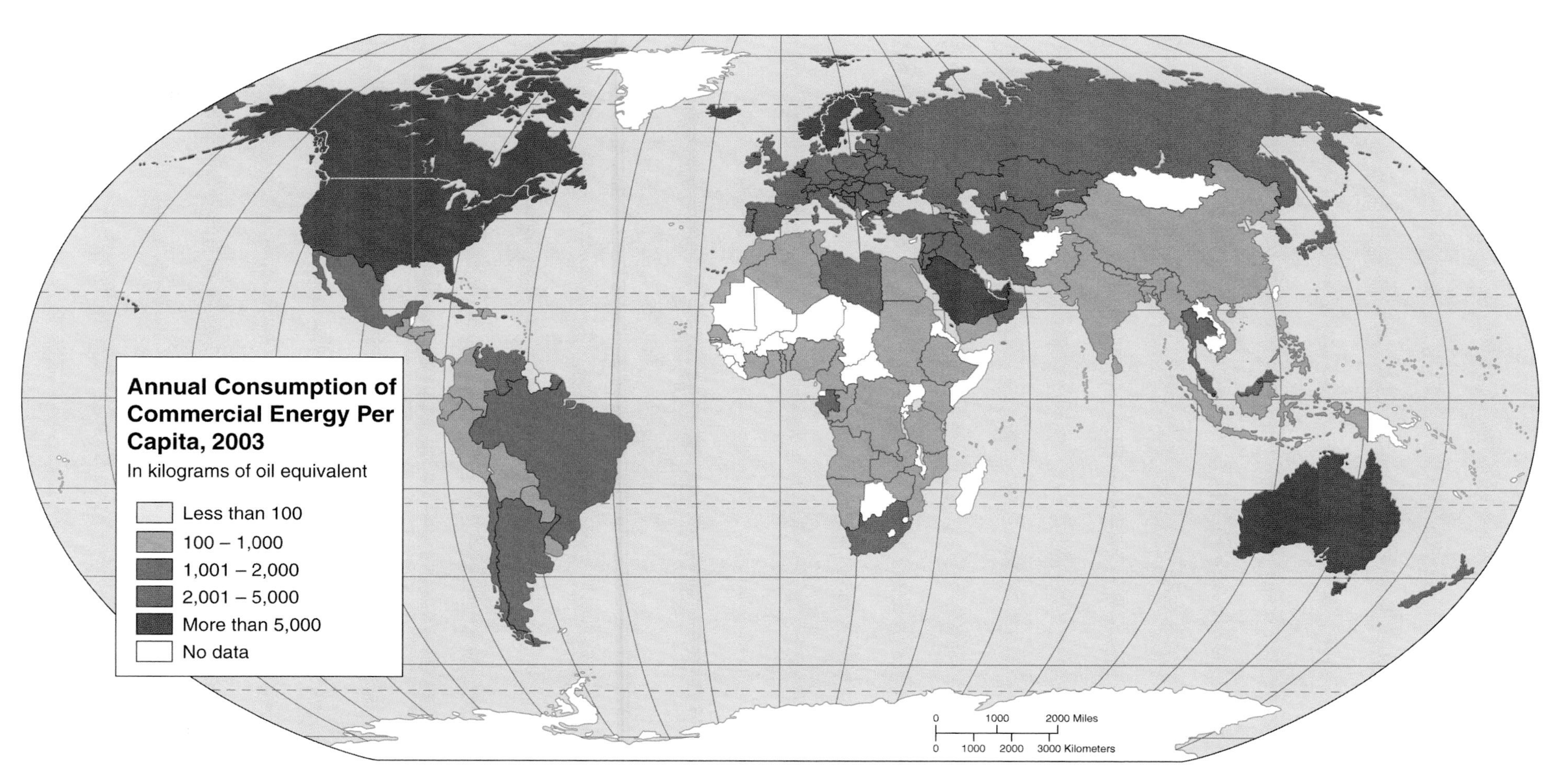

Of all the quantitative measures of economic well-being, energy consumption per capita may be the most expressive. All of the countries defined by the World Bank as having high incomes consume at least 100 gigajoules of commercial energy (the equivalent of about 3.5 metric tons of coal) per person per year, with some, such as the United States and Canada, having consumption rates in the 300 gigajoule range (the equivalent of more than 10 metric tons of coal per person per year). With the exception of the oil-rich Persian Gulf states, where consumption figures include the costly "burning off" of excess energy in the form of natural gas flares at wellheads, most of the highest-consuming countries are in the Northern Hemisphere, concentrated in North America and Western Europe. At the other end of the scale are low-income countries, whose consumption rates are often less than 1 percent of those of the United States and other high consumers. These figures do not, of course, include the consumption of noncommercial energy—the traditional fuels of firewood, animal dung, and other organic matter—widely used in the less developed parts of the world.

Map 64 Energy Dependency

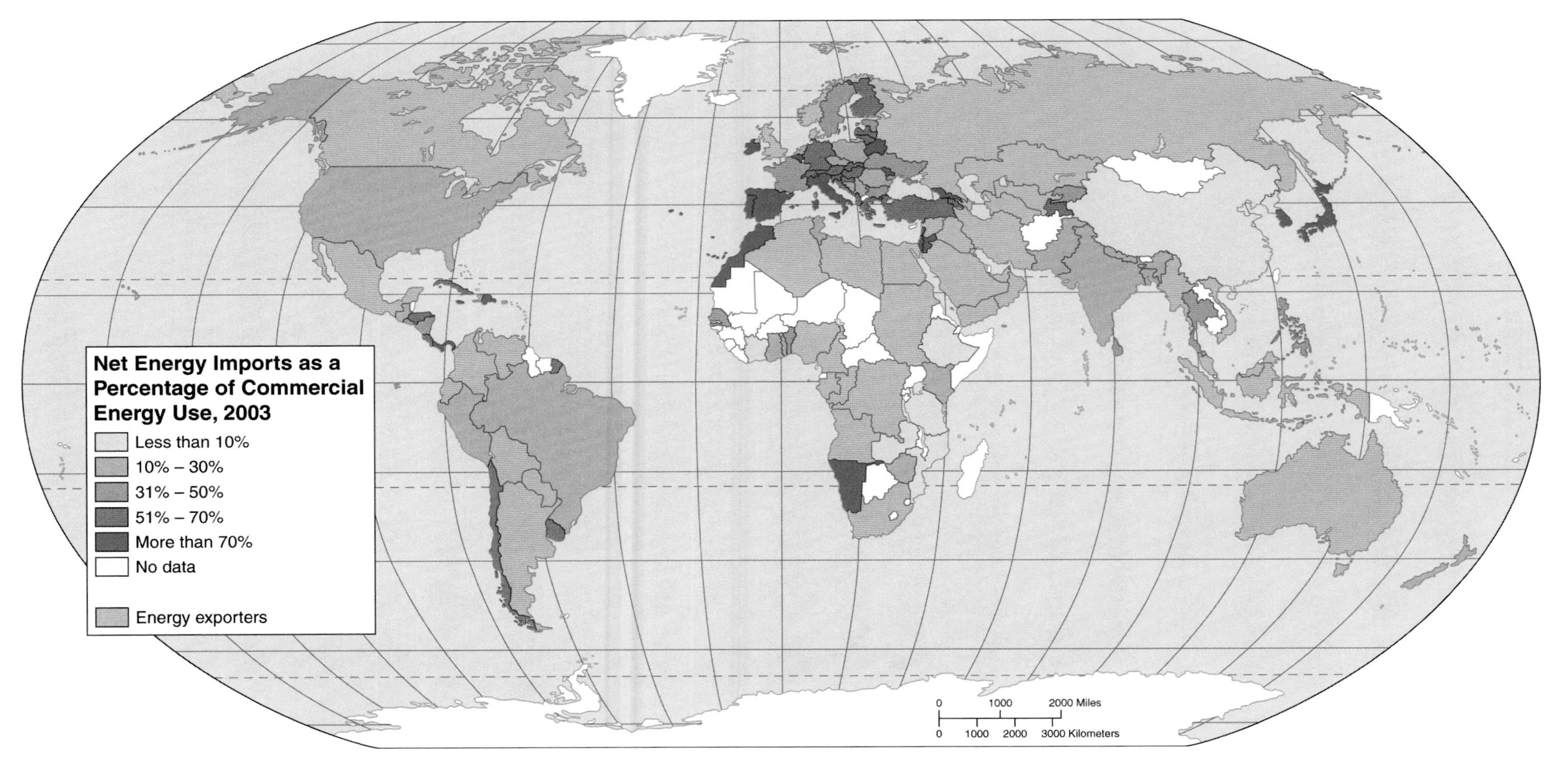

The patterns on the map show dependence on commercial energy before transformation to other end-use fuels such as electricity or refined petroleum products; energy from traditional sources such as fuelwood or dried animal dung is not included. Energy dependency is the difference between domestic consumption and domestic production of commercial energy and is most often expressed as a net energy import or export. A few of the world's countries are net exporters of energy: most are importers. The growth in global commercial energy use over the last decade indicates growth in the modern sectors of the economy—industry, transportation, and urbanization—particularly in the lesser developed countries. Still, the primary consumers of energy—and those having the greatest dependence on foreign sources of energy—are the more highly developed countries of Europe, North America, and Japan.

Map 65 Flows of Oil

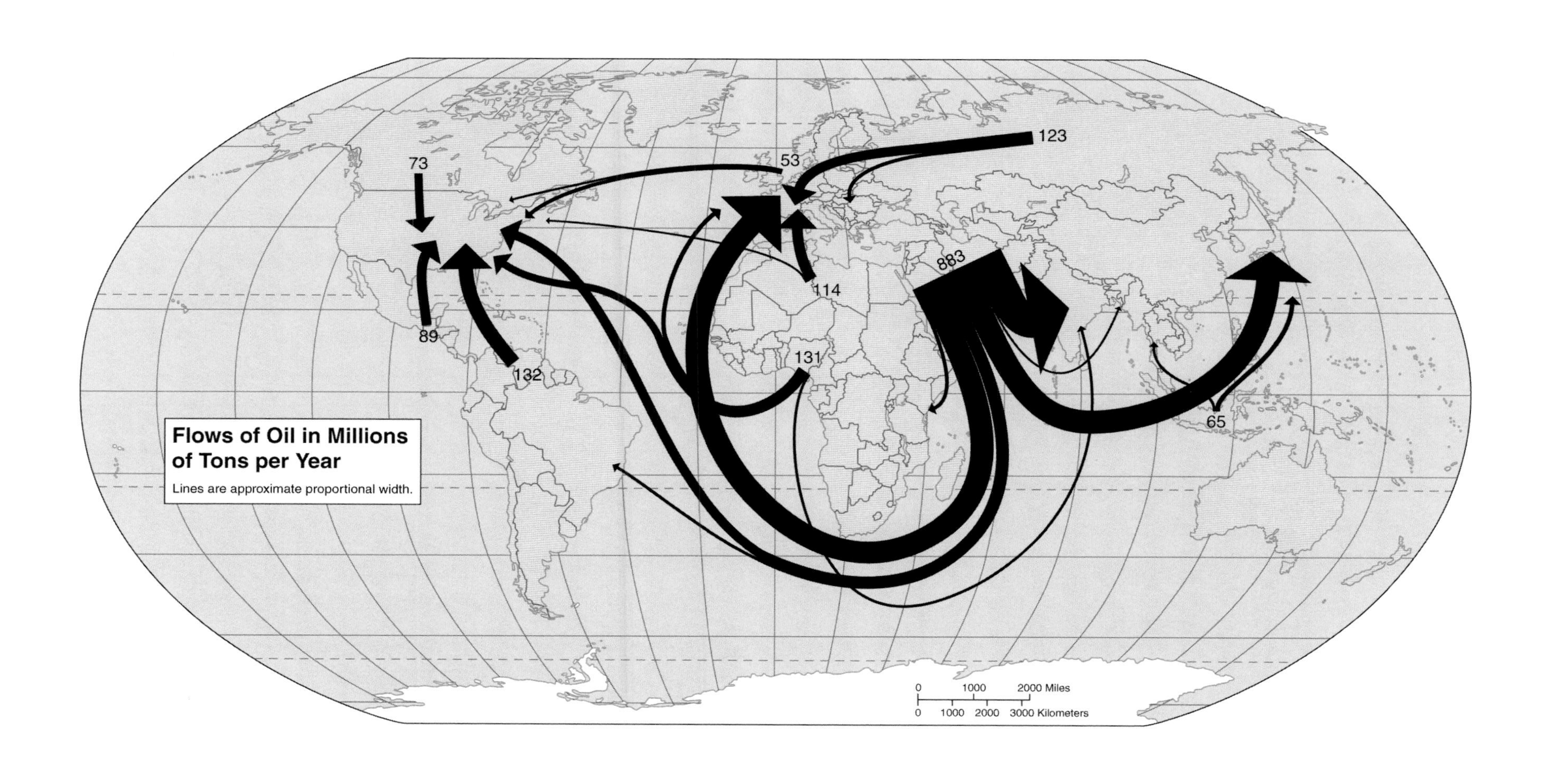

The pattern of oil movements from producing region to consuming region is one of the dominant facts of contemporary international maritime trade. Supertankers carry a million tons of crude oil and charge rates in excess of $0.10 per ton per mile, making the transportation of oil not only a necessity for the world's energy-hungry countries, but also an enormously profitable proposition. One of the major negatives of these massive oil flows is the damage done to the oceanic ecosystems—not just from the well-publicized and dramatic events like the wrecking of the *Exxon Valdez* but from the incalculable amounts of oil from leakage, scrubbings, purgings, and so on, which are a part of the oil transport technology. It is clear from the map that the primary recipients of these oil flows are the world's most highly developed economies.

Map 66 Production of Crucial Materials

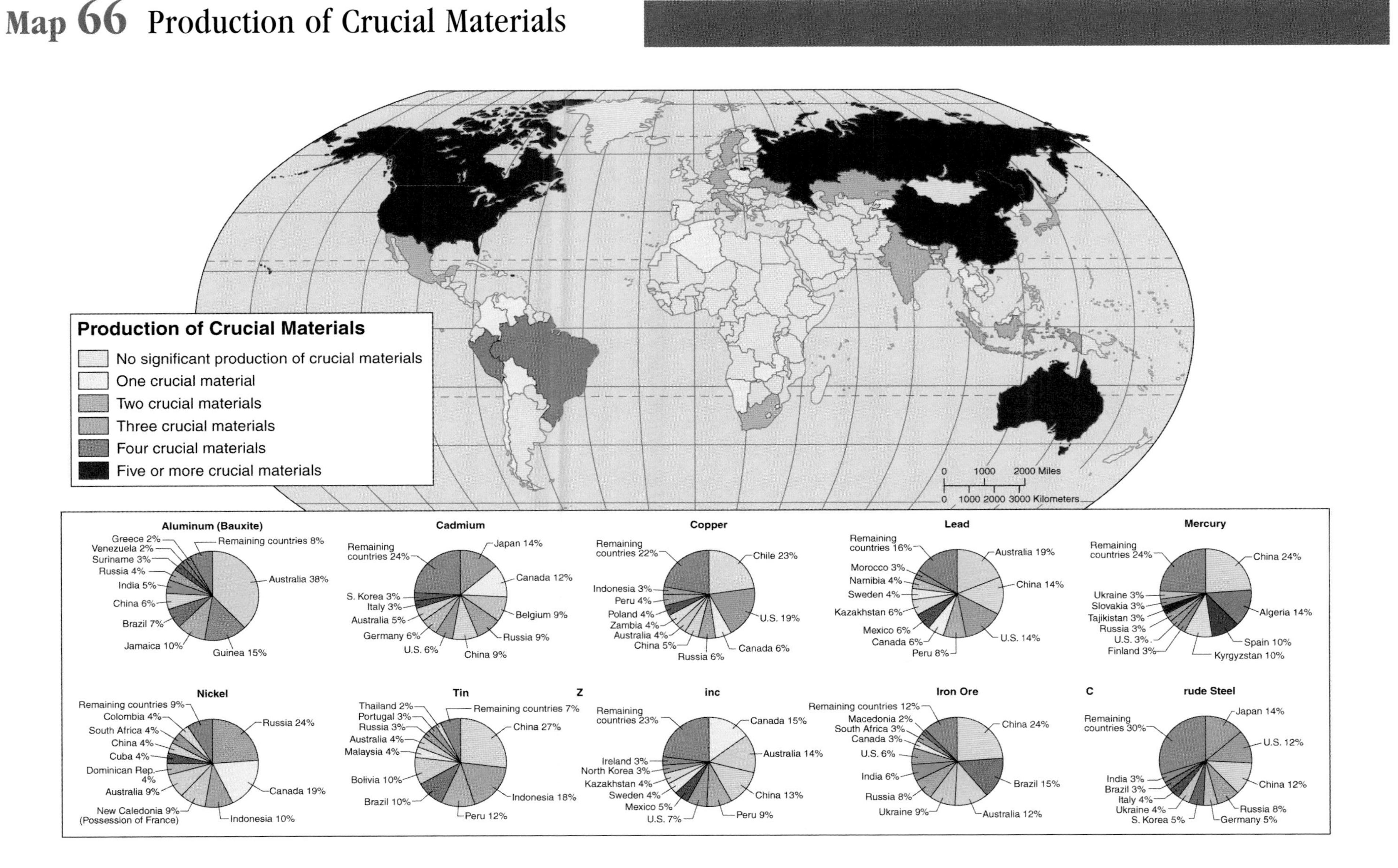

The data on this map portray world production of the metals most important for the operation of a modern industrial economy. The sector graphs across the bottom of the map show the percentage of production of crucial materials by the 10 leading countries for each of 10 materials. For copper, lead, mercury, nickel, tin, and zinc, the annual production data reflect the metal content of the ore mined. Aluminum (or bauxite ore) and iron ore production are expressed in gross weight of ore mined. Cadmium production refers to the refined metal, and crude steel production to usable ingots, cast products, and liquid steel. By comparing this map with Map 66, you will discover that some of the world's top producer nations of these critical materials are also among the world's top consumer nations.

Map 67 Consumption of Crucial Materials

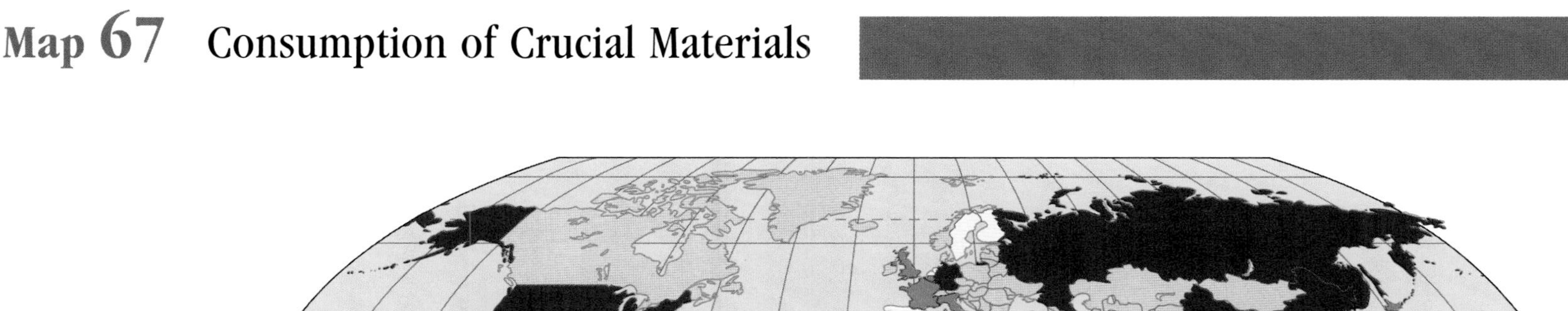

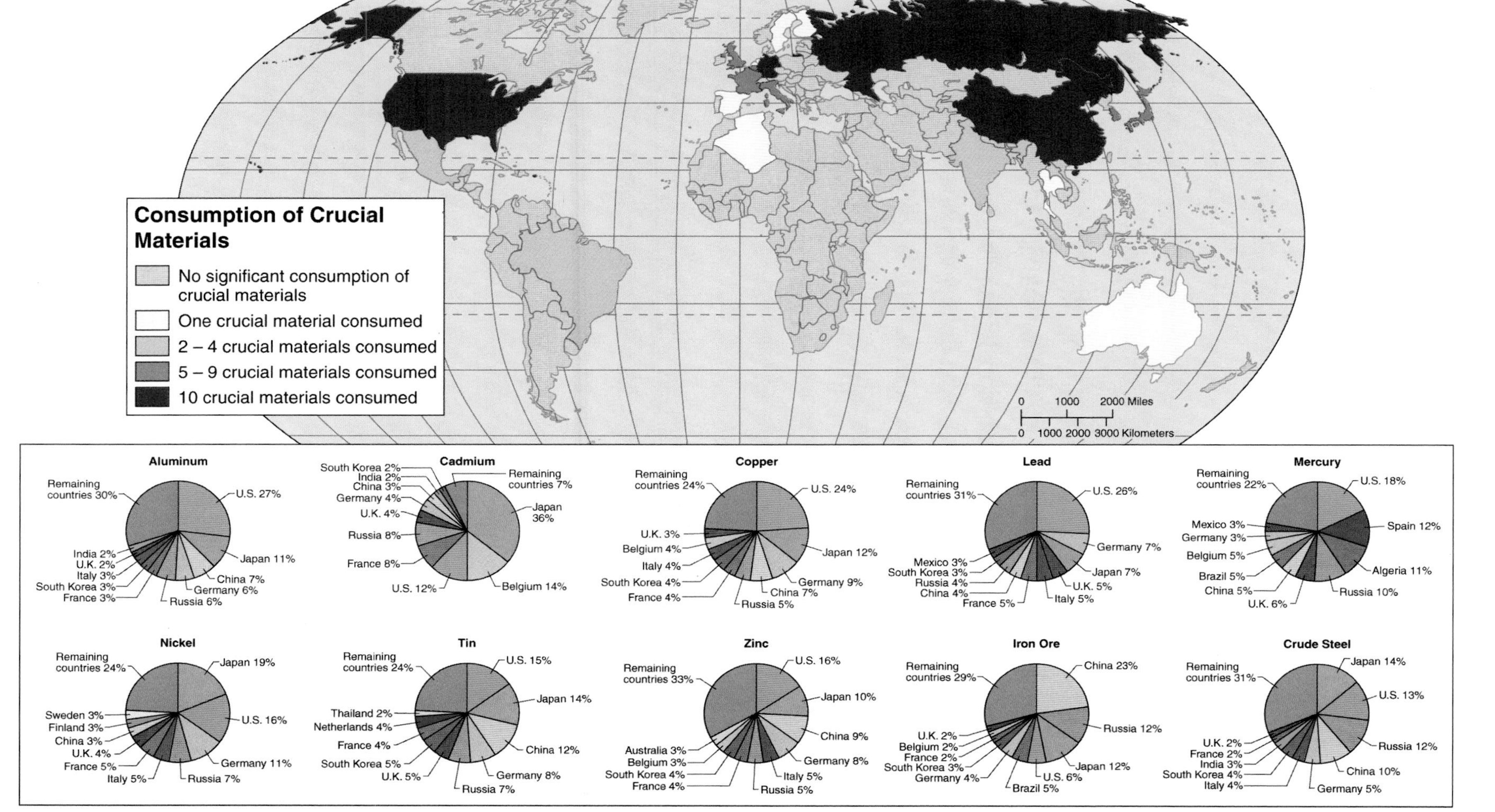

Consumption data refer to the domestic use of refined metals (for example, the tons of steel used in the manufacture of automobiles). Some countries rank among the top in both production and consumption, and those that do are among the most highly developed nations. The United States, for example, ranks in the top 4 consumers for each metal; but the United States also ranks in the top 10 producer countries for 7 of the metals. Many countries that rank high as producers but not as consumers have colonial dependency economies, producing raw materials for an export market, often at the mercy of the marketplace. Jamaica and Suriname, for example, depend extremely heavily upon the sale of bauxite ore (crude aluminum). When the United States, Japan, or Russia cuts its use of aluminum, the economies of Jamaica and Suriname crash.

Part VI

Environmental Conditions

Map 68 Deforestation and Desertification

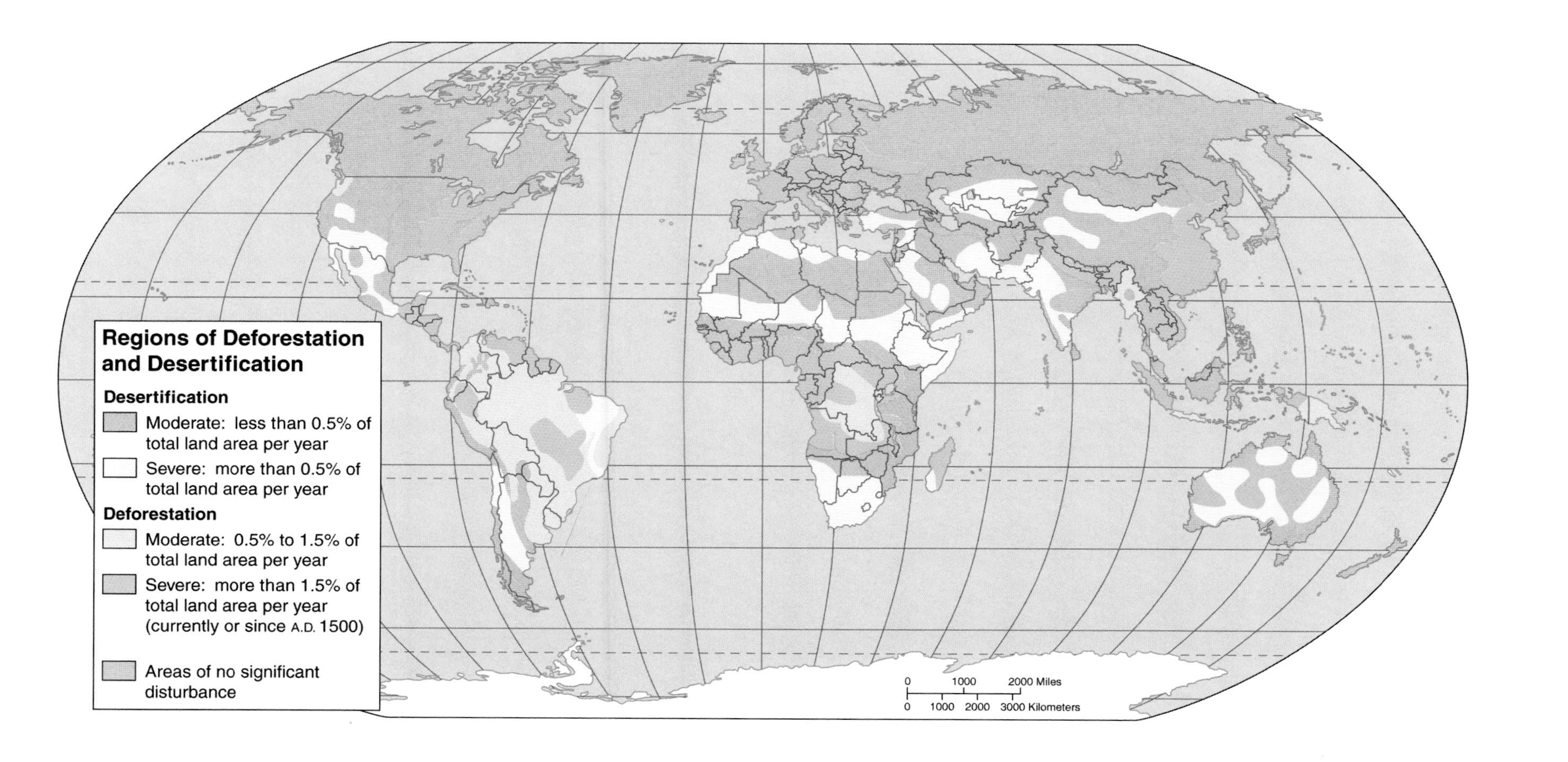

While those of us in the developed countries of the world tend to think of environmental deterioration as the consequence of our heavily industrialized economies, in fact the worst examples of current environmental degradation are found within the world's less developed regions. There, high population growth rates and economies limited primarily to farming have forced the increasing use of more marginal (less suited to cultivation) land. In the world's grassland and arid environments, which occupy approximately 40 percent of the world's total land area, increasing cultivation pressures are turning vulnerable areas into deserts incapable of sustaining agricultural productivity. In the world's forested regions, particularly in the tropical forests of Middle and South America, Africa, and Asia, a similar process is occurring: increasing pressure for more farmland is creating a process of deforestation or forest clearing that destroys the soil, reduces the biological diversity of the forest regions, and ultimately may have the capacity to alter the global climate by contributing to an increase in carbon dioxide in the atmosphere. This increases the heat trapped in the atmosphere and enhances the greenhouse effect.

Map 69 Soil Degradation

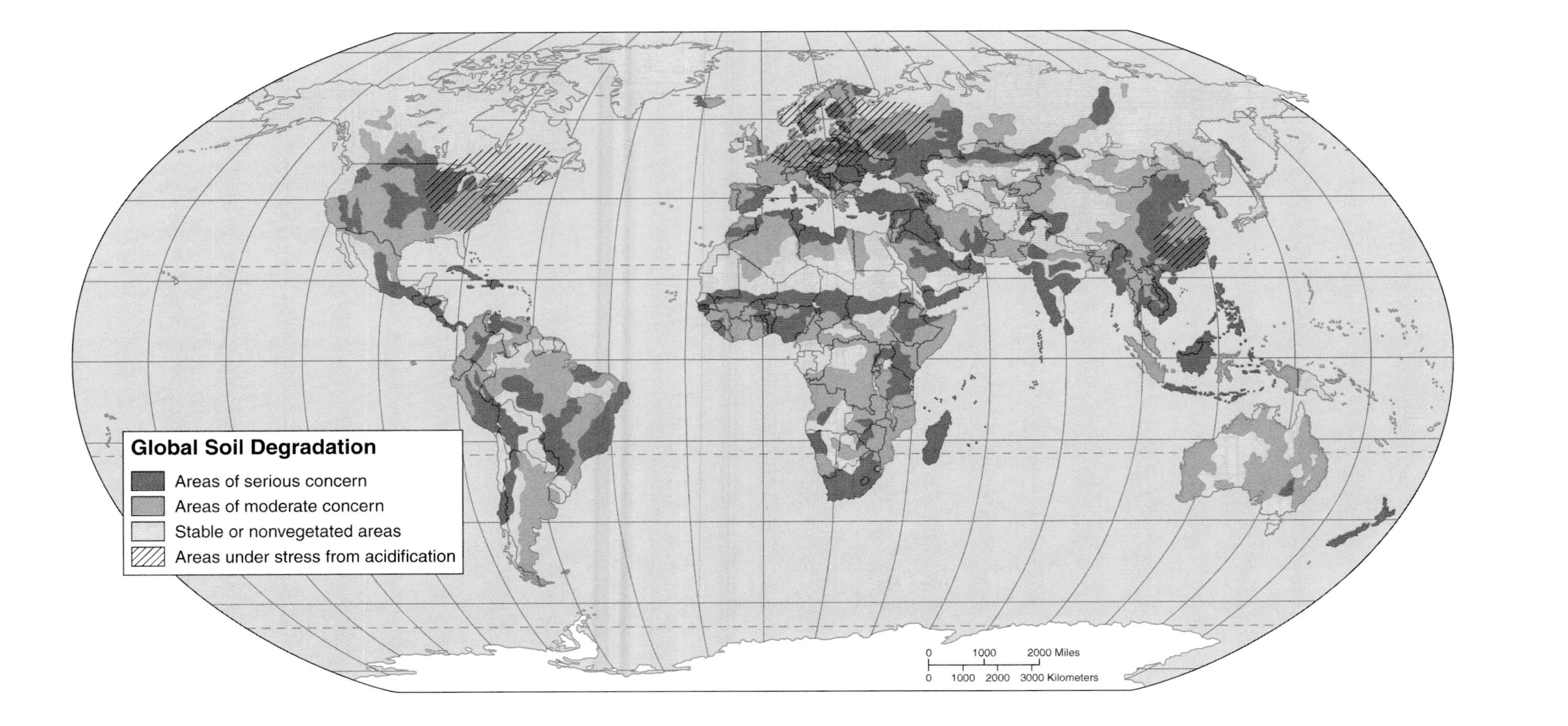

Recent research has shown that more than 3 billion acres of the world's surface suffer from serious soil degradation, with more than 22 million acres so severely eroded or poisoned with chemicals that they can no longer support productive crop agriculture. Most of this soil damage has been caused by poor farming practices, overgrazing of domestic livestock, and deforestation. These activities strip away the protective cover of natural vegetation—forests and grasslands—allowing wind and water erosion to remove the topsoil that contains the necessary nutrients and soil microbes for plant growth. But millions of acres of topsoil have been degraded by chemicals as well. In some instances these chemicals are the result of overapplication of fertilizers, herbicides, pesticides, and other agricultural chemicals. In other instances, chemical deposition from industrial and urban wastes and from acid precipitation has poisoned millions of acres of soil. As the map shows, soil erosion and pollution are problems not just in developing countries with high population densities and increasing use of marginal lands but in the more highly developed regions of mechanized, industrial agriculture as well. While many methods for preventing or reducing soil degradation exist, they are seldom used because of ignorance, cost, or perceived economic inefficiency.

Map 70 Air and Water Quality

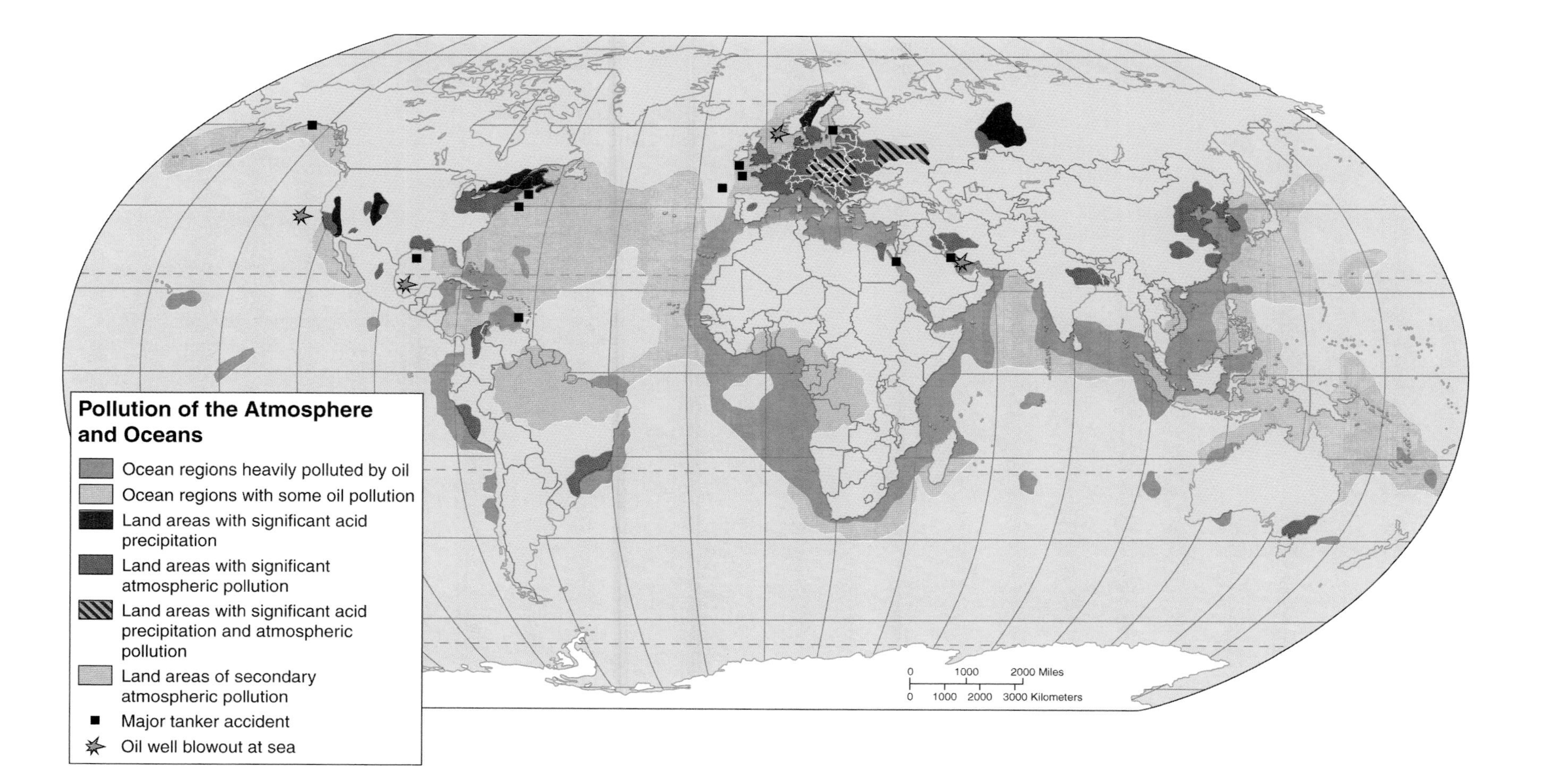

The pollution of the world's oceans and atmosphere has long been a matter of concern to environmental scientists. The great circulation systems of ocean and air are the controlling factors of the earth's natural environment, and modifications to those systems have unknown consequences. This map is based on what we can measure: (1) areas of oceans where oil pollution has been proven to have inflicted significant damage to ocean ecosystems and lifeforms (including phytoplankton—the oceans' primary food producers, the equivalent of land vegetation); (2) areas of oceans where unusually high concentrations of hydrocarbons from oil spills may have inflicted some damage to the oceans' biota; (3) land areas where the combination of sulphur and nitrogen oxides with atmospheric water vapor has produced acid precipitation at high enough levels to have produced significant damage to terrestrial vegetation systems; (4) land areas where the emissions from industrial, transportation, commercial, residential, and other uses of fossil fuels have produced concentrations of atmospheric pollutants high enough to be damaging to human health; and (5) land areas of secondary air pollution where the primary pollutant is smoke from forest clearance. A glance at the map shows that there are few areas of the world where some form of oceanic or atmospheric pollution is not a part of our environmental system. Scientists are still debating the long-range implications of this pollution, but nearly all agree that the consequences, whatever they may be, will not be good.

Map 71 Per Capita Carbon Dioxide (CO_2) Emissions

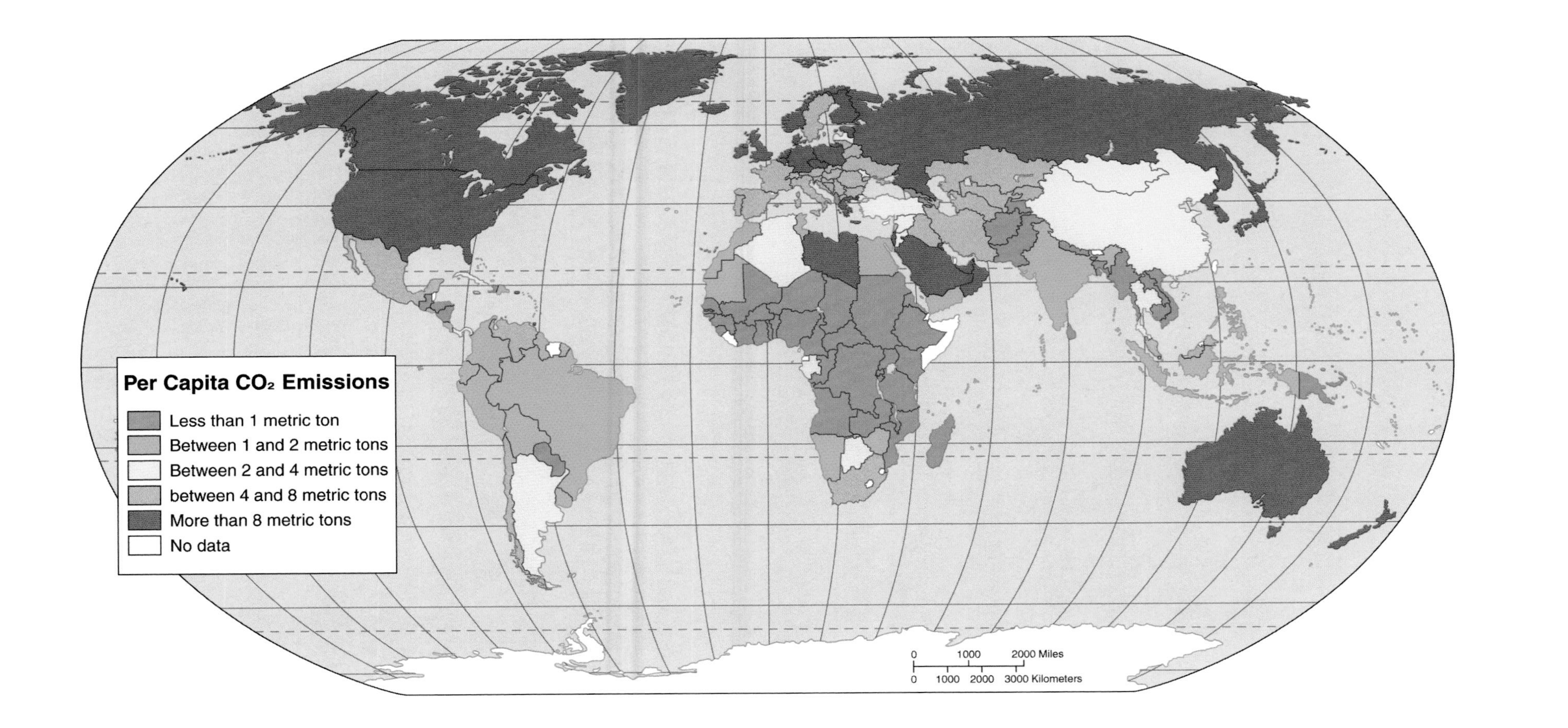

Carbon dioxide emissions are a major indicator of economic development, since they are generated largely by burning of fossil fuels for electrical power generation, for industrial processes, for domestic and commercial heating, and for the internal combustion engines of automobiles, trucks, buses, planes, and trains. Scientists have long known that carbon dioxide in the atmosphere increases the ability of atmosphere to retain heat, a phenomenon known as the greenhouse effect. While the greenhouse effect is a natural process (and life on earth as we know it would not be possible without it), many scientists are concerned that an increase in carbon dioxide in the atmosphere will augment this process, creating a global warming trend and a potential worldwide change of climate patterns. These climatological changes threaten disaster for many regions and their peoples in both the developed and less developed areas of the world. You will note from the map that the countries of the midlatitude regions generate extremely high levels of carbon dioxide per capita.

Map 72 Potential Global Temperature Change

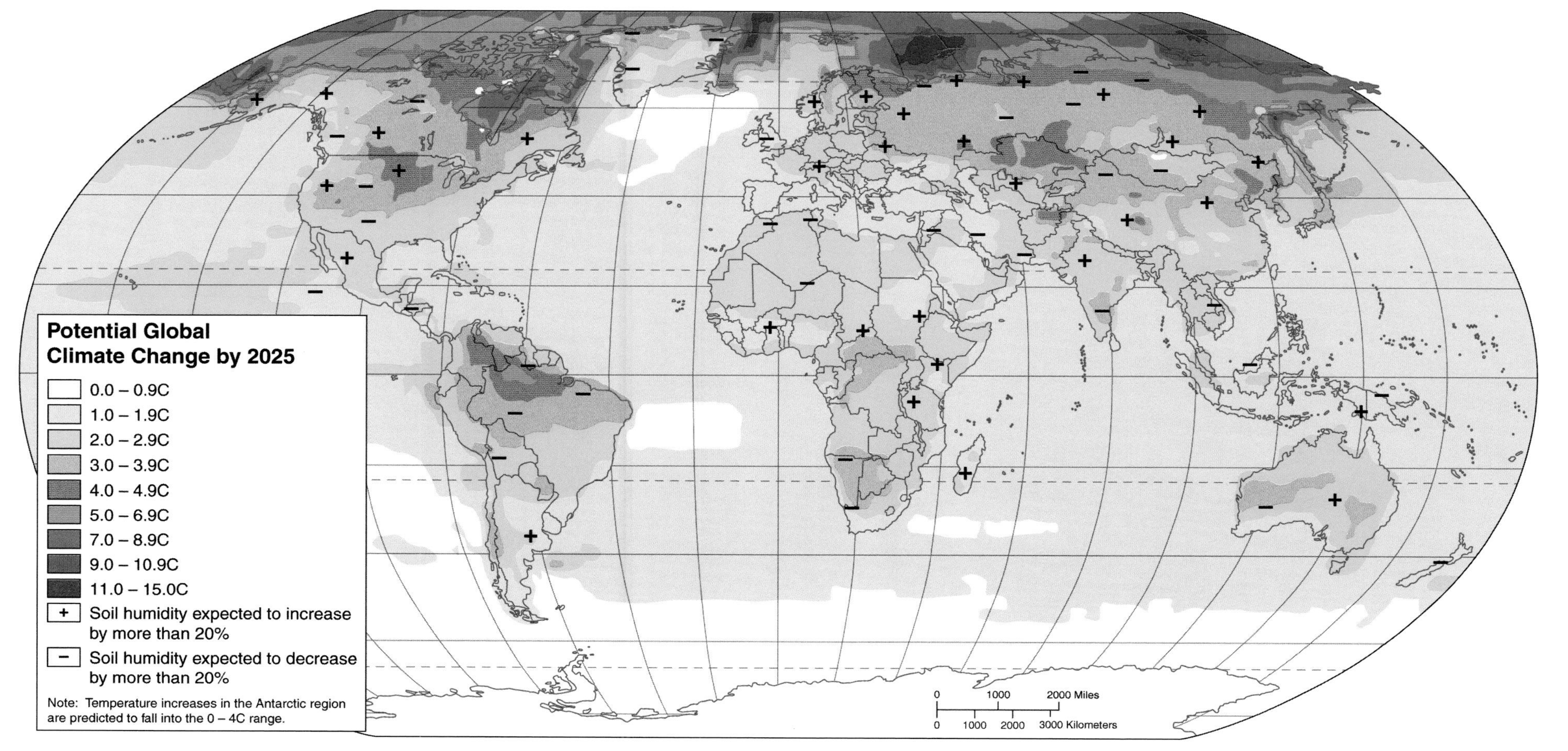

According to atmospheric scientists, one of the major problems of the twenty-first century will be "global warming," produced as the atmosphere's natural ability to trap and retain heat is enhanced by increased percentages of carbon dioxide, methane, chlorinated fluorocarbons or "CFCs," and other "greenhouse gases" in the earth's atmosphere. Computer models based on atmospheric percentages of carbon dioxide resulting from present use of fossil fuels show that warming is not just a possibility but a probability. Increased temperatures would cause precipitation patterns to alter significantly as well and would produce a number of other harmful effects, including a rise in the level of the world's oceans that could flood most coastal cities. International conferences on the topic of the enhanced greenhouse effect have resulted in several international agreements to reduce the emission of carbon dioxide or to maintain it at present levels. Unfortunately, the solution is not that simple since reduction of carbon dioxide emissions is, in the short run, expensive—particularly as long as the world's energy systems continue to be based on fossil fuels. Chief among the countries that could be hit by serious international mandates to reduce emissions are those highest on the development scale who use the highest levels of fossil fuels and, therefore, produce the highest emissions, and those on the lowest end of the development scale whose efforts to industrialize could be severely impeded by the more expensive energy systems that would replace fossil fuels.

Map 73 The Loss of Biodiversity: Globally Threatened Animal Species

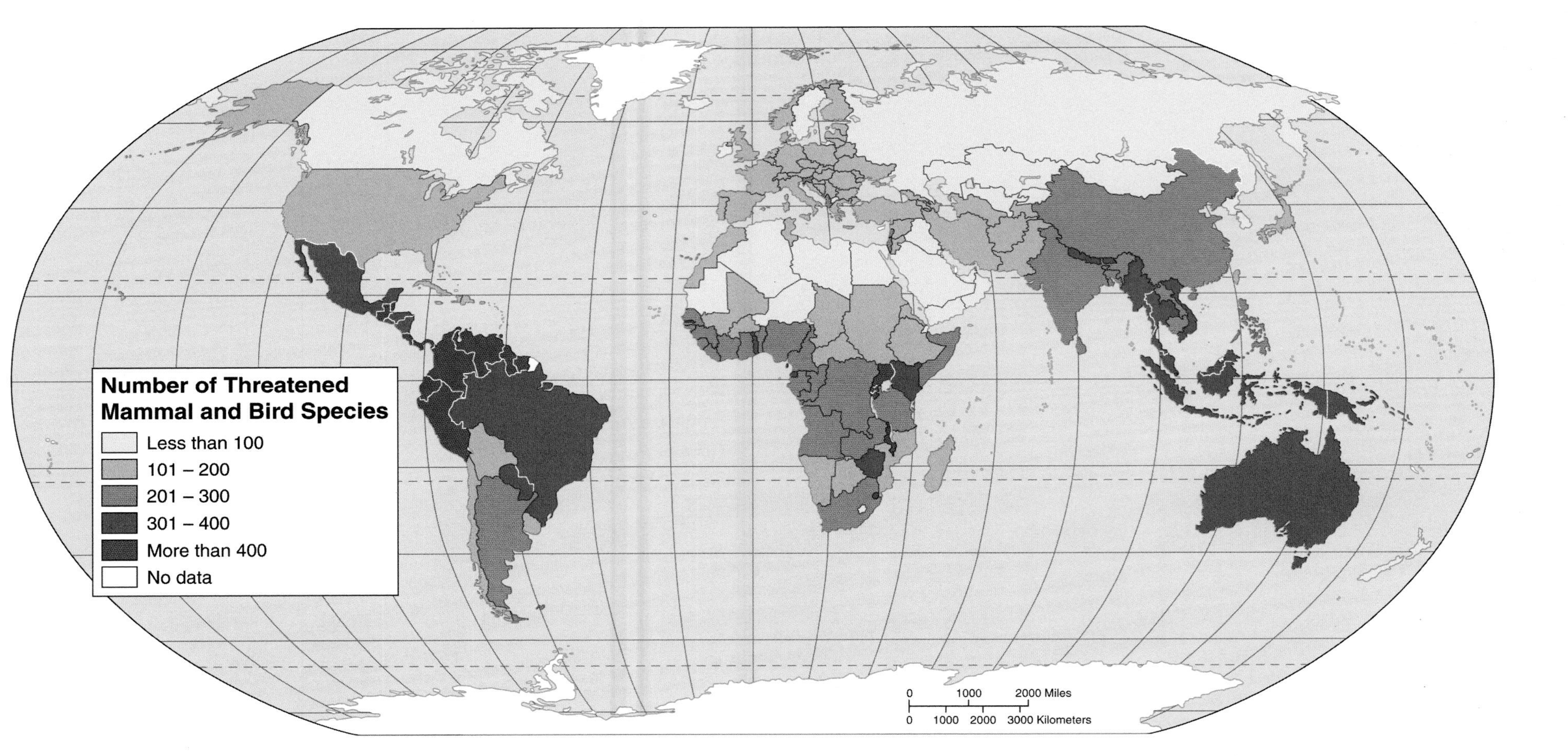

Threatened species are those in grave danger of going extinct. Their populations are becoming restricted in range, and the size of the populations required for sustained breeding is nearing a critical minimum. *Endangered species* are in immediate danger of becoming extinct. Their range is already so reduced that the animals may no longer be able to move freely within an ecozone, and their populations are at the level where the species may no longer be able to sustain breeding. Most species become threatened first and then endangered as their range and numbers continue to decrease. When people think of animal extinction, they think of large herbivorous species like the rhinoceros or fierce carnivores like lions, tigers, or grizzly bears. Certainly these animals make almost any list of endangered or threatened species. But there are literally hundreds of less conspicuous animals that are equally threatened. Extinction is normally nature's way of informing a species that it is inefficient. But conditions in the late twentieth century are controlled more by human activities than by natural evolutionary processes. Species that are endangered or threatened fall into that category because, somehow, they are competing with us or with our domesticated livestock for space and food. And in that competition the animals are always going to lose.

Map 74 The Loss of Biodiversity: Globally Threatened Plant Species

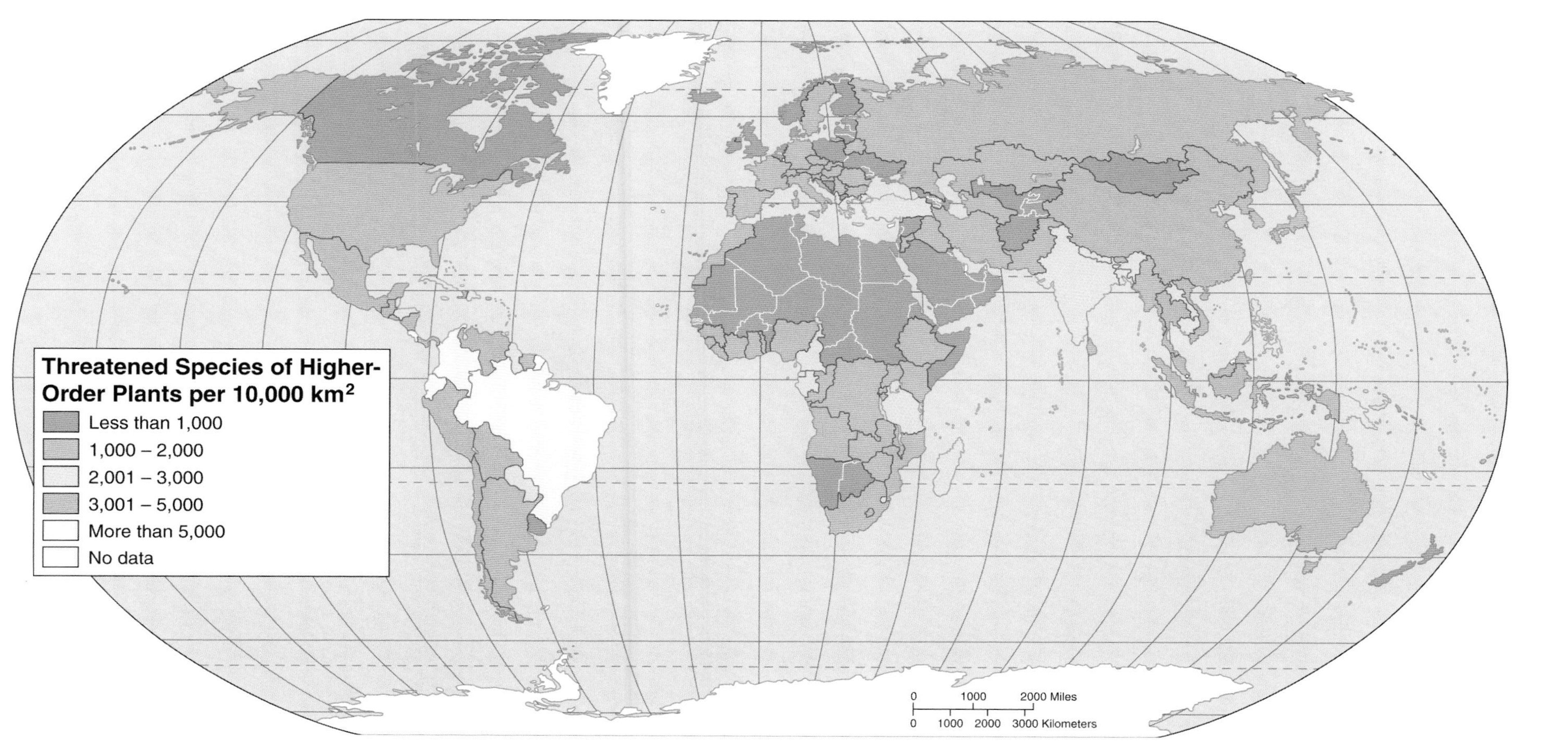

While most people tend to be more concerned about the animals on threatened and endangered species lists, the fact is that many more plants are in jeopardy, and the loss of plant life is, in all ecological regions, a more critical occurrence than the loss of animal populations. Plants are the primary producers in the ecosystem; that is, plants produce the food upon which all other species in the food web, including human beings, depend for sustenance. It is plants from which many of our critical medicines come, and it is plants that maintain the delicate balance between soil and water in most of the world's regions. When environmental scientists speak of a loss of biodiversity, what they are most often describing is a loss of the richness and complexity of plant life that lends stability to ecosystems. Systems with more plant life tend to be more stable than those with less. For these and other reasons, the scientific concern over extinction is greater when applied to plants than to animals. It is difficult for people to become as emotional over a teak tree as they would over an elephant. But as great a tragedy as the loss of the elephant would be, the loss of the teak would be greater.

Map 75 Global Hotspots of Biodiversity

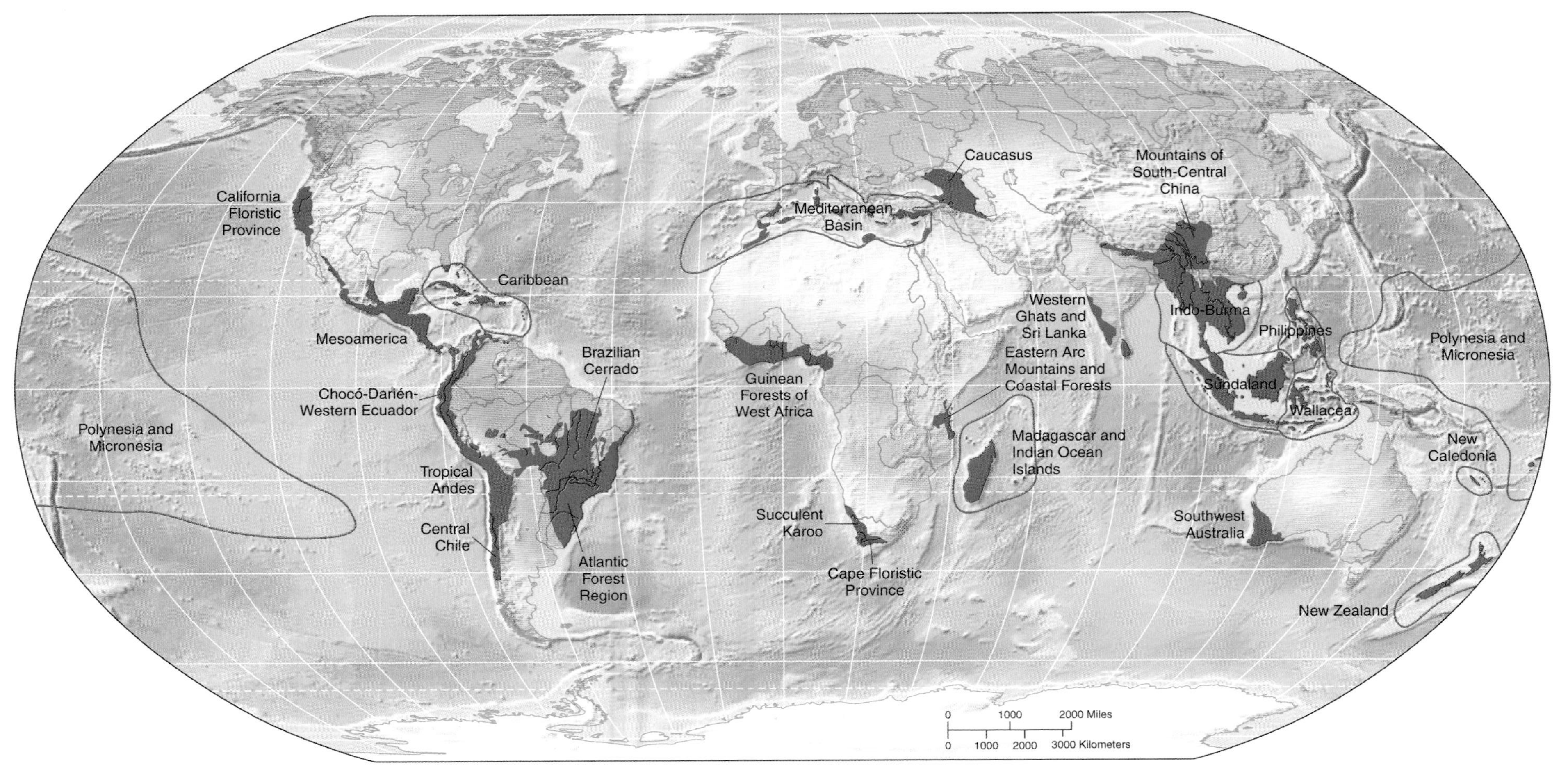

Where we have normally thought of tropical forest basins such as Amazonia as the world's most biologically diverse ecosystems, recent research has discovered the surprising fact that a number of hotspots of biological diversity exist outside the major tropical forest regions. These hotspot regions contain slightly less than 2 percent of the world's total land area but may contain up to 60 percent of the total world's terrestrial species of plants and animals. Geographically, the hotspot areas are characterized by vertical zonation (that is, they tend to be hilly to mountainous regions), long known to be a factor in biological complexity. They are also in coastal locations or near large bodies of water, locations that stimulate climatic variability and, hence, biological complexity. Although some of the hotspots are sparsely populated, others, such as "Sundaland," are among the world's most densely populated areas. Protection of the rich biodiversity of these hotspots is, most biologists feel, of crucial importance to the preservation of the world's biological heritage.

Map 76 Degree of Human Disturbance

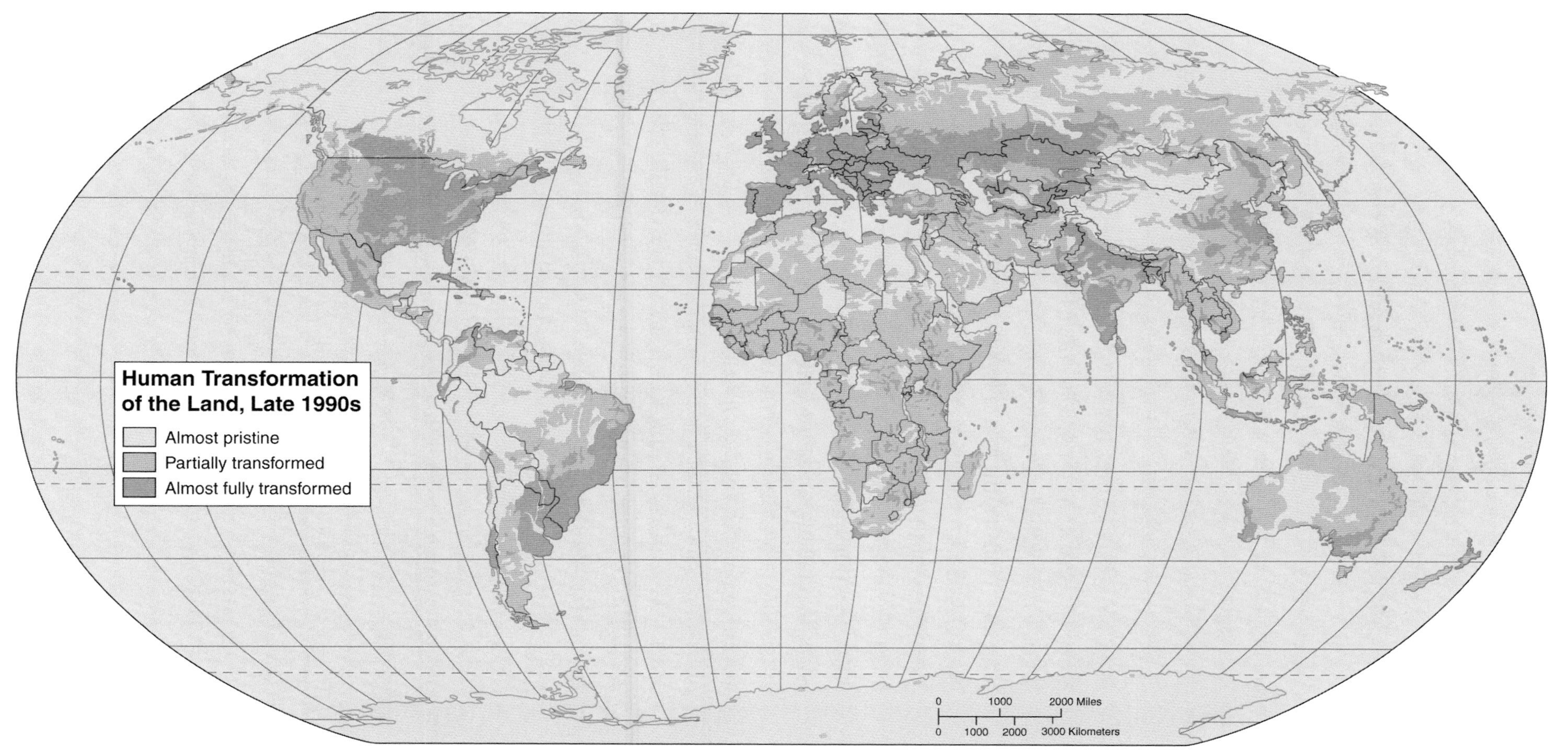

The data on human disturbance have been gathered from a wide variety of sources, some of them conflicting and not all of them reliable. Nevertheless, at a global scale this map fairly depicts the state of the world in terms of the degree to which humans have modified its surface. The almost pristine areas, covered with natural vegetation, generally have population densities under 10 persons per square mile. These areas are, for the most part, in the most inhospitable parts of the world: too high, too dry, too cold for permanent human habitation in large numbers. The partially transformed areas are normally agricultural areas, either subsistence (such as shifting cultivation) or extensive (such as livestock grazing). They often contain areas of secondary vegetation, regrown after removal of original vegetation by humans. They are also often marked by a density of livestock in excess of carrying capacity, leading to overgrazing, which further alters the condition of the vegetation. The almost fully transformed areas are those of permanent and intensive agriculture and urban settlement. The primary vegetation of these regions has been removed, with no evidence of regrowth or with current vegetation that is quite different from natural (potential) vegetation. Soils are in a state of depletion and degradation, and, in drier lands, desertification is a factor of human occupation. The disturbed areas match closely those areas of the world with the densest human populations.

Part VII

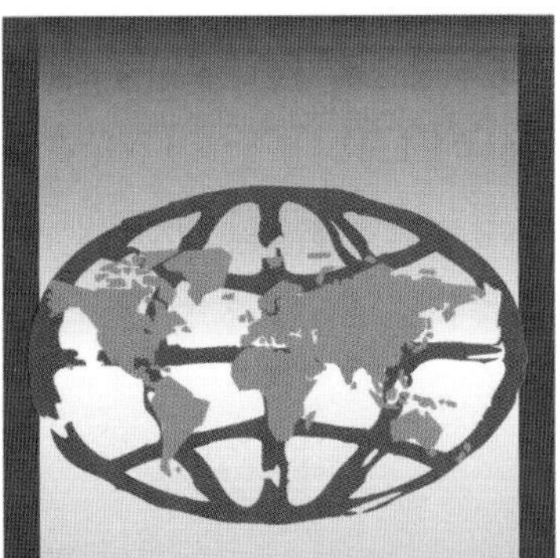

Regions of the World

Map 77 North America: Physical Features

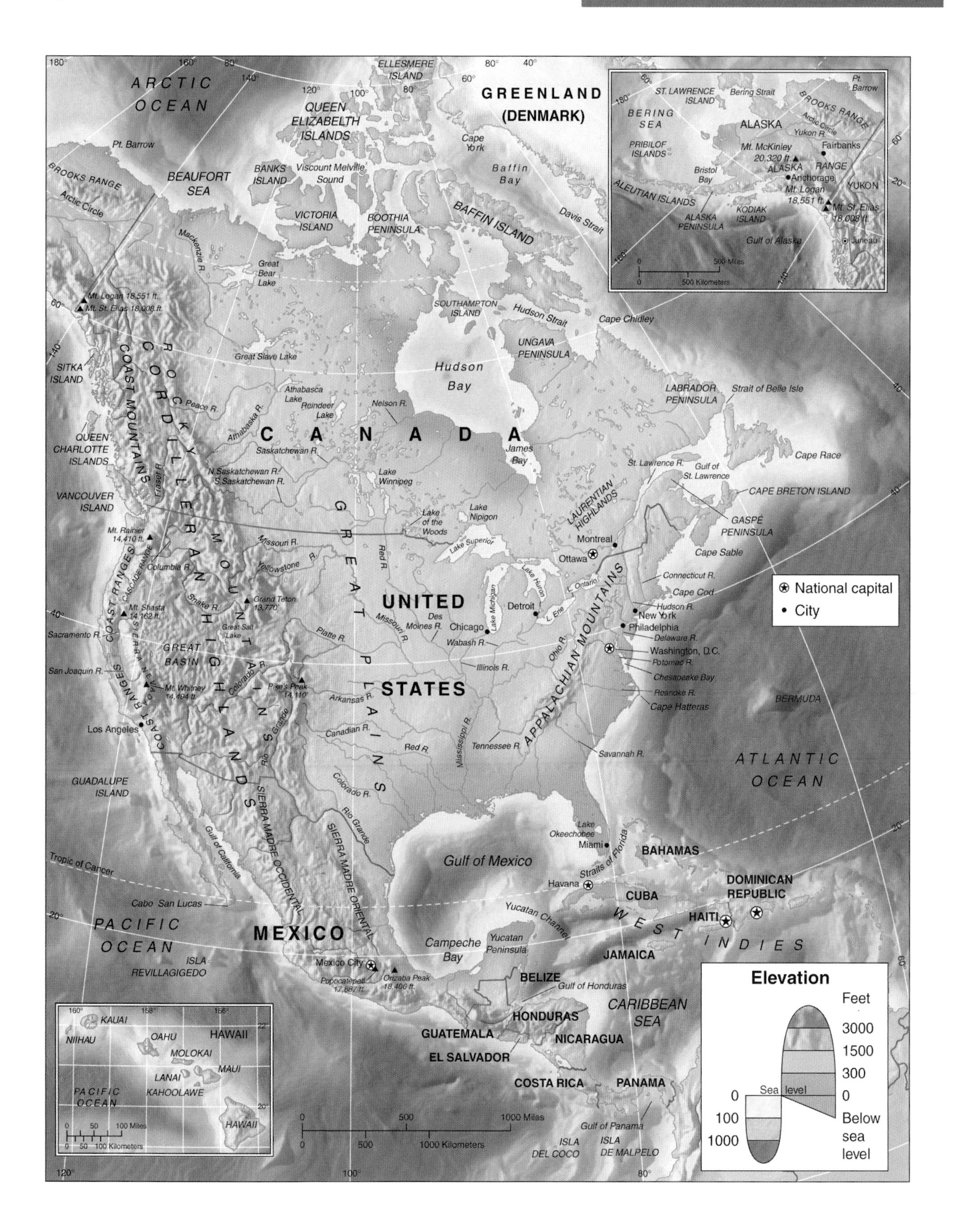

Map 78 North America: Political Divisions

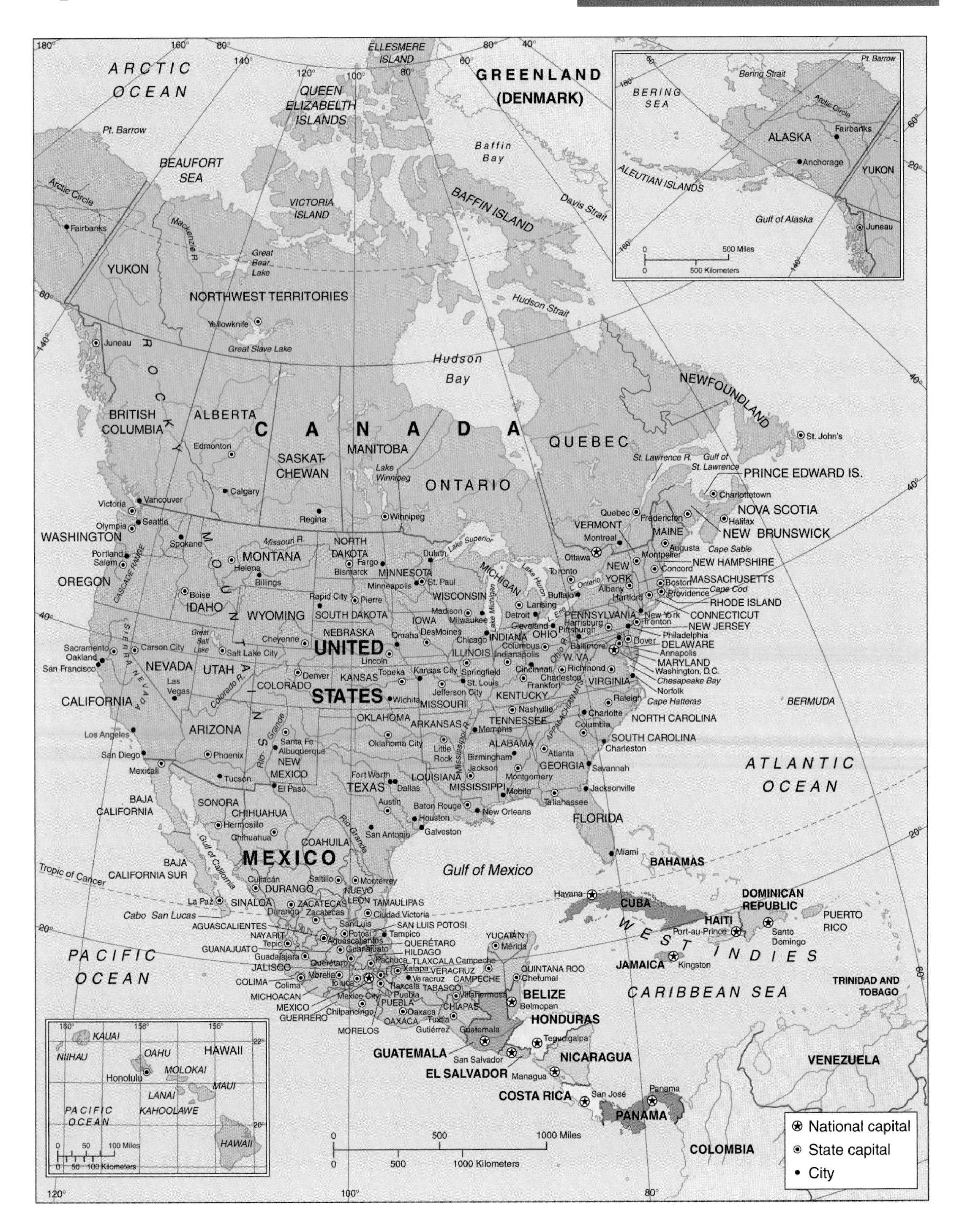

Map 79 North American Land Use

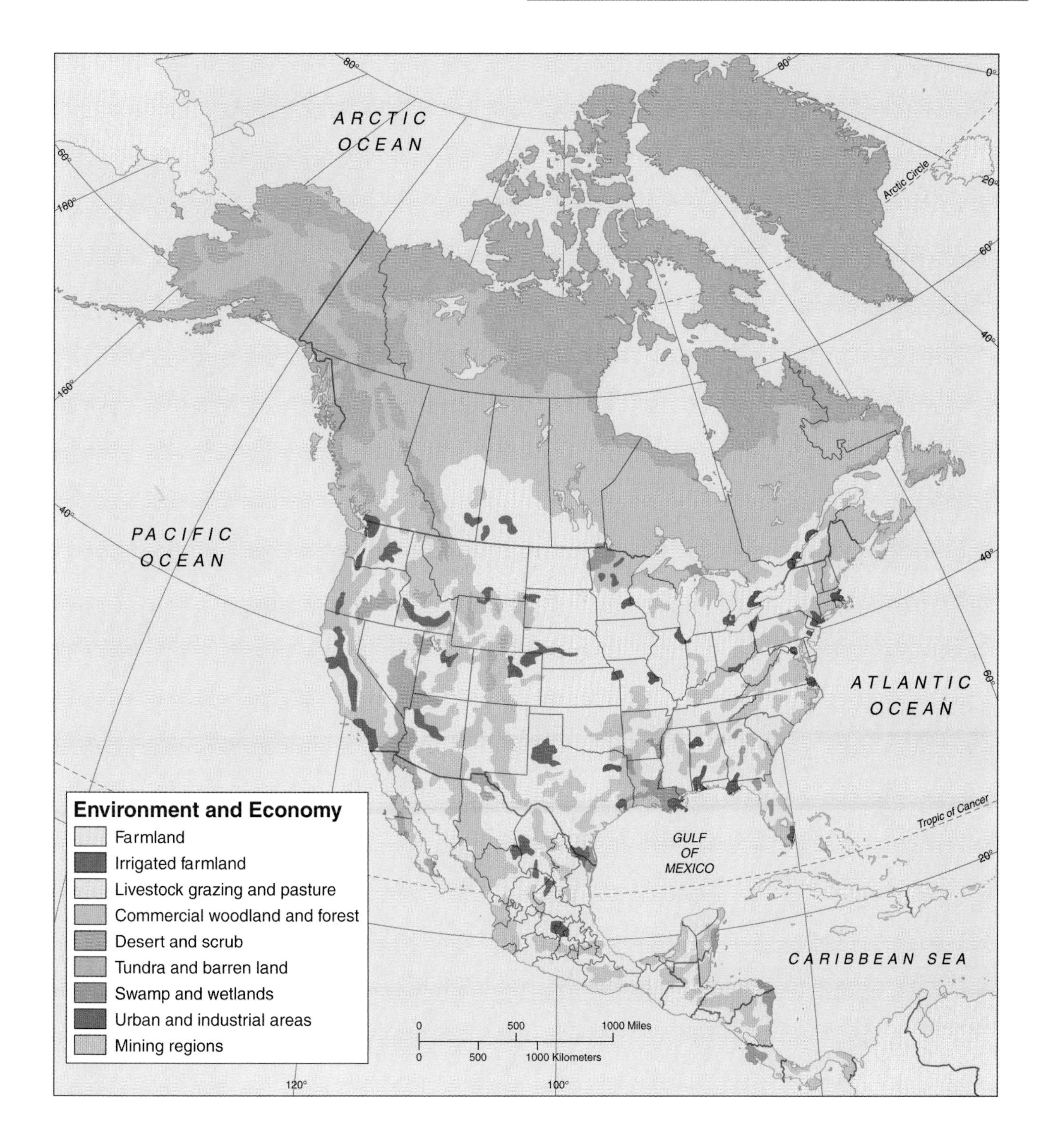

The use of land in North America represents a balance between agriculture, resource extraction, and manufacturing that is unmatched. The United States, as the world's leading industrial power, is also the world's leader in commercial agricultural production. Canada, despite its small population, is a ranking producer of both agricultural and industrial products and Mexico has begun to emerge from its developing nation status to become an important industrial and agricultural nation as well. The countries of Middle America and the Caribbean are just beginning the transition from agriculture to modern industrial economies. Part of the basis for the high levels of economic productivity in North America is environmental: a superb blend of soil, climate, and raw materials. But just as important is the cultural and social mix of the plural societies of North America, a mix that historically aided the growth of the economic diversity necessary for developed economies.

Map 80 South America: Physical Features

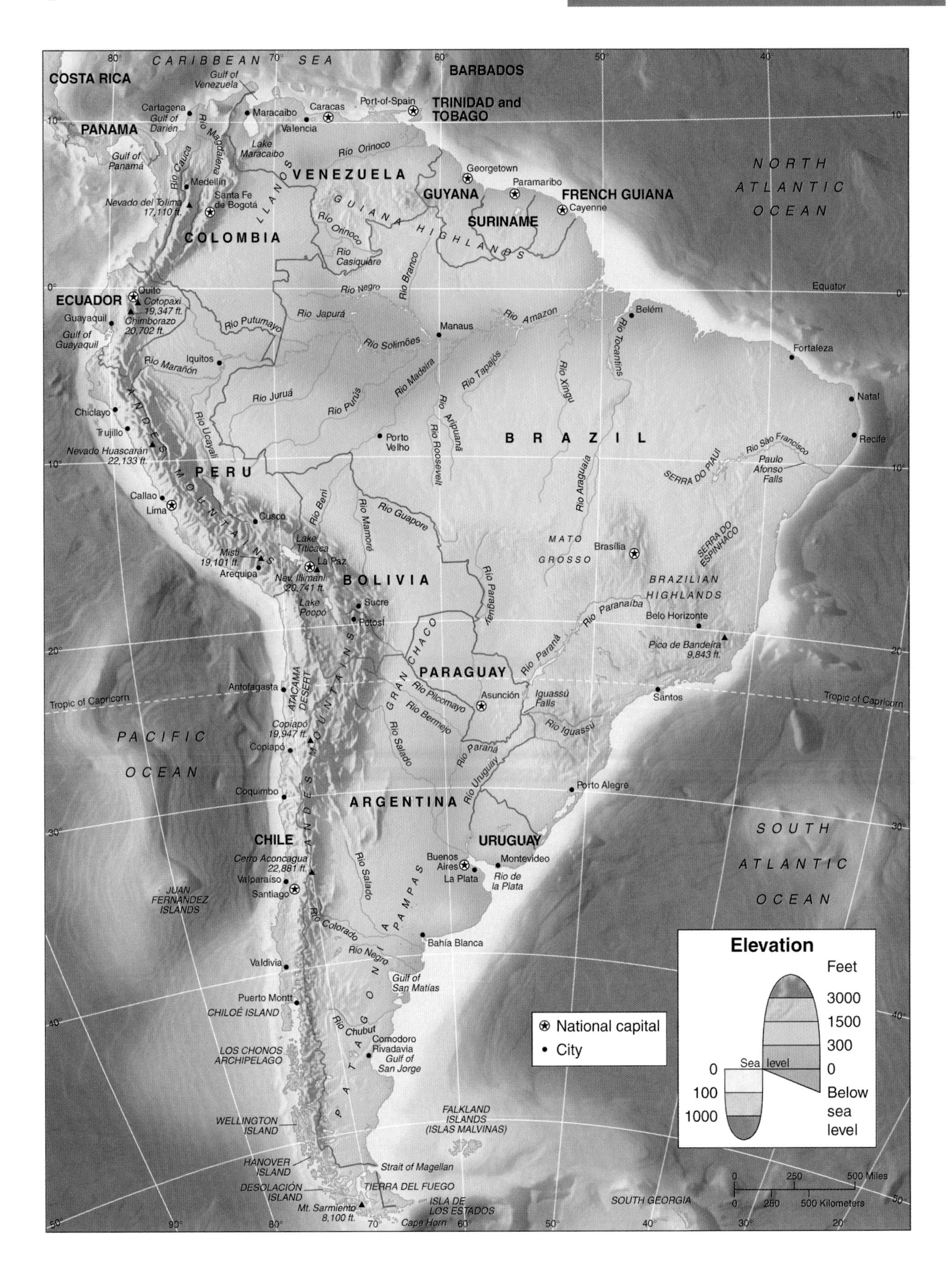

Map 81 South America: Political Divisions

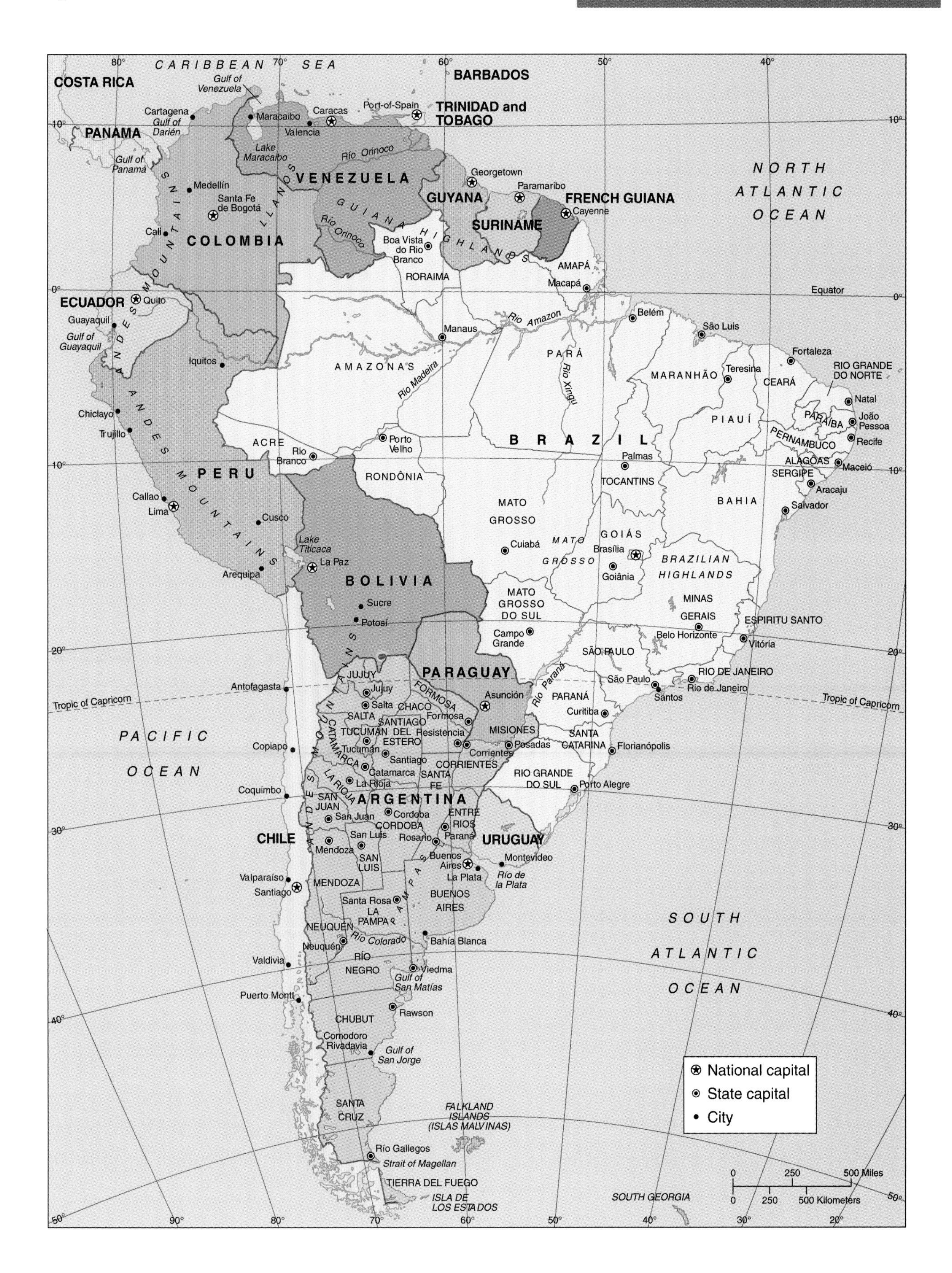

Map 82 South American Land Use

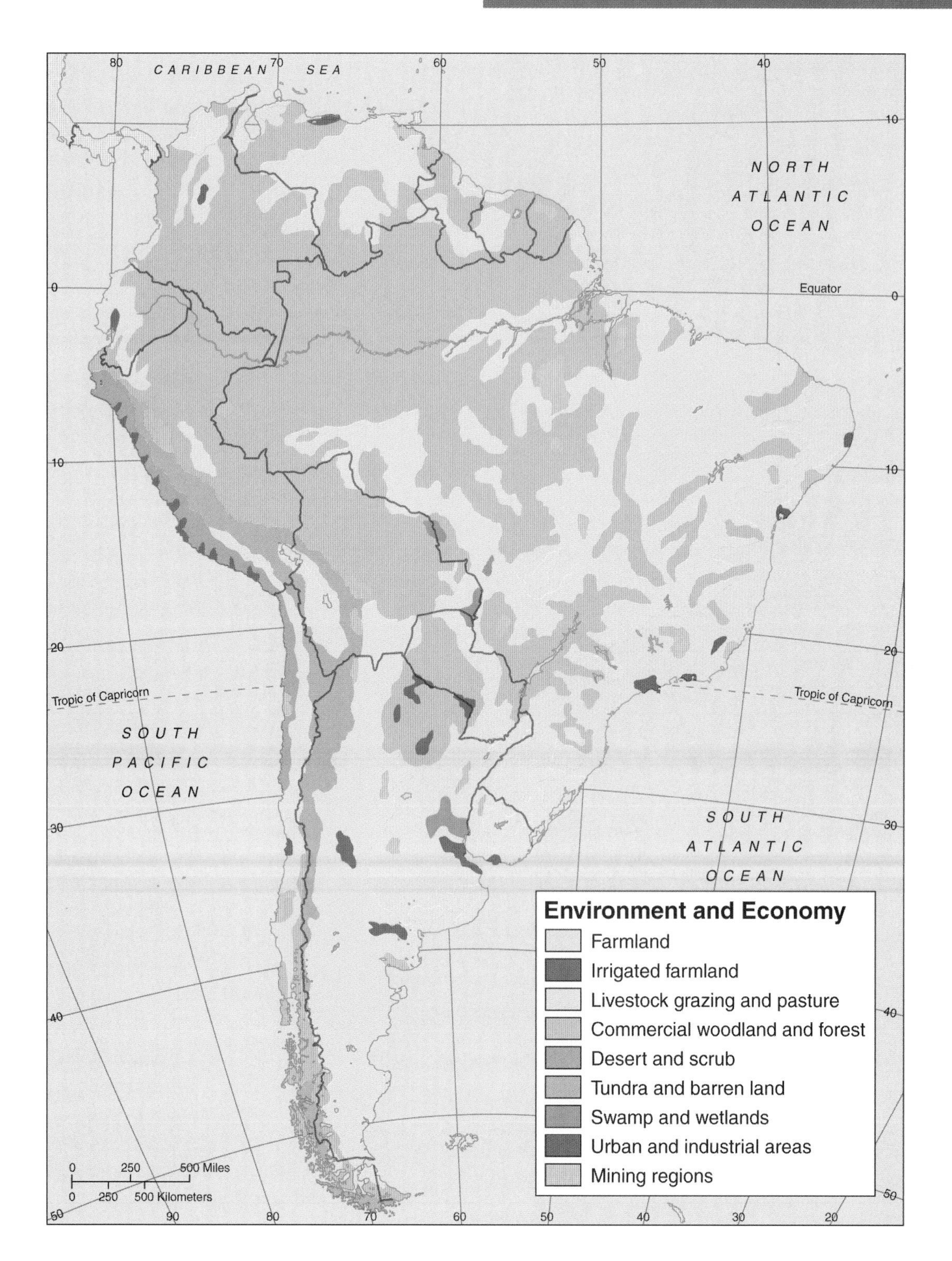

South America is a region just beginning to emerge from a colonial-dependency economy in which raw materials flowed from the continent to more highly developed economic regions. With the exception of Brazil, Argentina, Chile, and Uruguay, most of the continent's countries still operate under the traditional mode of exporting raw materials in exchange for capital that tends to accumulate in the pockets of a small percentage of the population. The land use patterns of the continent are, therefore, still dominated by resource extraction and agriculture. A problem posed by these patterns is that little of the continent's land area is actually suitable for either commercial forestry or commercial crop agriculture without extremely high environmental costs. Much of the agriculture, then, is based on high value tropical crops that can be grown in small areas profitably, or on extensive livestock grazing. Even within the forested areas of the Amazon Basin where forest clearance is taking place at unprecedented rates, much of the land use that replaces forest is grazing.

Map 83 Europe: Physical Features

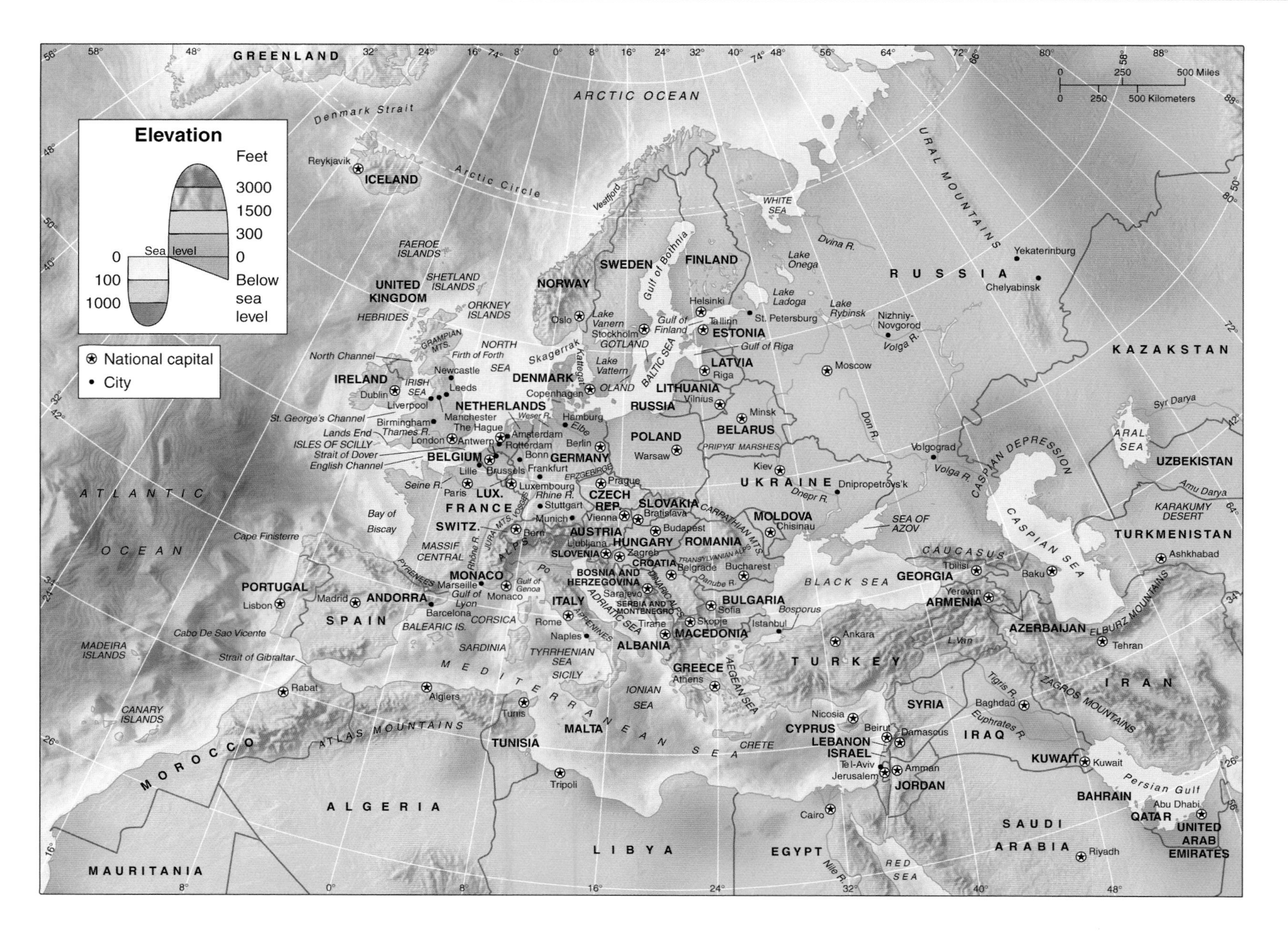

Map 84 Europe: Political Divisions

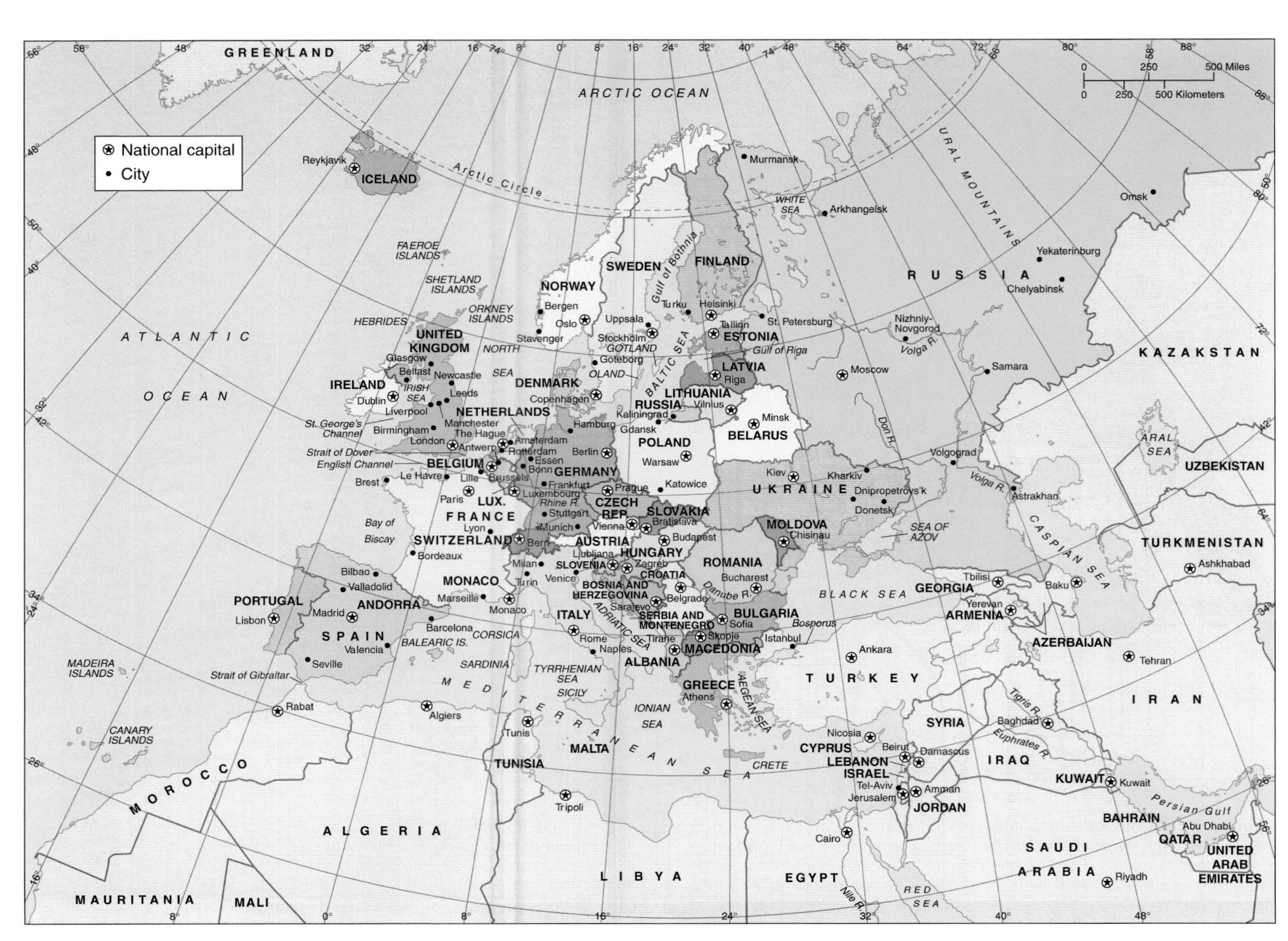

Map 85 European Land Use

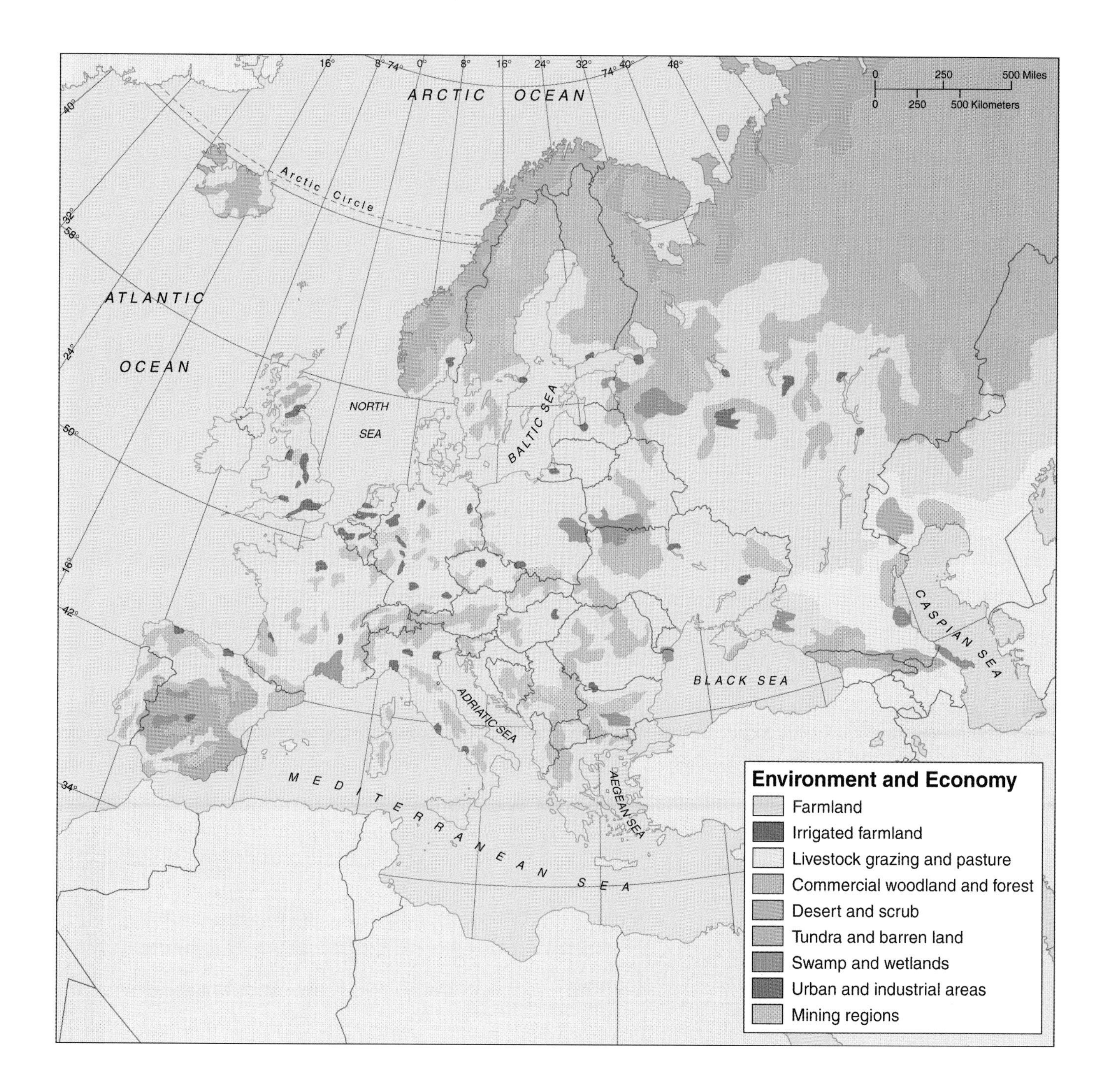

More than any other continent, Europe bears the imprint of human activity—mining, forestry, agriculture, industry, and urbanization. Virtually all of western and central Europe's natural forest vegetation is gone, lost to clearing for agriculture beginning in prehistory, to lumbering that began in earnest during the Middle Ages, or more recently, to disease and destruction brought about by acid precipitation. Only in the far north and the east do some natural stands remain. The region is the world's most heavily industrialized and the industrial areas on the map represent only the largest and most significant. Not shown are the industries that are found in virtually every small town and village and smaller city throughout the industrial countries for Europe. Europe also possesses abundant raw materials and a very productive agricultural base. The mineral resources have long been in a state of active exploitation and the mining regions shown on the map are, for the most part, old regions in upland areas that are somewhat less significant now than they may have been in the past. Agriculturally, the northern European plain is one of the world's great agricultural regions but most of Europe contains decent land for agriculture.

Map 86 Asia: Physical Features

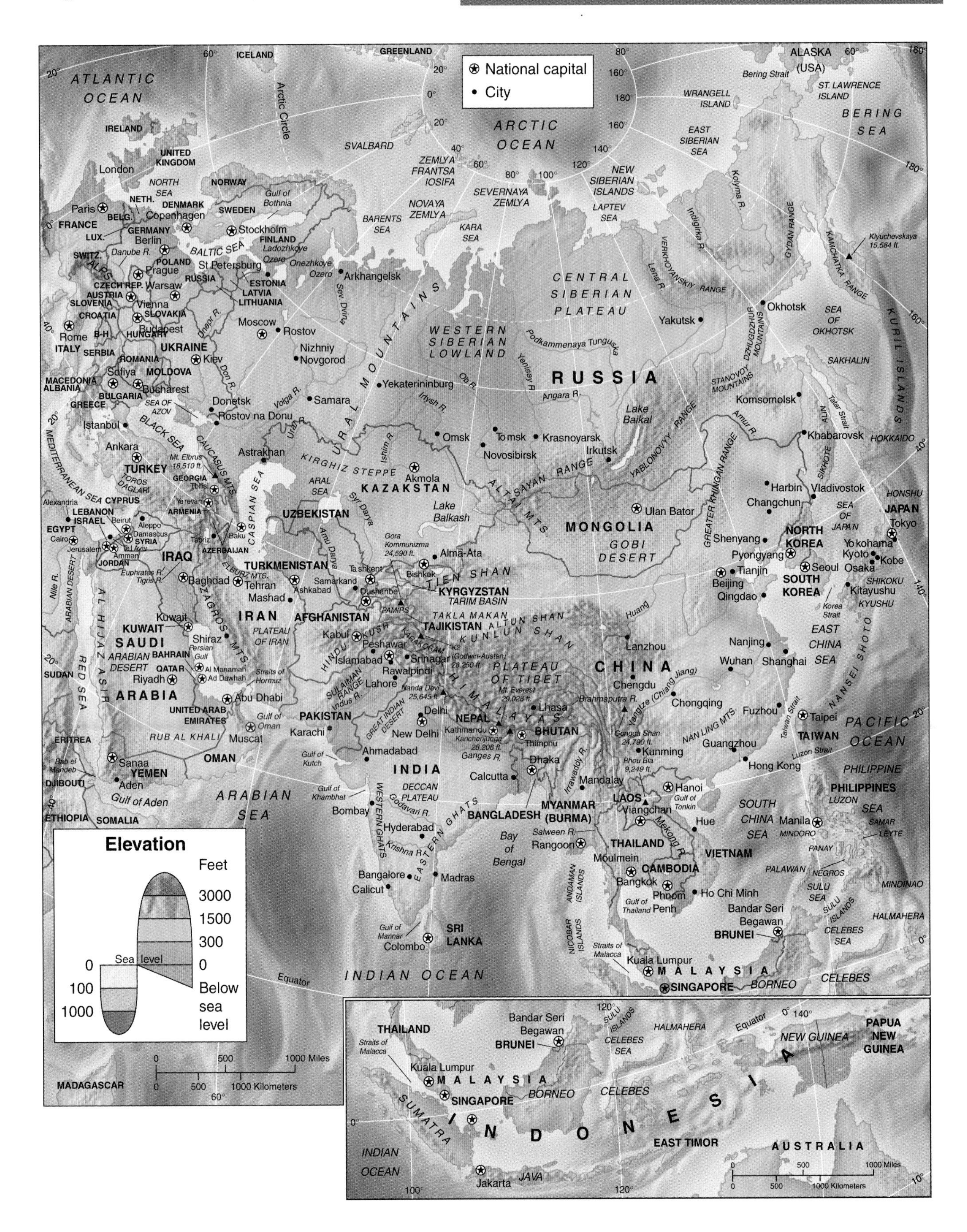

Map 87 Asia: Political Divisions

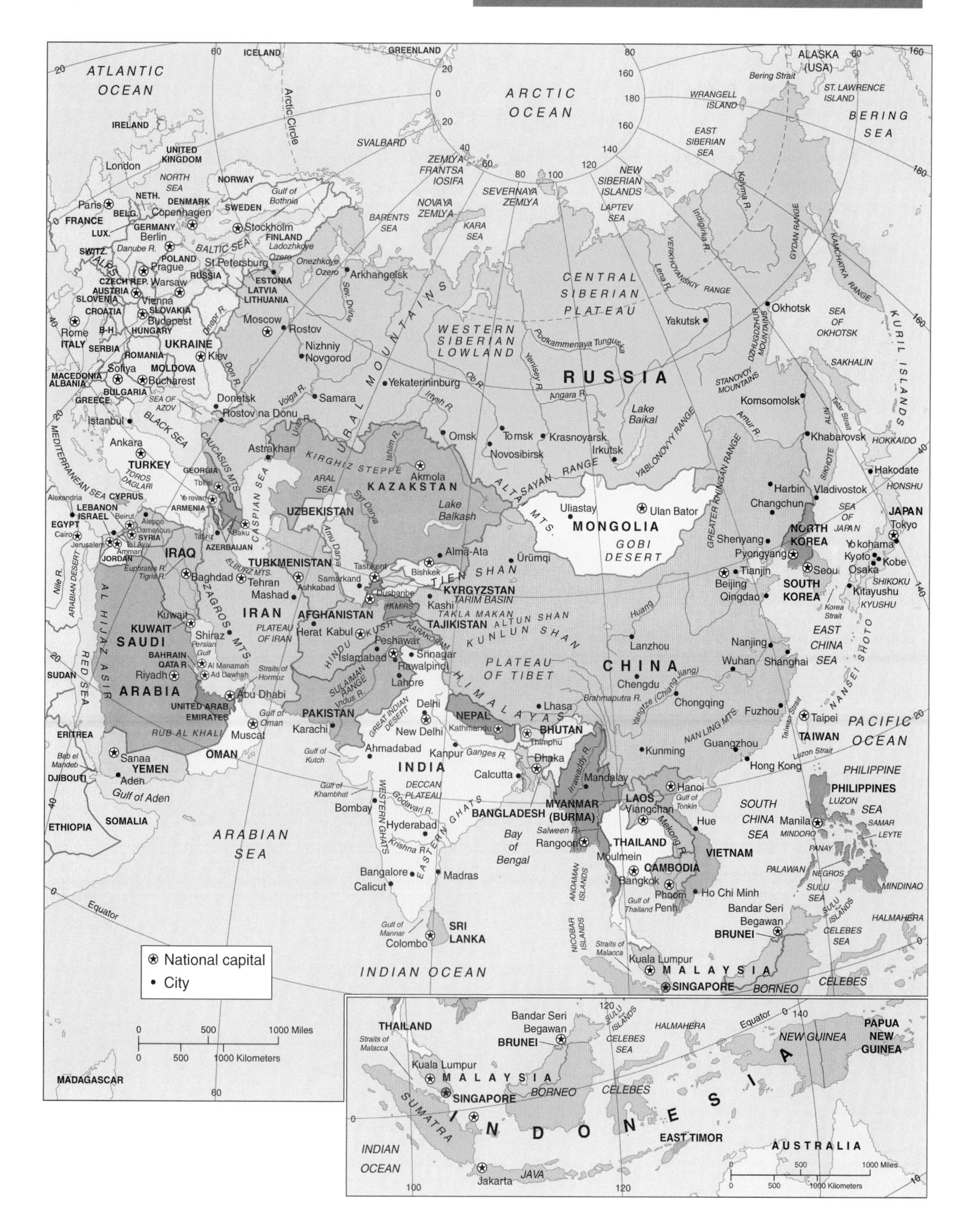

Map 88 Asian Land Use

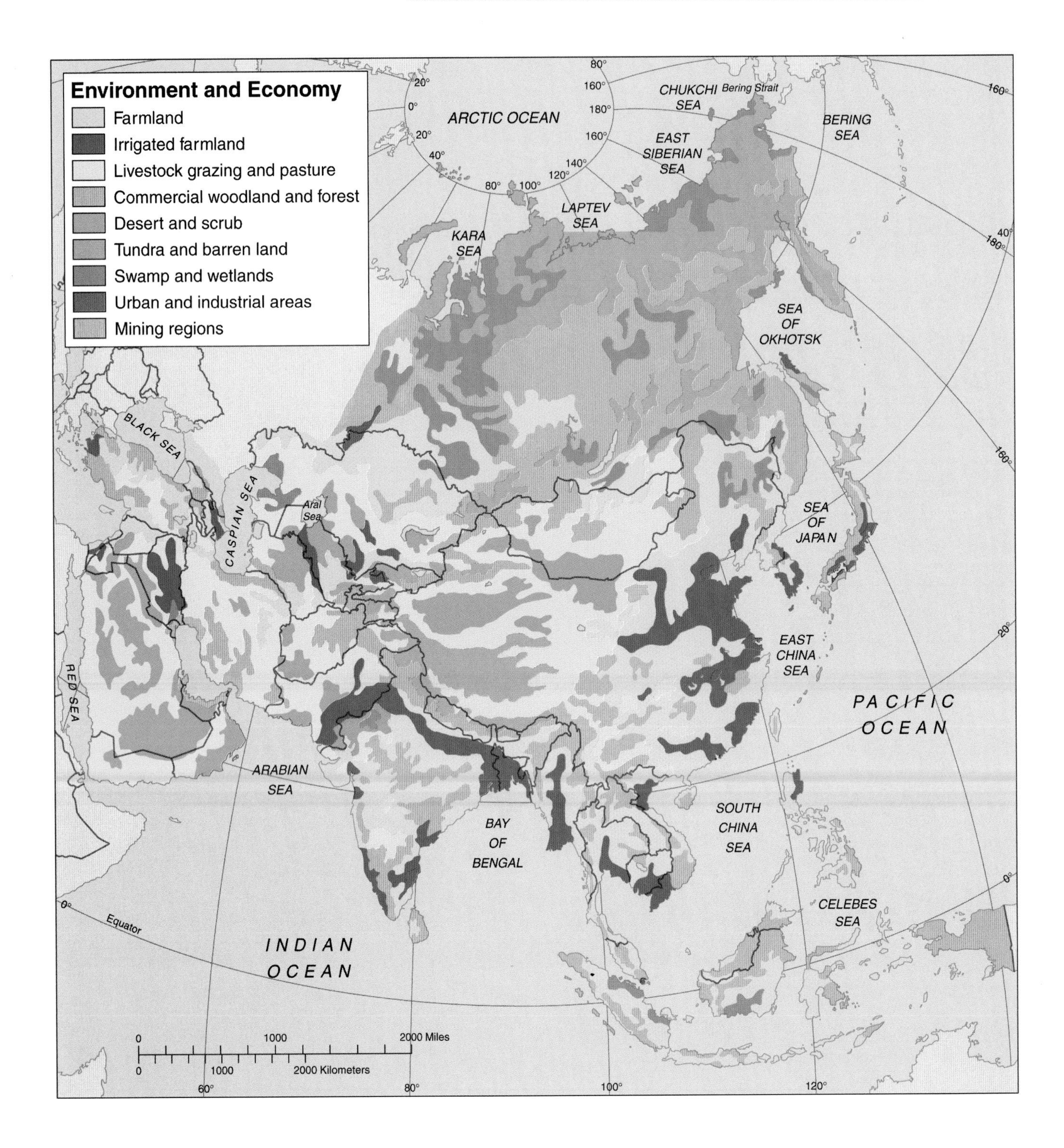

Asia is a land of extremes of land use with some of the world's most heavily industrialized regions, barren and empty areas, and productive and densely populated farm regions. Asia is a region of rapid industrial growth. Yet Asia remains an agricultural region with three out of every four workers engaged in agriculture. Asian commercial agriculture and intensive subsistence agriculture is characterized by irrigation. Some of Asia's irrigated lands are desert requiring additional water. But most of the Asian irrigated regions have sufficient precipitation for crop agriculture and irrigation is a way of coping with seasonal drought—the wet-and-dry cycle of the monsoon—often gaining more than one crop per year on irrigated farms. Agricultural yields per unit area in many areas of Asia are among the world's highest. Because the Asian population is so large and the demands for agricultural land so great, Asia is undergoing rapid deforestation and some areas of the continent have only small remnants of a once-abundant forest reserve.

Map 89 Southwest and Central Asia: Physical Features

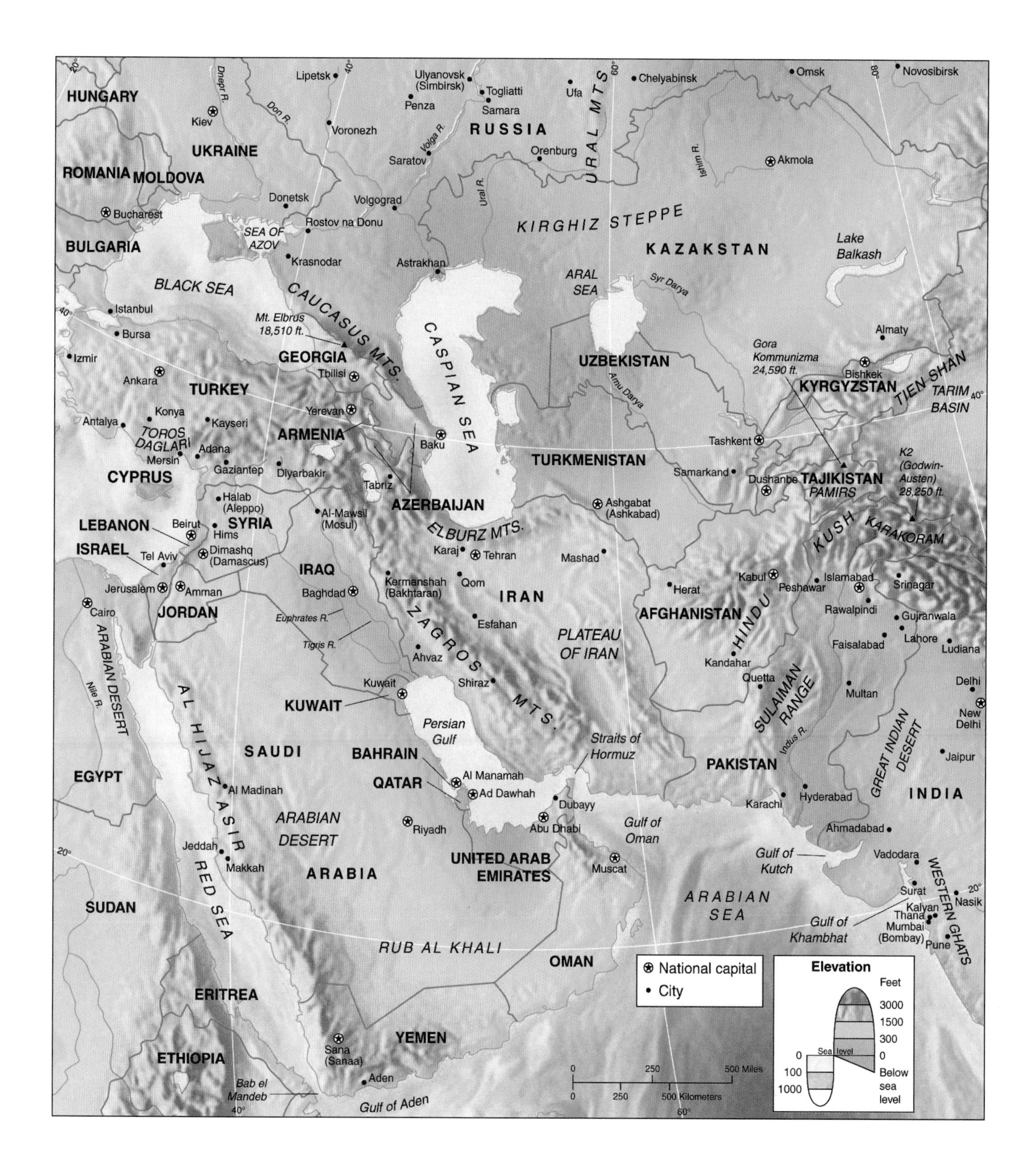

Map 90 Southwest and Central Asia: Political Divisions

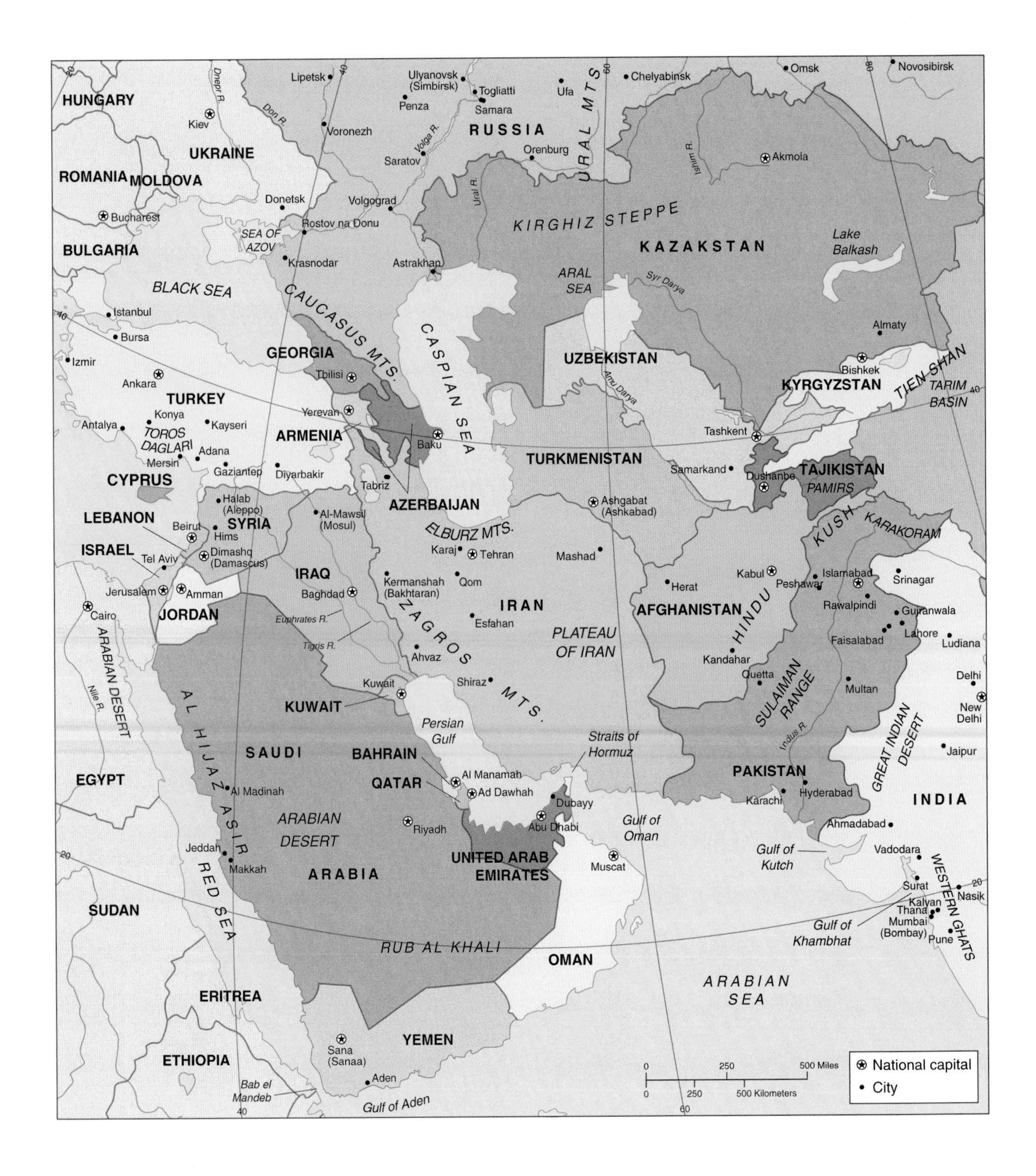

Map 91 South and Southeast Asia: Physical Features

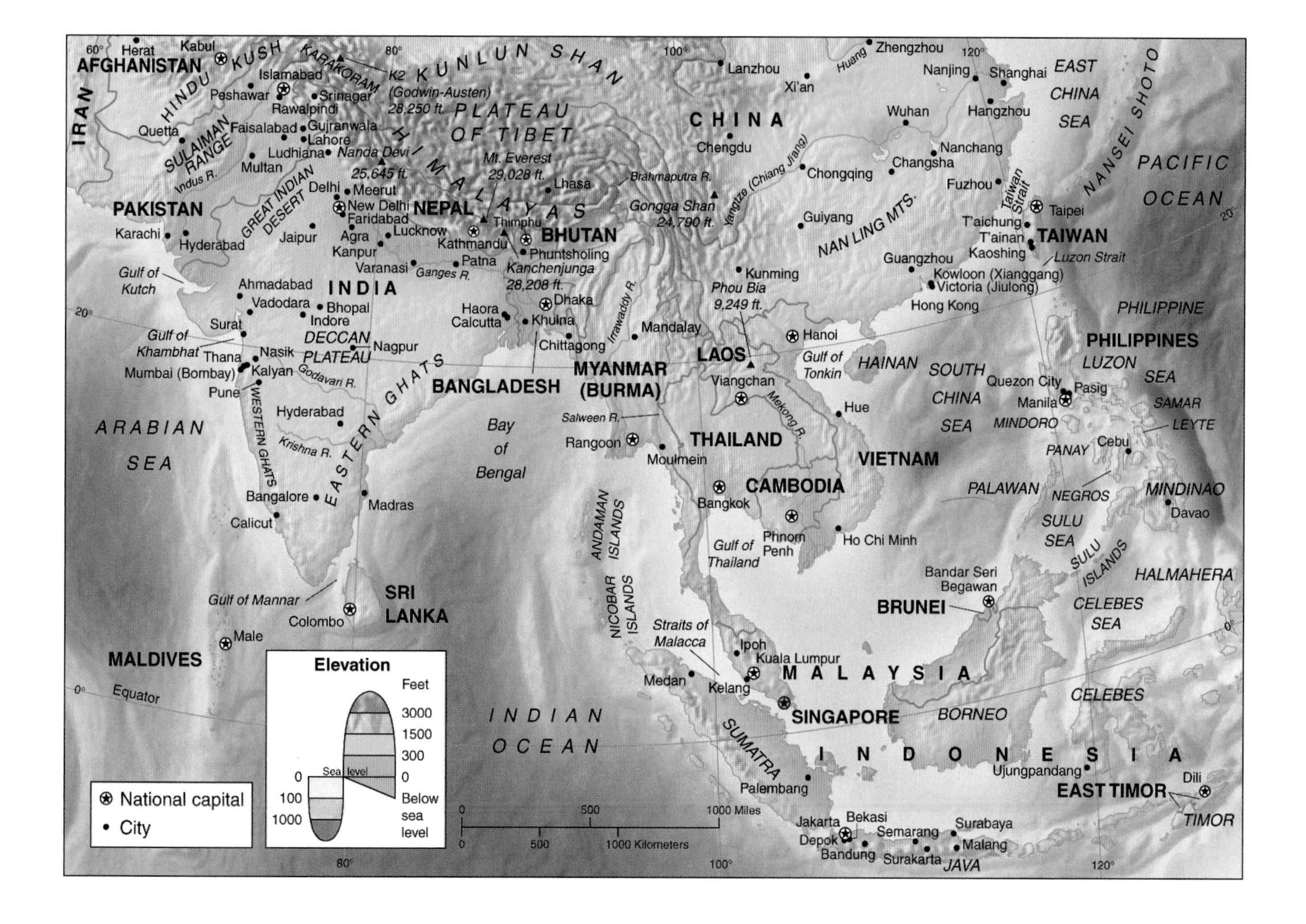

Map 92 South and Southeast Asia: Political Divisions

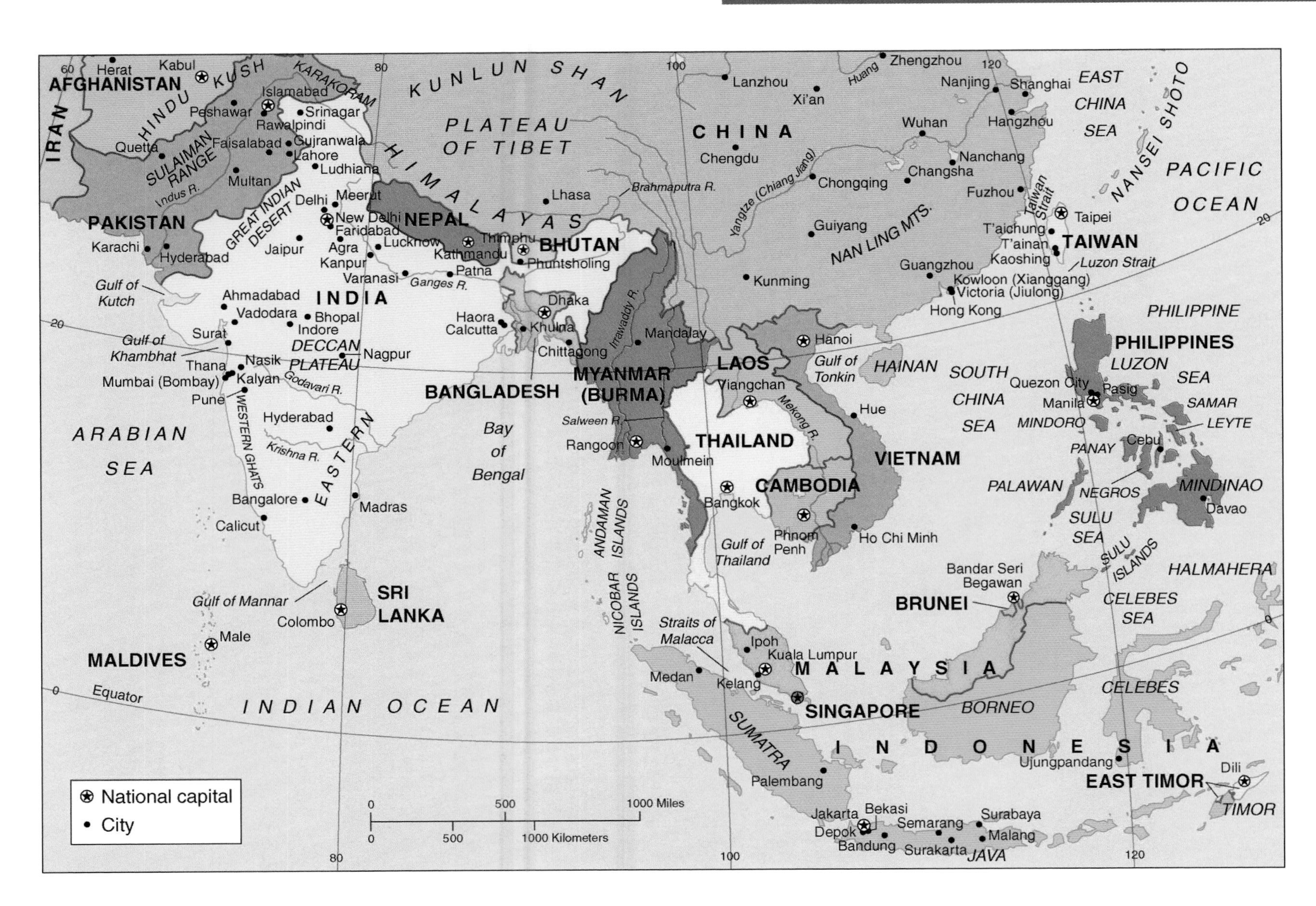

Map 93 North and East Asia: Physical Features

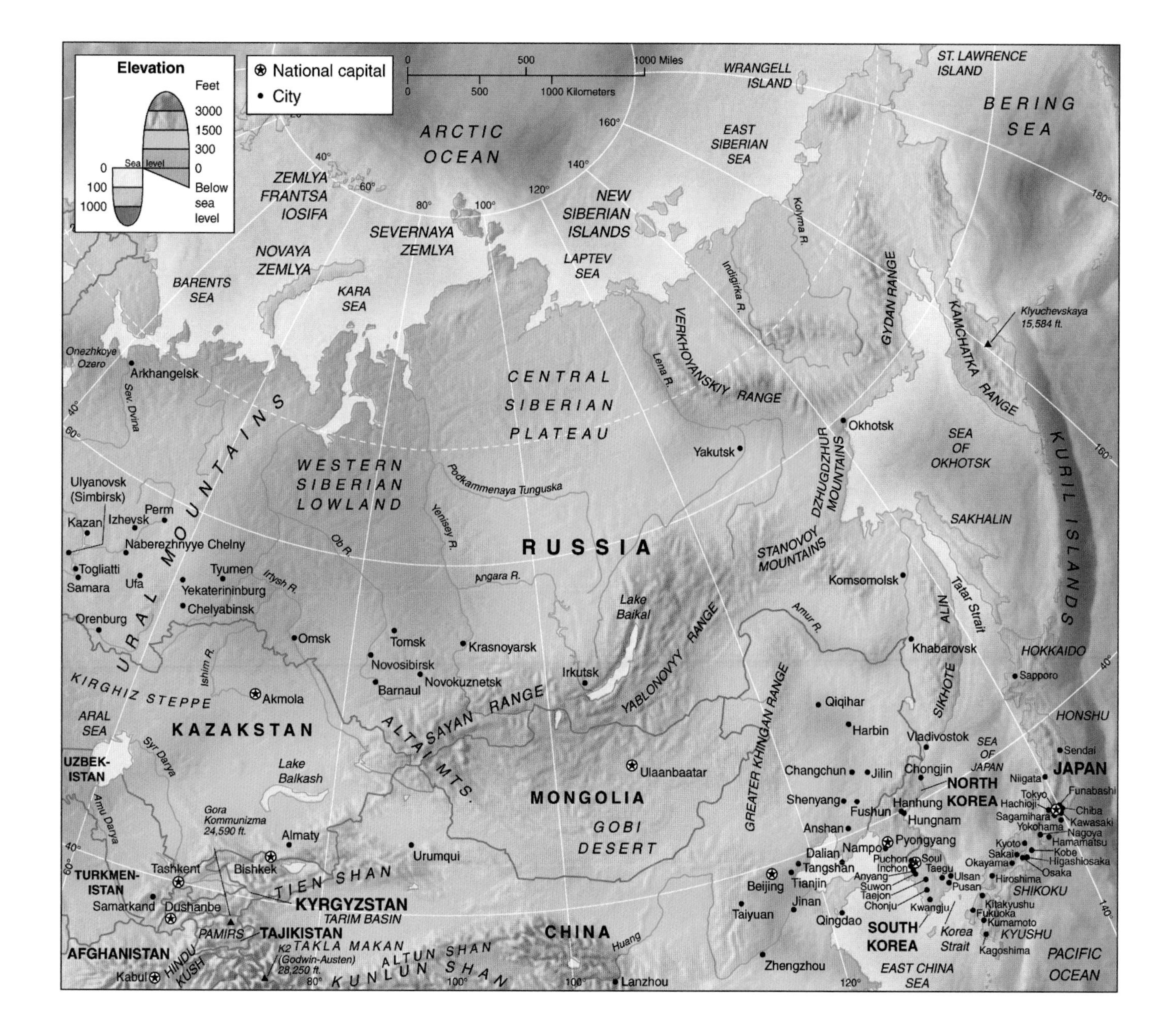

Map 94 North and East Asia: Political Divisions

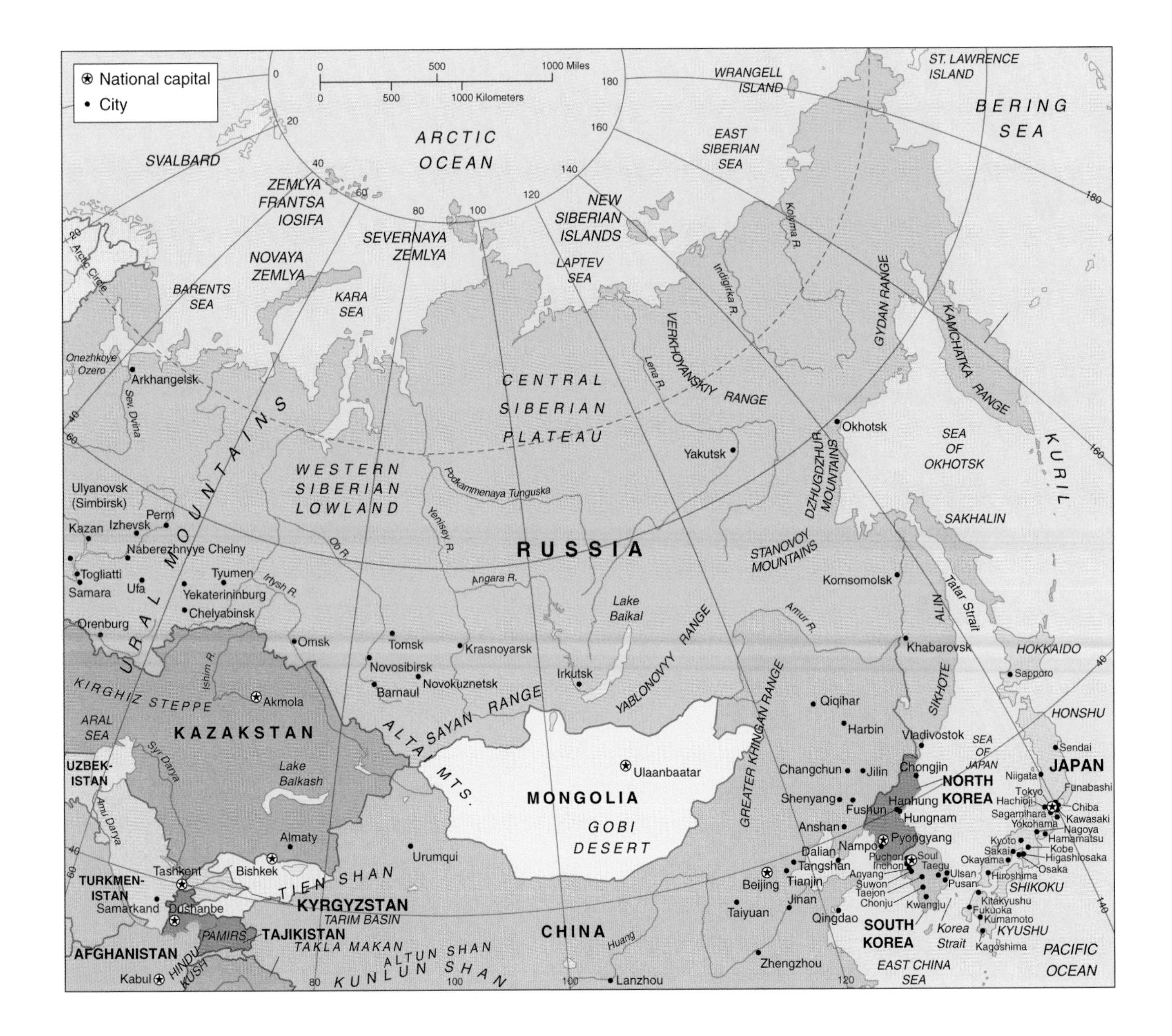

Map 95 Africa: Physical Features

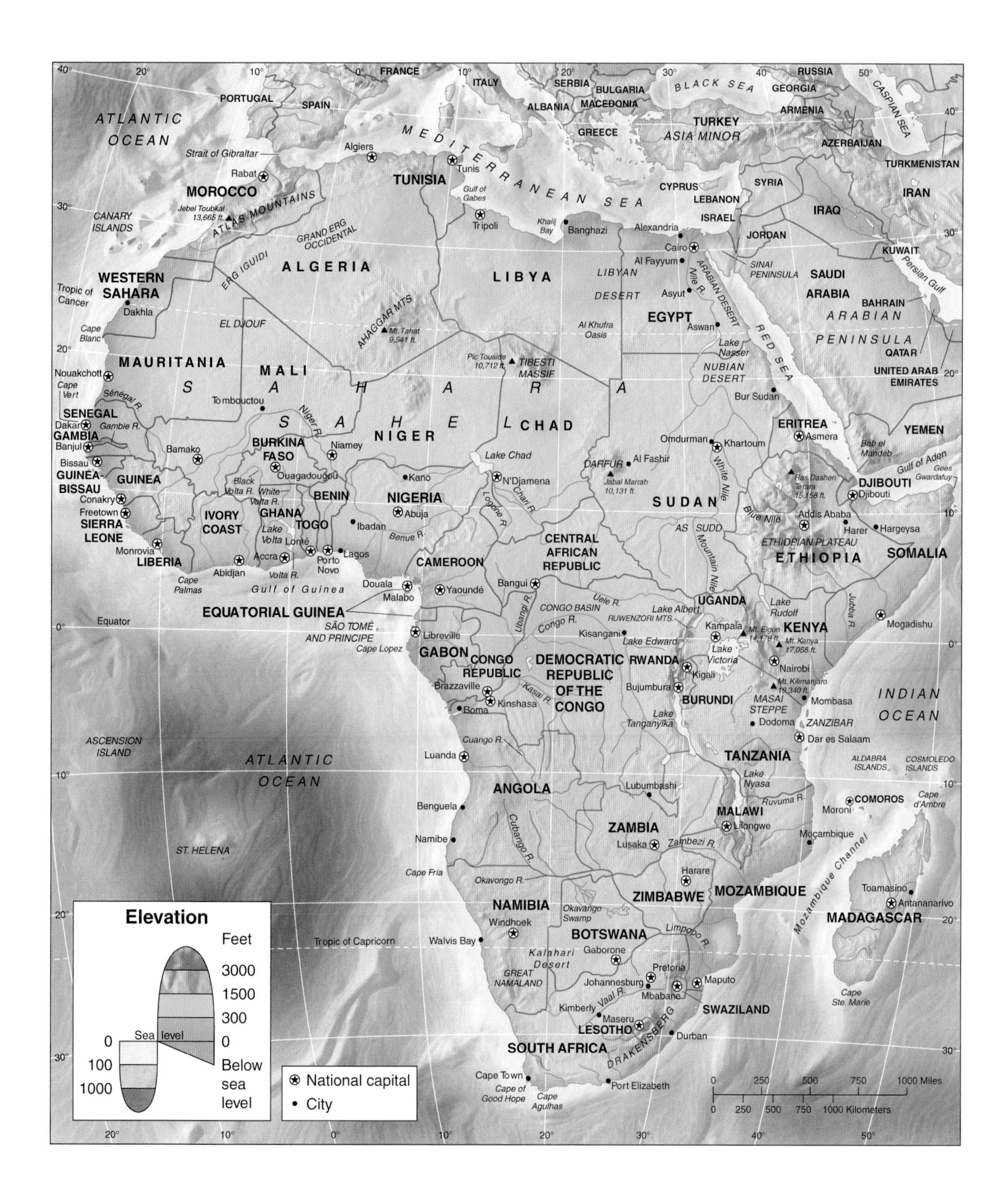

Map 96 Africa: Political Divisions

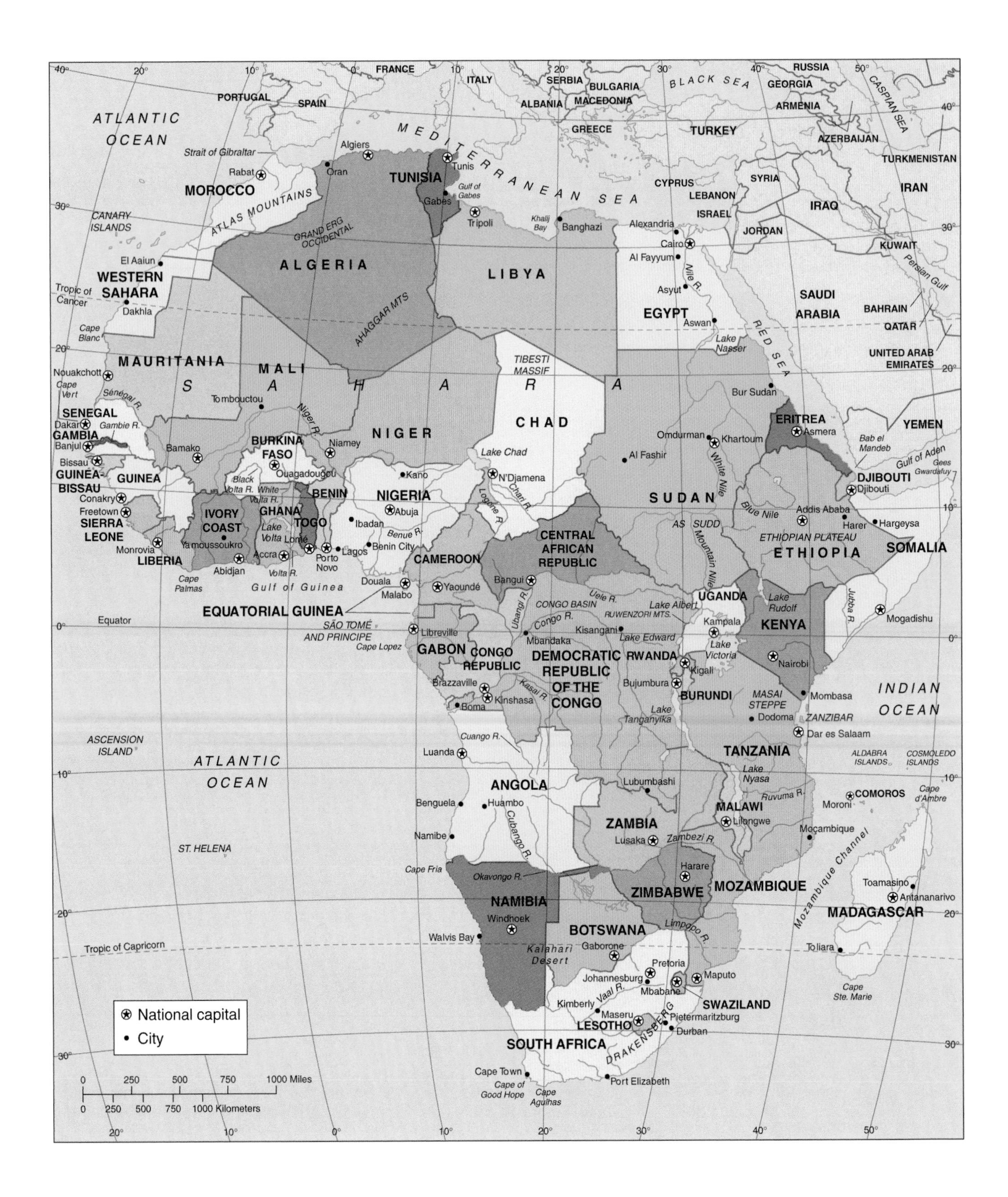

African Land Use

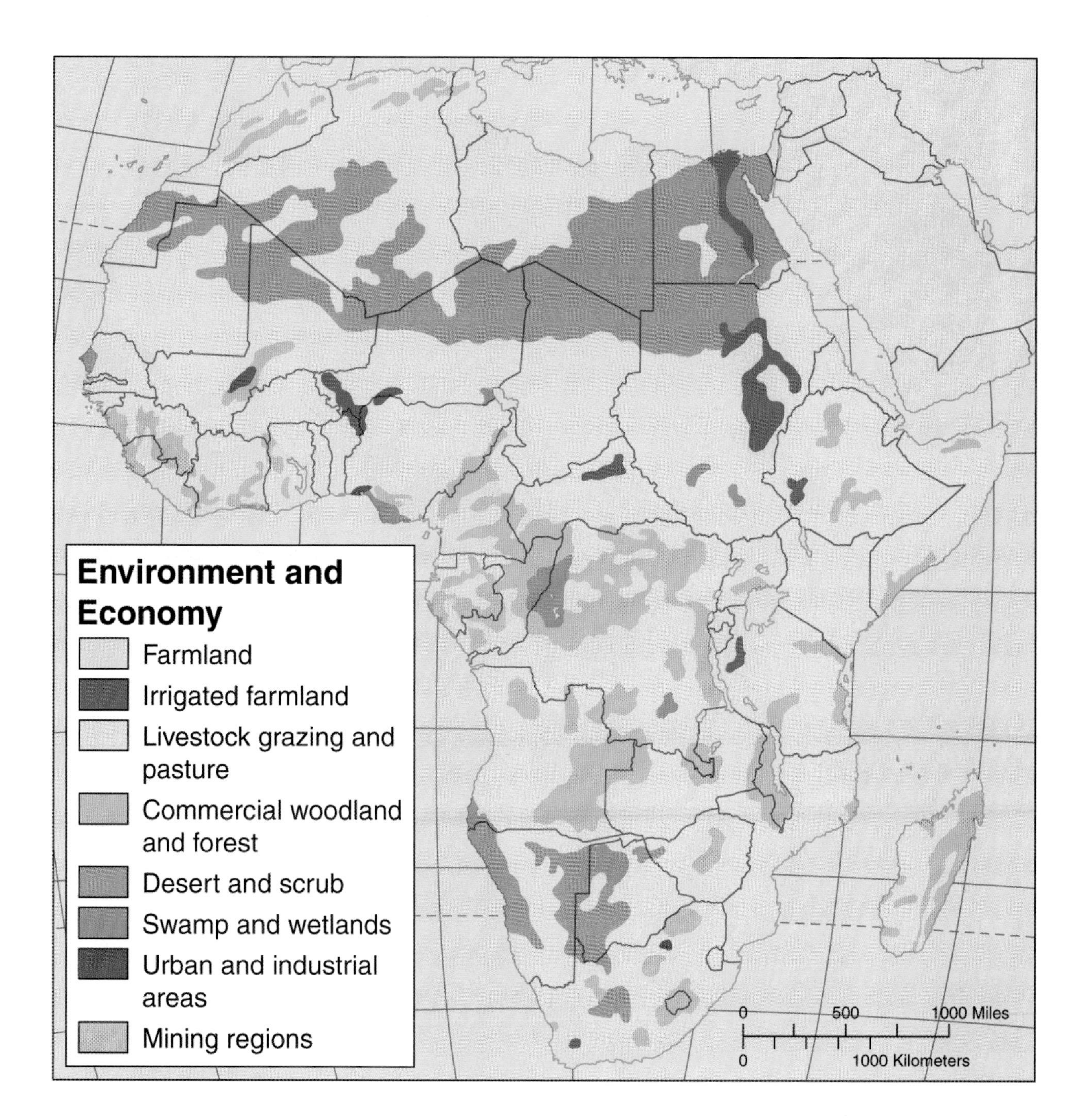

Africa's economic landscape is dominated by subsistence, or marginally-commercial agricultural activities and raw material extraction, engaging three-fourths of Africa's workers. Much of this grazing land is very poor desert scrub and bunch grass that is easily impacted by cattle, sheep, and goats. Growing human and livestock populations place enormous stress on this fragile support capacity and the result is desertification: the conversion of even the most minimal of grazing environments to a small quantity of land suitable for crop farming. Although the continent has approximately 20 percent of the world's total land area, the proportion of Africa's arable land is small. The agricultural environment is also uncertain; unpredictable precipitation and poor soils hamper crop agriculture.

Map 98 Australia and Oceania: Physical Features

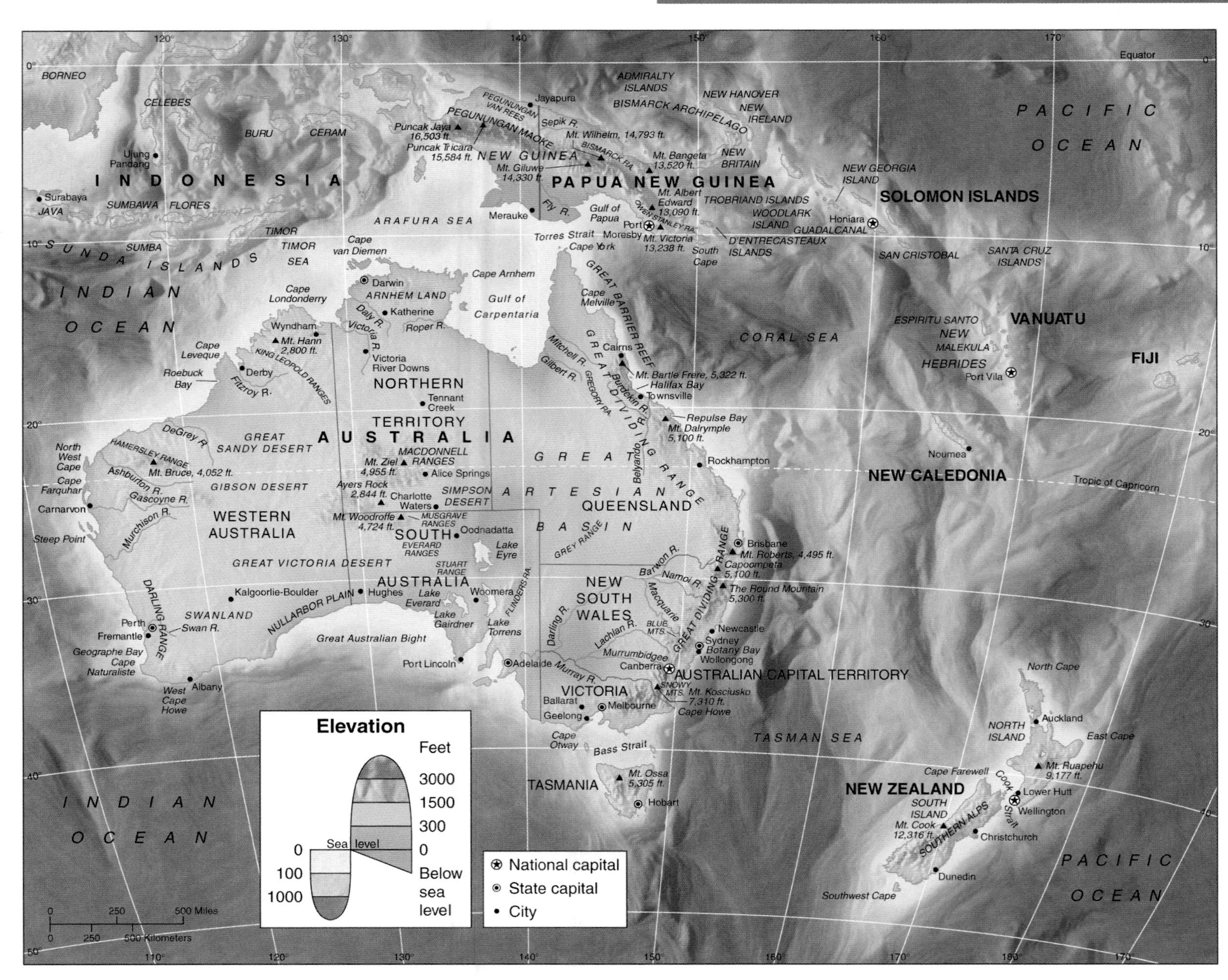

Map 99 Australia and Oceania: Political Divisions

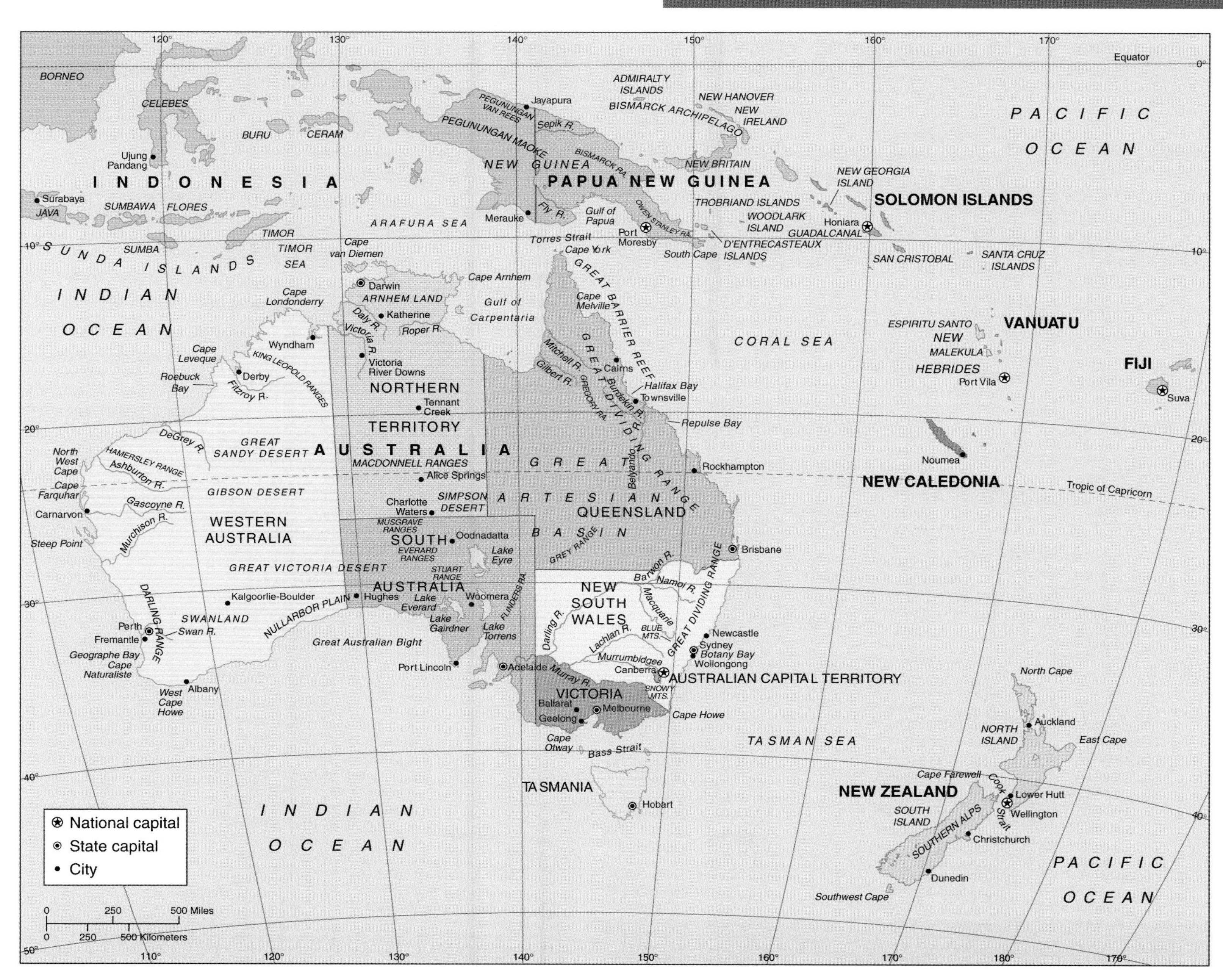

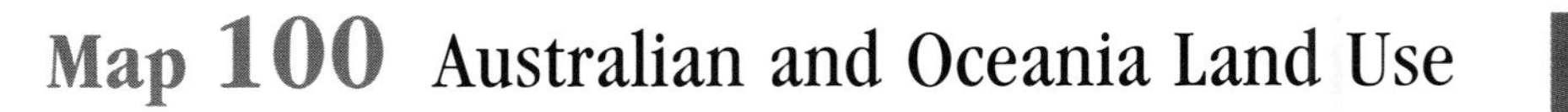

Map 100 Australian and Oceania Land Use

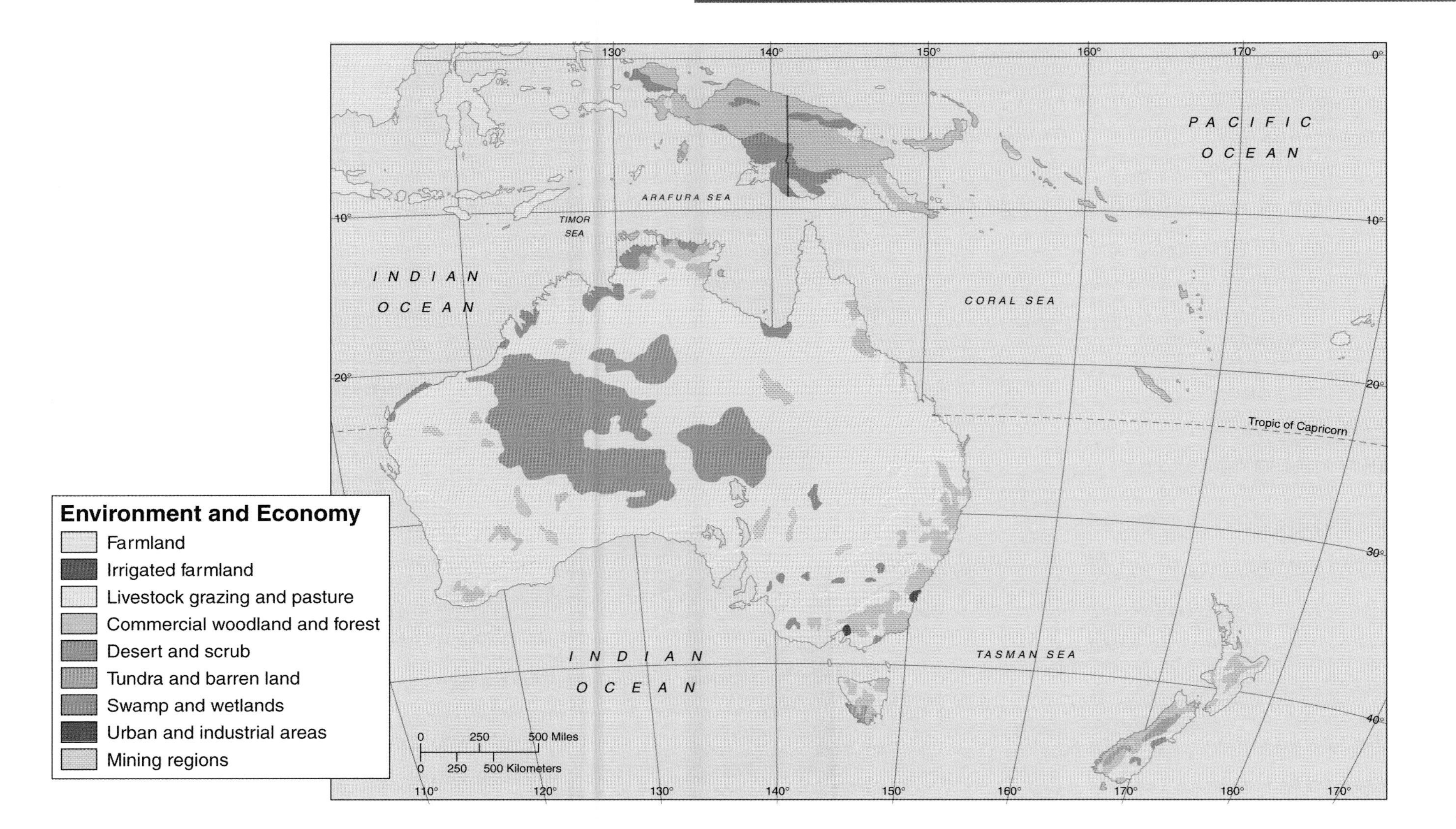

Australasia is dominated by the world's smallest and most uniform continent. Flat, dry, and mostly hot, Australia has the simplest of land use patterns: where rainfall exists so does agricultural activity. Two agricultural patterns dominate the map: livestock grazing, primarily sheep, and wheat farming, although some sugar cane production exists in the north and some cotton is grown elsewhere. Only about 6 percent of the continent consists of arable land so the areas of wheat farming, dominant as they may be in the context of Australian agriculture, are small. Australia also supports a healthy mineral resource economy, with iron and copper and precious metals making up the bulk of the extraction. Elsewhere in the region, tropical forests dominate Papua New Guinea, with some subsistence agriculture and livestock. New Zealand's temperate climate with abundant precipitation supports a productive livestock industry and little else besides tourism—which is an important economic element throughout the remainder of the region as well.

Map 101 The Pacific Rim

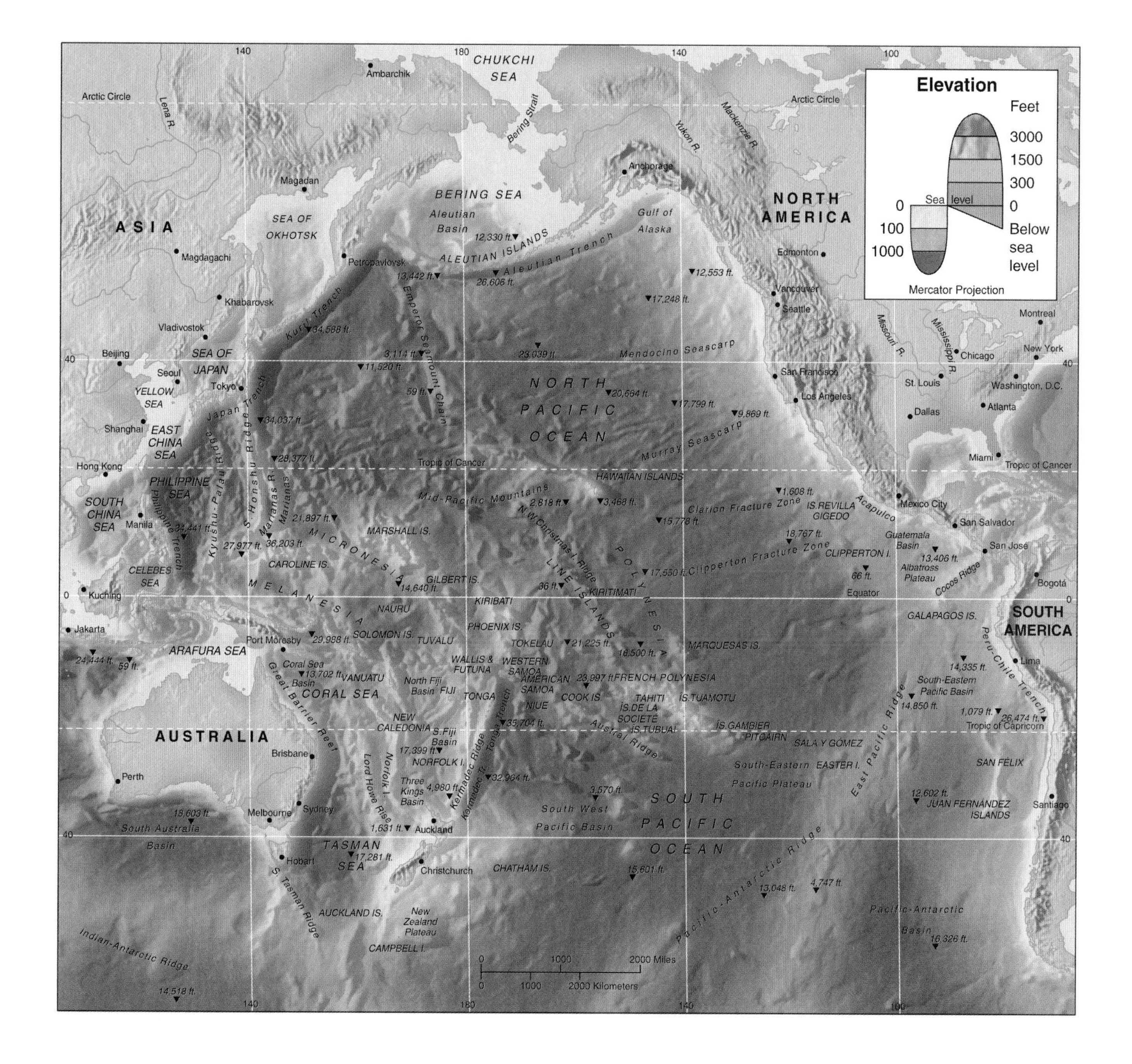

Map 102 The Atlantic Ocean

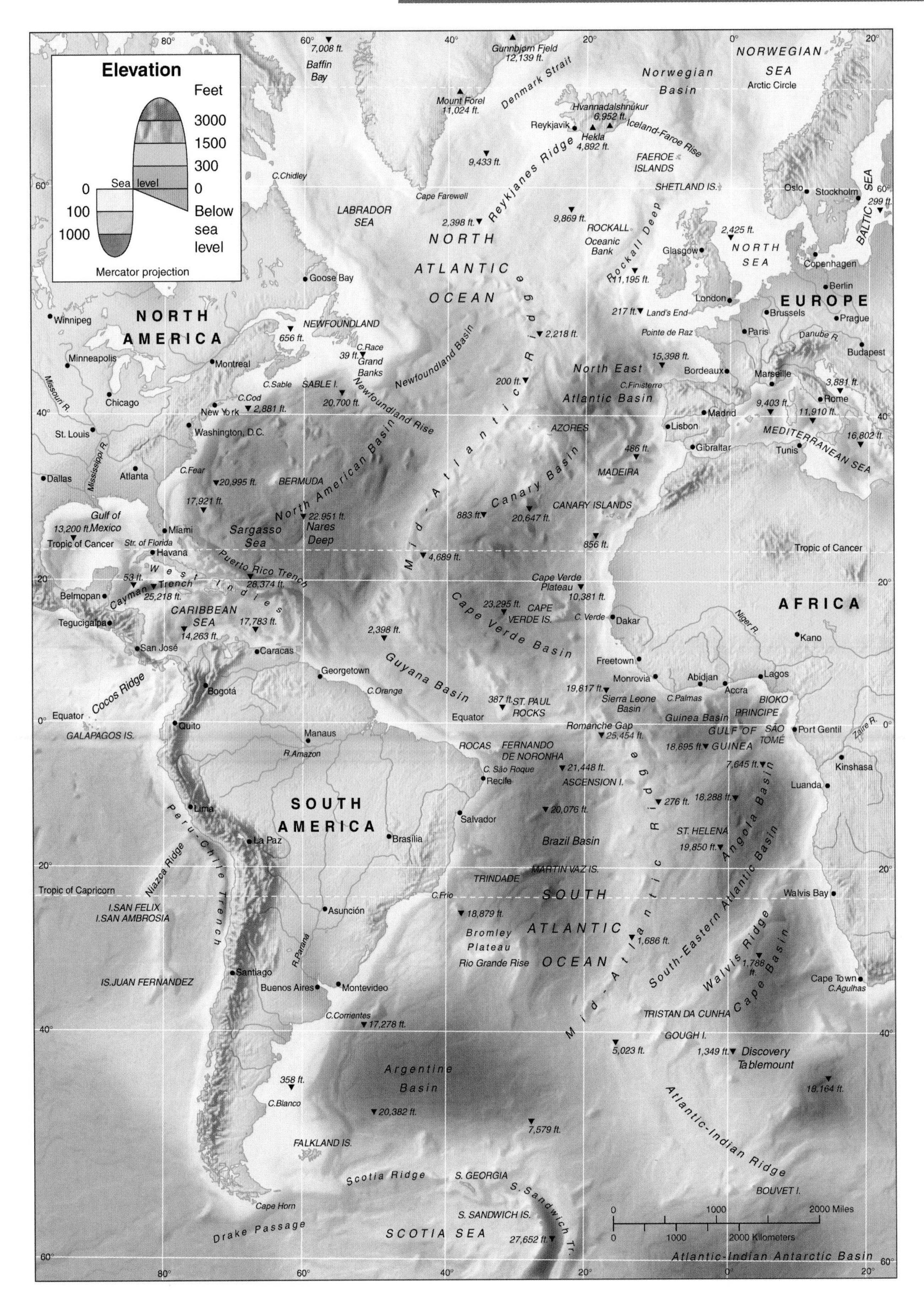

Map 103 The Indian Ocean

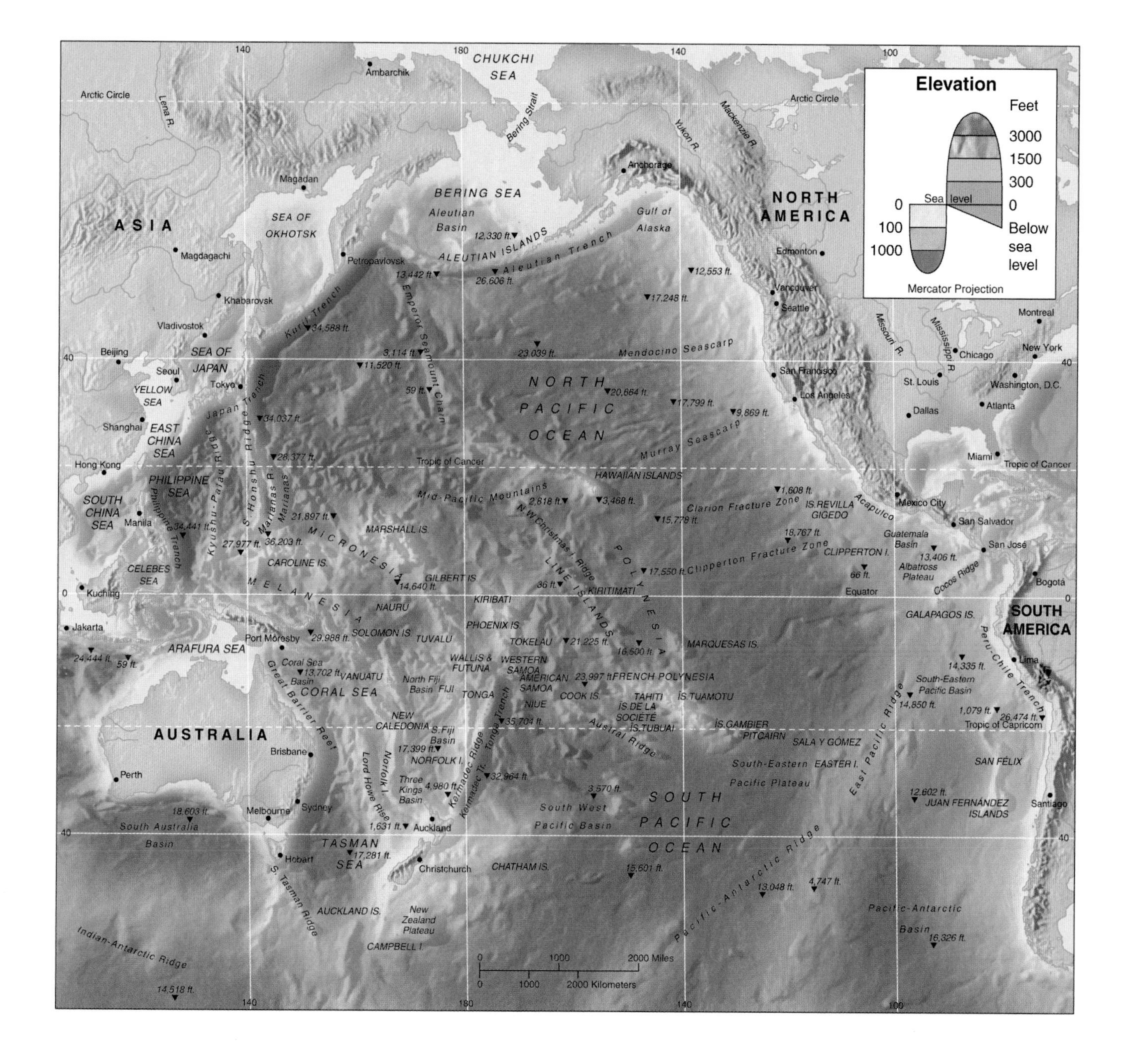

Map 104 The Arctic

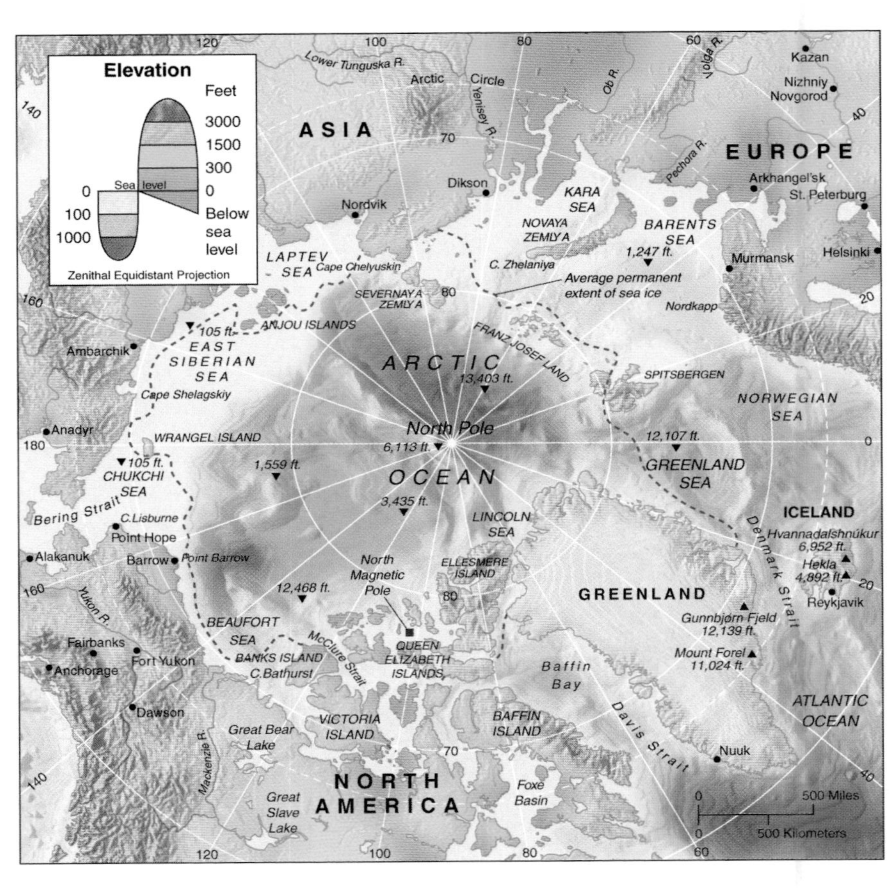

Map 105 Antarctica

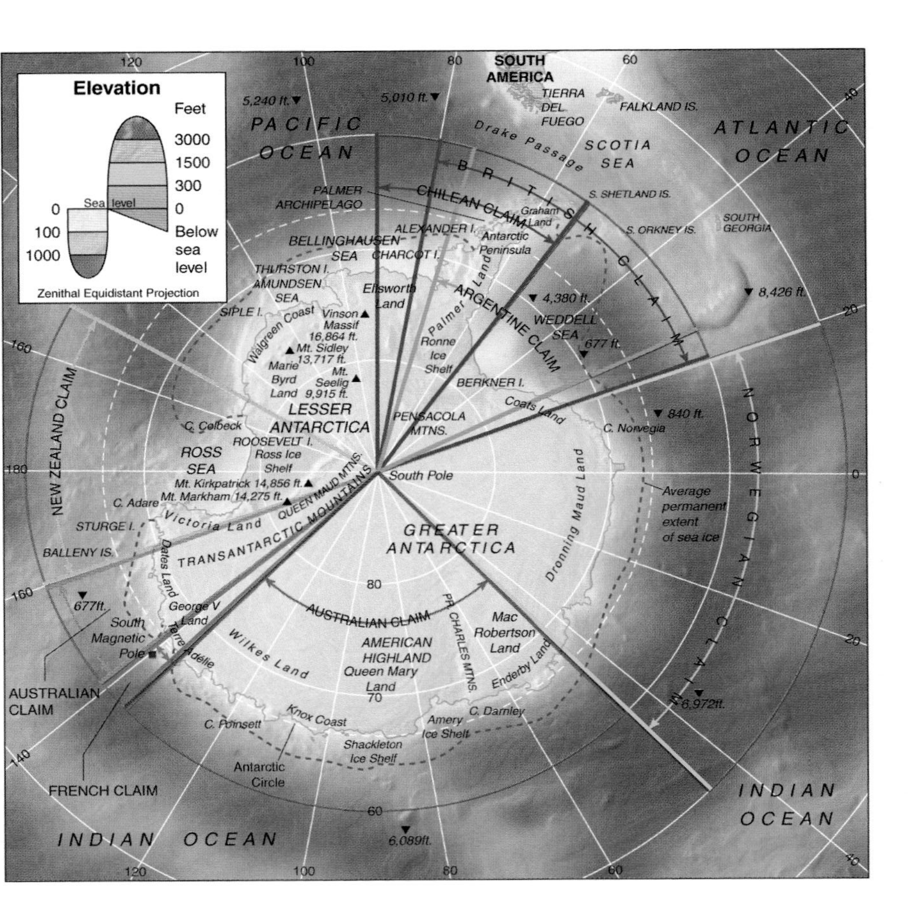

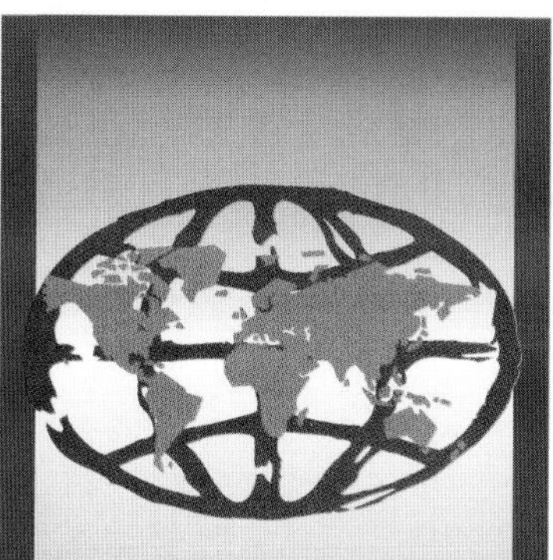

Part VIII

World Countries: Data Tables

Table A
World Countries: Area, Population, and Population Density, 2004

COUNTRY	AREA		POPULATION	DENSITY	
	(Mi^2)	(Km^2)	(Estimated 12/2004)	(Pop/Mi^2)	(Pop/Km^2)
Afghanistan	251,826	652,090	29,928,987	118.85	45.90
Albania	11,100	28,750	3,169,064	285.50	115.66
Algeria	919,595	2,381,740	31,832,610	34.62	13.37
Angola	481,354	1,246,700	13,522,110	28.09	10.85
Antigua and Barbuda	171	440	78,580	459.53	178.59
Argentina	1,073,400	2,780,400	36,771,840	34.26	13.44
Armenia	11,506	29,800	3,055,630	265.57	108.36
Australia	2,966,155	7,741,220	19,881,000	6.70	2.59
Austria	32,377	83,860	8,090,000	249.87	97.79
Azerbaijan	33,436	86,600	8,233,000	246.23	99.67
Bahamas, The	5,382	13,880	317,413	58.98	31.71
Bahrain	267	710	711,662	2665.40	1,002.34
Bangladesh	55,598	144,000	138,066,400	2483.30	1,060.66
Barbados	166	430	270,584	1630.02	629.27
Belarus	80,155	207,600	9,880,963	123.27	47.62
Belgium	11,783	30,510	10,376,000	880.59	343.24
Belize	8,866	22,960	273,700	30.87	12.00
Benin	43,475	112,620	6,720,250	154.58	60.75
Bhutan	18,200	47,000	873,663	48.00	18.59
Bolivia	424,165	1,098,580	8,814,158	20.78	8.13
Bosnia and Herzegovina	19,776	51,210	4,139,835	209.34	80.86
Botswana	231,803	581,730	1,722,468	7.43	3.04
Brazil	3,286,488	8,514,880	176,596,300	53.73	20.88
Brunei	2,228	5,770	356,447	159.99	67.64
Bulgaria	42,823	110,990	7,823,000	182.68	70.71
Burkina Faso	105,869	274,000	12,109,230	114.38	44.26
Burundi	10,745	27,830	7,205,982	670.64	280.61
Cambodia	69,898	181,040	13,403,640	191.76	75.93
Cameroon	183,569	475,440	16,087,470	87.64	34.57
Canada	3,849,674	9,970,610	31,630,000	8.22	3.43
Cape Verde	1,557	4,030	469,681	301.66	116.55
Central African Republic	240,535	622,980	3,880,847	16.13	6.23
Chad	495,755	1,284,000	8,581,741	17.31	6.82
Chile	292,259	756,630	15,774,000	53.97	21.07
China	3,705,392	9,598,050	1,288,400,000	347.71	138.13
Colombia	439,734	1,138,910	44,584,000	101.39	42.92
Comoros	838	2,230	600,142	716.16	269.12
Congo, Democratic Republic	905,564	2,344,860	53,153,362	58.70	23.45
Congo, Republic	132,047	342,000	3,757,263	28.45	11.00
Costa Rica	19,730	51,100	4,004,680	202.97	78.43
Côte d'Ivoire	124,502	322,460	16,835,420	135.22	52.94
Croatia	21,824	56,540	4,444,653	203.66	79.48
Cuba	42,804	110,860	11,326,000	264.60	103.13
Cyprus	3,571	9,250	769,954	215.61	83.33
Czech Republic	30,387	78,870	10,202,000	335.74	132.01

Table A *(Continued)*
World Countries: Area, Population, and Population Density, 2004

COUNTRY	AREA		POPULATION	DENSITY	
	(Mi^2)	(Km^2)	(Estimated 12/2004)	(Pop/Mi^2)	(Pop/Km^2)
Denmark	16,629	43,090	5,387,200	323.96	126.97
Djibouti	8,494	23,200	705,480	83.06	30.43
Dominica	290	750	71,213	254.56	94.95
Dominican Republic	18,815	48,730	8,738,639	464.45	180.63
Ecuador	109,484	283,560	13,007,940	118.81	46.99
Egypt, Arab Republic	386,662	1,001,450	67,559,040	174.72	67.87
El Salvador	8,124	21,040	6,533,215	804.19	315.31
Equatorial Guinea	10,831	28,050	494,000	45.61	17.61
Eritrea	46,842	117,600	4,389,500	93.71	43.46
Estonia	17,413	45,230	1,353,000	77.70	31.92
Ethiopia	435,184	1,104,300	68,613,470	157.67	68.61
Fiji	7,054	18,270	835,000	118.37	45.70
Finland	130,127	338,150	5,212,000	40.05	17.11
France	176,460	551,500	59,762,000	338.67	108.64
Gabon	103,347	267,670	1,344,433	13.01	5.22
Gambia, The	4,363	11,300	1,420,895	325.67	142.09
Georgia	26,911	69,700	5,126,000	190.48	73.77
Germany	137,803	357,030	82,541,000	598.98	236.54
Ghana	92,098	238,540	20,669,260	224.43	90.84
Greece	50,942	131,960	11,033,000	216.58	85.59
Grenada	131	340	104,600	798.47	307.65
Guatemala	42,042	108,890	12,307,090	292.73	113.50
Guinea	94,926	245,860	7,908,905	83.32	32.19
Guinea-Bissau	13,948	36,120	1,489,209	106.77	52.96
Guyana	83,000	214,970	768,888	9.26	3.91
Haiti	10,714	27,750	8,439,799	787.74	306.23
Honduras	43,277	112,090	6,968,512	161.02	62.28
Hungary	35,920	93,030	10,128,000	281.96	109.97
Iceland	39,768	103,000	289,000	7.27	2.88
India	1,269,340	3,287,260	1,064,399,000	838.55	358.00
Indonesia	741,097	1,904,570	214,674,200	289.67	118.50
Iran, Islamic Republic	636,294	1,648,200	66,392,020	104.34	40.58
Iraq	168,754	438,320	24,699,540	146.36	56.47
Ireland	27,137	70,270	3,994,000	147.18	57.98
Israel	8,019	22,140	6,688,000	834.02	308.06
Italy	116,305	301,340	57,646,270	495.65	196.00
Jamaica	4,244	10,990	2,642,628	622.67	244.01
Japan	145,882	377,890	127,573,000	874.49	349.99
Jordan	35,445	89,210	5,307,895	149.75	59.69
Kazakhstan	1,049,156	2,724,900	14,878,100	14.18	5.51
Kenya	224,961	580,370	31,915,850	141.87	56.08
Kiribati	277	730	96,377	347.93	132.02
Korea, Democratic Republic	46,540	120,540	22,612,280	485.87	187.79
Korea, Republic	38,023	99,260	47,911,730	1,260.07	485.28
Kuwait	6,880	17,820	2,396,417	348.32	134.48

Table A *(Continued)*
World Countries: Area, Population, and Population Density, 2004

COUNTRY	AREA		POPULATION	DENSITY	
	(Mi^2)	(Km^2)	(Estimated 12/2004)	(Pop/Mi^2)	(Pop/Km^2)
Kyrgyz Republic	76,641	199,900	5,052,000	65.92	26.34
Lao PDR	91,429	236,800	5,659,834	61.90	24.52
Latvia	24,749	64,600	2,321,000	93.78	37.41
Lebanon	4,015	10,400	4,497,669	1,120.22	439.65
Lesotho	11,720	30,350	1,792,744	152.96	59.07
Liberia	43,000	111,370	3,373,542	78.45	35.02
Libya	679,362	1,759,540	5,559,289	8.18	3.16
Liechtenstein	62	160	33,000	532.26	206.00
Lithuania	25,174	65,300	3,454,000	137.21	55.11
Luxembourg	998	2,586	448,000	448.90	173.24
Macedonia, FYR	9,781	25,710	2,049,000	209.49	80.57
Madagascar	226,658	587,040	16,893,900	74.53	29.05
Malawi	45,747	118,480	10,962,010	239.62	116.52
Malaysia	127,317	329,750	24,774,250	194.59	75.40
Maldives	115	300	293,080	2,548.52	976.93
Mali	478,767	1,240,190	11,651,500	24.34	9.55
Malta	124	320	399,000	3,217.74	1,246.88
Marshall Islands	70	181	52,500	750.00	289.58
Mauritania	397,954	1,025,520	2,847,869	7.16	2.78
Mauritius	718	2,040	1,222,188	1,702.21	602.06
Mexico	761,603	1,958,200	102,291,000	134.31	53.59
Micronesia, Federal States	271	702	124,560	459.63	177.44
Moldova	13,012	33,840	4,237,600	325.67	128.88
Monaco	1	2	33,000	27,272.73	16,923.00
Mongolia	604,427	1,566,500	2,479,568	4.10	1.58
Morocco	172,413	446,550	30,112,640	174.65	67.47
Mozambique	309,494	801,590	18,791,420	60.72	23.97
Myanmar	261,969	676,580	49,362,500	188.43	75.07
Namibia	318,259	824,290	2,014,546	6.33	2.45
Nepal	54,363	147,180	24,659,960	453.62	172.45
Netherlands	14,413	41,530	16,221,800	1,125.50	478.80
New Zealand	103,738	270,530	4,009,200	38.65	14.96
Nicaragua	49,998	130,000	5,480,000	109.60	45.14
Niger	489,191	1,267,000	11,762,250	24.04	9.29
Nigeria	356,669	923,770	136,461,000	382.60	149.83
Norway	125,182	323,760	4,562,000	36.44	14.90
Oman	82,030	309,500	2,598,832	31.68	8.40
Pakistan	310,402	796,100	148,438,800	478.21	192.56
Palau	177	460	20,000	112.99	43.00
Panama	30,193	75,520	2,984,022	98.83	40.09
Papua New Guinea	178,259	462,840	5,501,871	30.86	12.15
Paraguay	157,048	406,750	5,643,097	35.93	14.20
Peru	496,225	1,285,220	27,148,000	54.71	21.21
Philippines	115,831	300,000	81,502,620	703.63	273.34
Poland	120,728	312,690	38,196,000	316.38	124.71

Table A *(Continued)*
World Countries: Area, Population, and Population Density, 2004

COUNTRY	AREA		POPULATION	DENSITY	
	(Mi^2)	(Km^2)	(Estimated 12/2004)	(Pop/Mi^2)	(Pop/Km^2)
Portugal	35,552	92,980	10,444,000	293.77	114.14
Qatar	4,247	11,000	623,703	146.86	56.70
Romania	91,699	238,390	21,744,000	237.12	94.59
Russian Federation	6,592,745	17,075,400	143,425,000	21.75	8.49
Rwanda	10,169	26,340	8,395,000	825.55	340.29
Samoa	1,104	2,840	178,000	161.23	62.90
San Marino	23	61	28,000	1,217.39	463.00
São Tomé and Principe	372	960	157,400	423.12	163.93
Saudi Arabia	756,982	2,149,690	22,528,300	29.76	10.48
Senegal	75,749	196,720	10,239,850	135.18	53.19
Serbia-Montenegro	39,517	102,170	8,104,000	205.08	79.45
Seychelles	175	450	83,639	477.94	185.86
Sierra Leone	27,699	71,740	5,336,568	192.66	74.51
Singapore	244	680	4,250,000	17,418.03	6,343.28
Slovak Republic	18,859	48,845	5,390,000	285.81	110.45
Slovenia	7,836	20,250	1,995,000	254.59	99.16
Solomon Islands	10,985	28,900	456,645	41.57	16.31
Somalia	246,201	637,660	9,625,918	39.10	15.34
South Africa	471,444	1,219,090	45,828,700	97.21	37.74
Spain	194,885	505,990	41,101,430	210.90	82.30
Sri Lanka	25,332	65,610	19,231,760	759.19	297.57
Sudan	967,500	2,505,810	33,545,730	34.67	14.12
Suriname	63,039	163,270	438,104	6.95	2.81
Swaziland	6,704	17,360	1,105,525	164.91	64.27
Sweden	173,732	449,960	8,956,000	51.55	21.76
Switzerland	15,943	41,290	7,350,000	461.02	185.84
Syrian Arab Republic	71,498	185,180	17,384,490	243.15	94.59
Taiwan	13,892	35,980	23,546,171	1,694.94	627.00
Tajikistan	55,251	143,100	6,304,700	114.11	44.84
Tanzania	364,900	945,090	35,888,960	98.35	40.62
Thailand	198,456	513,120	62,014,220	312.48	121.38
Timor-Leste	5,741	14,870	877,000	152.76	58.98
Togo	21,925	56,790	4,861,493	221.73	89.38
Tonga	290	750	101,524	350.08	141.01
Trinidad and Tobago	1,980	5,130	1,312,664	662.96	255.88
Tunisia	63,170	163,610	9,895,201	156.64	63.69
Turkey	301,382	774,820	70,712,000	234.63	91.88
Turkmenistan	188,456	488,100	4,863,500	25.81	10.35
Uganda	93,135	241,040	25,280,000	271.43	128.26
Ukraine	233,090	603,700	48,355,700	207.46	83.47
United Arab Emirates	31,696	83,600	4,041,000	127.49	48.34
United Kingdom	94,525	242,910	59,329,000	627.65	246.30
United States	3,717,797	9,629,090	290,810,000	78.22	31.75
Uruguay	68,039	176,220	3,380,177	49.68	19.31
Uzbekistan	172,742	447,400	25,590,000	148.14	61.78

Table A *(Continued)*
World Countries: Area, Population, and Population Density, 2004

COUNTRY	AREA		POPULATION	DENSITY	
	(Mi^2)	(Km^2)	(Estimated 12/2004)	(Pop/Mi^2)	(Pop/Km^2)
Vanuatu	5,699	12,190	210,164	36.88	17.24
Venezuela, RB	352,145	912,050	25,674,000	72.91	29.11
Vietnam	127,243	331,690	81,314,240	639.05	249.82
West Bank and Gaza	2,401	6,220	3,366,702	1,402.21	541.27
Yemen, Republic	203,850	527,970	19,173,160	94.06	48.13
Zambia	290,586	752,610	10,402,960	35.80	36.31
Zimbabwe	150,803	390,760	13,101,750	86.88	13.99
World	51,715,141	133,941,600	6,272,522,000	121.29	33.87

Sources: World Development Indicators 2003 (The World Bank); *World Almanac and Book of Facts; World Population Prospects: The 2000 Revision* (United Nations Population Information Network, 2001).

Table B
World Countries: Form of Government, Capital City, Major Languages

Notes: Unless indicated otherwise, republics are multi-party. Theocratic normally refers to fundamentalist Islamic rule. Transitional governments are those still in the process of change from a previous form (eg. Single-party communist state to multi-party republic).

COUNTRY	GOVERNMENT	CAPITAL	MAJOR LANGUAGES
Afghanistan	Transitional	Kabul	Dari, Pashtu, Uzbek, Turkmen
Albania	Multi-party democracy	Tiranë	Albanian, Greek
Algeria	Republic	Algiers	Arabic, Berber, dialects, French
Andorra	Parliamentary democracy	Andorra	Cataln, French, Spanish, Portuguese
Angola	Multi-party republic	Luanda	Portugese; Bantu and other African
Antigua and Barbuda	Parliamentary democracy	St. John's	English, local dialects
Argentina	Federal republic	Buenos Aires	Spanish, English, Italian, German, other
Armenia	Republic	Yerevan	Armenian, Russian, other
Australia	Federal parliamentary democracy	Canberra	English, indigenous
Austria	Federal republic	Vienna	German
Azerbaijan	Republic	Baku	Azerbaijani, Russian, Armenian, other
Bahamas	Parliamentary democracy; independent Commonwealth	Nassau	English, Creole
Bahrain	Constitutional hereditary monarchy	Al Manamah	Arabic, English, Farsi, Urdu
Bangladesh	Parliamentary democracy	Dhaka	Bangla, English
Barbados	Parliamentary democracy	Bridgetown	English
Belarus	Republic, authoritarian rule	Minsk	Byelorussian, Russian, other
Belgium	Constitutional monarchy	Brussels	Dutch (Flemish), French, German
Belize	Parliamentary democracy	Belmopan	English, Spanish, Garifuna, Mayan
Benin	Multi-party republic	Porto-Novo	French, Fon, Yoruba
Bhutan	Monarchy; special treaty relationship with India	Timphu	Dzongkha, Tibetan, Nepalese
Bolivia	Republic	La Paz, Sucre	Spanish, Quechua, Aymara
Bosnia-Herzegovina	Emerging federal democratic republic	Sarajevo	Croatian, Serbian, Bosnian
Botswana	Parliamentary republic	Gaborone	English, Setswana
Brazil	Federal republic	Brasilia	Portugese, Spanish, English, French
Brunei	Constitutional monarchy	Bandar Seri Begawan	Malay, English, Chinese
Bulgaria	Parliamentary democracy	Sofia	Bulgarian
Burkina Faso	Parliamentary republic	Ouagadougou	French, indigenous
Burundi	Republic	Bujumbura	French, Kirundi, Swahili
Cambodia	Multi-party democracy (under UN supervision)	Phnom Penh	Khmer, French, English
Cameroon	Multi-party republic	Yaoundé	English, French, indigenous
Canada	Federal parliamentary	Ottawa	English, French, other
Cape Verde	Republic	Cidade de Praia	Portugese, Crioulu
Central African Republic	Republic	Bangui	French, Sangho
Chad	Republic	N'Djamena	French, Arabic, Sara, other indigenous
Chile	Republic	Santiago	Spanish
China	Single-party communist state	Beijing	Various Chinese dialects
Colombia	Republic	Bogotá	Spanish
Comoros	Republic	Moroni	Arabic, French, Shikomoro
Democratic Republic of the Congo (formerly Zaire)	Republic/transitional from military dictatorship	Kinshasa	French, Lingala, Kingwana, Kikongo, Tshilluba
Congo Republic	Multi-party republic	Brazzaville	French, Lingala, Monokutuba, Kikongo
Costa Rica	Democratic republic	San José	Spanish

Table B *(Continued)*
World Countries: Form of Government, Capital City, Major Languages

Notes: Unless indicated otherwise, republics are multi-party. Theocratic normally refers to fundamentalist Islamic rule. Transitional governments are those still in the process of change from a previous form (eg. Single-party communist state to multi-party republic).

COUNTRY	GOVERNMENT	CAPITAL	MAJOR LANGUAGES
Côte d'Ivoire	Multi-party republic	Abidjan, Yamoussoukro	French, indigenous
Croatia	Parliamentary democracy	Zagreb	Croatian
Cuba	Single-party communist state	Havana	Spanish
Cyprus	Republic	Nicosia	Greek, Turkish, English
Czech Republic	Parliamentary democracy	Prague	Czech
Denmark	Constitutional monarchy	Copenhagen	Danish, Faroese, Greenlandic, German
Djibouti	Republic	Djibouti	French, Somali, Afar, Arabic
Dominica	Parliamentary democracy, republic within Commonwealth	Roseau	English, French
Dominican Republic	Republic	Santo Domingo	Spanish
East Timor	Republic	Dili	Tetum, Portuguese, Indonesian, English
Ecuador	Republic	Quito	Spanish, Quechua, indigenous
Egypt	Republic	Cairo	Arabic
El Salvador	Republic	San Salvador	Spanish, Nahua
Equatorial Guinea	Republic	Malabo	Spanish, French, indigenous, English
Eritrea	Transitional government	Asmara	Afar, Amharic, Arabic, Tigre, other indigenous
Estonia	Parliamentary republic	Tallinn	Estonian, Russian, Ukranian, Finnish
Ethiopia	Federal republic	Addis Ababa	Amharic, Tigrinya, Orominga, Somali, Arabic, English
Fiji	Republic	Suva	English, Fijian, Hindustani
Finland	Republic	Helsinki	Finnish, Swedish
France	Republic	Paris	French
Gabon	Multi-party republic	Libreville	French, Fang, indigenous
The Gambia	Multi-party democratic republic	Banjul	English, Mandinka, Wolof, Fula
Georgia	Republic	Tbilisi	Georgian, Russian, Armenian
Germany	Federal republic	Berlin	German
Ghana	Parliamentary democracy	Accra	English, indigenous
Greece	Parliamentary republic	Athens	Greek
Grenada	Parliamentary democracy	St. George's	English, French
Guatemala	Republic	Guatemala City	Spanish, Quiche, Cakchiquel, other indigenous
Guinea	Republic	Conakry	French, indigenous
Guinea-Bissau	Multi-party republic	Bissau	Portugese, Crioulo, indigenous
Guyana	Republic within Commonwealth	Georgetown	English, Creole, Hindi, Urdu, indigenous
Haiti	Republic	Port-au-Prince	Creole, French
Honduras	Republic	Tegucigalpa	Spanish, indigenous
Hungary	Parliamentary democracy	Budapest	Hungarian
Iceland	Republic	Reykjavk	Icelandic
India	Federal republic	New Delhi	English, Hindi, 14 other official
Indonesia	Republic	Jakarta	Bahasa Indonesian, English, Dutch, Javanese
Iran	Theocratic republic	Tehran	Farsi, Turkish, Kurdish
Iraq	In transition following US-led invasion	Baghdad	Arabic, Kurdish, Assyrian, Armenian
Ireland	Republic	Dublin	English, Irish Gaelic
Israel	Parliamentary democracy	Jerusalem	Hebrew, Arabic, English
Italy	Republic	Rome	Italian
Jamaica	Parliamentary democracy	Kingston	English, Creole

Table B *(Continued)*
World Countries: Form of Government, Capital City, Major Languages

Notes: Unless indicated otherwise, republics are multi-party. Theocratic normally refers to fundamentalist Islamic rule. Transitional governments are those still in the process of change from a previous form (eg. Single-party communist state to multi-party republic).

COUNTRY	GOVERNMENT	CAPITAL	MAJOR LANGUAGES
Japan	Constitutional monarchy	Tokyo	Japanese
Jordan	Constitutional monarchy	Amman	Arabic
Kazakhstan	Republic	Astana	Kazakh, Russian
Kenya	Republic	Nairobi	English, Swahili, indigenous
Kiribati	Republic	Tarawa	English, I-Kiribati
Korea, North	Single-party communist state	Pyongyang	Korean
Korea, South	Republic	Seoul	Korean
Kuwait	Constitutional monarchy	Kuwait	Arabic, English
Kyrgyzstan	Republic	Bishkek	Kirghiz, Russian
Laos	Single-party communist state	Vientiane	Lao, French, English, indigenous
Latvia	Parliamentary democracy	Riga	Latvian, Lithuanian, Russian
Lebanon	Republic	Beirut	Arabic, French, Armenian, English
Lesotho	Constitutional monarchy	Maseru	English, Sesotho, Zulu, Xhosa
Liberia	Republic	Monrovia	English, indigenous
Libya	Single party/military dictatorship	Tripoli	Arabic
Liechtenstein	Constitutional monarchy	Vaduz	German
Lithuania	Parliamentary democracy	Vilnius	Lithuanian, Russian, Polish
Luxembourg	Constitutional monarchy	Luxembourg	French, Luxembourgian, German
Macedonia	Parliamentary democracy	Skopje	Macedonian, Albanian, Turkish, Serbo-Croatian
Madagascar	Republic	Antananarivo	Malagasy, French
Malawi	Multi-party democracy	Lilongwe	Chichewa, English, Tombuka
Malaysia	Constitutional monarchy	Kuala Lumpur	Malay, Chinese, English, indigenous
Maldives	Republic	Male	Dhivehi
Mali	Republic	Bamako	French, Bambara, indigenous
Malta	Parliamentary democracy	Valletta	English, Maltese
Marshall Islands	Constitutional government (free association with U.S.)	Majuro	English, Polynesian dialects, Japanese
Mauritania	Republic	Nouakchott	Arabic, Wolof, Pular, French, Solinke
Mauritius	Parliamentary democracy	Port Louis	English, Creole, French, Hindi, Urdu, Bojpoori, Hakka
Mexico	Federal republic	Mexico City	Spanish, indigenous
Micronesia	Constitutional government (free association with U.S.)	Palikir	English, Trukese, Pohnpeian, Yapese, others
Moldova	Republic	Chisinau	Moldavian, Russian, Gagauz
Monaco	Constitutional monarchy	Monaco	French, English, Italian, Monegasque
Mongolia	Republic	Ulaanbaatar	Khalkha Mongol, Turkic, Russian
Morocco	Constitutional monarchy	Rabat	Arabic, Berber dialects, French
Mozambique	Republic	Maputo	Portuguese, indigenous
Myanmar (Burma)	Military regime	Rangoon	Burmese, indigenous
Namibia	Republic	Windhoek	Afrikaans, English, German, indigenous
Nauru	Republic	Yaren district	Nauruan, English
Nepal	Constitutional monarchy	Kathmandu	Nepali, indigenous
Netherlands	Constitutional monarchy	Amsterdam	Dutch
New Zealand	Parliamentary democracy	Wellington	English, Maori
Nicaragua	Republic	Managua	Spanish, English, indigenous

Table B *(Continued)*
World Countries: Form of Government, Capital City, Major Languages

Notes: Unless indicated otherwise, republics are multi-party. Theocratic normally refers to fundamentalist Islamic rule. Transitional governments are those still in the process of change from a previous form (eg. Single-party communist state to multi-party republic).

COUNTRY	GOVERNMENT	CAPITAL	MAJOR LANGUAGES
Niger	Provisional military	Niamey	French, Hausa, Djerma
Nigeria	Military/transitional	Abuja	English, Hausa, Fulani, Yorbua, Ibo
Norway	Constitutional monarchy	Oslo	Norwegian, Sami
Oman	Monarchy	Muscat	Arabic, English, Baluchi, Urdu
Pakistan	Federal republic	Islamabad	Punjabi, Sindhi, Siraiki, Pashtu, Urbu, English, others
Palau	Constitutional government (free association with U.S.)	Koror	English, Palauan, Sonsolorese, Tobi, Angaur, Japanese
Panama	Constitutional democracy	Panama	Spanish, English
Papua New Guinea	Parliamentary democracy	Port Moresby	Various indigenous, English, Motu
Paraguay	Constitutional republic	Asunción	Spanish, Guarani
Peru	Republic	Lima	Quechua, Spanish, Aymara
Philippines	Republic	Manila	English, Filipino, 8 major dialects
Poland	Republic	Warsaw	Polish
Portugal	Parliamentary democracy	Lisbon	Portuguese
Qatar	Traditional monarchy	Doha	Arabic, English
Romania	Republic	Bucharest	Romanian, Hungarian, German
Russia	Federation	Moscow	Russian, numerous other
Rwanda	Republic	Kigali	French, Kinyarwanda, English, Kiswahili
St. Kitts and Nevis	Constitutional monarchy	Basseterre	English
St. Lucia	Parliamentary democracy	Castries	English, French
St. Vincent/Grenadines	Parliamentary monarchy independent within Commonwealth	Kingstown	English, French
Samoa	Constitutional monarchy	Apia	Samoan, English
San Marino	Republic	San Marino	Italian
São Tomé and Principe	Republic	São Tome	Portuguese
Saudi Arabia	Monarchy	Riyadh	Arabic
Senegal	Republic	Dakar	French, Wolof, indigenous
Serbia and Montenegro	Republic	Belgrade	Serbian, Albanian
Seychelles	Republic	Victoria	English, French, Creole
Sierra Leone	Constitutional democracy	Freetown	English, Krio, Mende, Temne
Singapore	Parliamentary republic	Singapore	Chinese, English, Malay,Tamil
Slovakia	Parliamentary democracy	Bratislava	Slovak, Hungarian
Slovenia	Republic	Ljubljana	Slovenian, Serbo-Croatian, other
Solomon Islands	Parliamentary democracy	Honiara	English, indigenous
Somalia	Disordered state, no permanent government	Mogadishu	Arabic, Somali, English, Italian
South Africa	Republic	Pretoria	Afrikaans, English, Zulu, Xhosa, other
Spain	Parliamentary monarchy	Madrid	Castilian Spanish, Catalan, Galician, Basque
Sri Lanka	Republic	Colombo	English, Sinhala, Tamil
Sudan	Provisional military	Khartoum	Arabic, Nubian, others
Suriname	Constitutional democracy	Paramaribo	Dutch, Sranang Tongo, English, Hindustani, Javanese
Swaziland	Monarchy within Commonwealth	Mbabane	English, siSwati
Sweden	Constitutional monarchy	Stockholm	Swedish
Switzerland	Federal republic	Bern	German, French, Italian, Romansch
Syria	Republic (under military regime)	Damascus	Arabic, Kurdish, Armenian, Aramaic

Table B *(Continued)*
World Countries: Form of Government, Capital City, Major Languages

Notes: Unless indicated otherwise, republics are multi-party. Theocratic normally refers to fundamentalist Islamic rule. Transitional governments are those still in the process of change from a previous form (eg. Single-party communist state to multi-party republic).

COUNTRY	GOVERNMENT	CAPITAL	MAJOR LANGUAGES
Taiwan	Multi-party democracy	Taipei	Mandarin Chinese, Taiwanese, Hakka
Tajikistan	Republic	Dushanbe	Tajik, Russian
Tanzania	Republic	Dar es Salaam	Kiswahili, English, Arabic, indigenous
Thailand	Constitutional monarchy	Bangkok	Thai, English
Togo	Republic/transitional	Lomé	French, indigenous
Tonga	Constitutional monarchy	Nuku'alofa	Tongan, English
Trinidad and Tobago	Parliamentary democracy	Port-of-Spain	English, Hindi, French, Spanish, Chinese
Tunisia	Republic	Tunis	Arabic, French
Turkey	Parliamentary republic	Ankara	Turkish, Kurdish, Arabic, Armenian, Greek
Turkmenistan	Republic	Ashkhabad	Turkmen, Russian, Uzbek, other
Tuvalu	Constitutional monarchy	Funafuti	Tuvaluan, English, Samoan, Kiribati
Uganda	Republic	Kampala	English, Luganda, Swahili, Arabic, indigenous
Ukraine	Republic	Kiev	Ukranian, Russian, Romanian, Polish, Hungarian
United Arab Emirates	Federated monarchy	Abu Dhabi	Arabic, English, Farsi, Hindi, Urdu
United Kingdom	Constitutional monarchy	London	English, Welsh, Scottish Gaelic
United States	Federal republic	Washington	English, Spanish
Uruguay	Republic	Montevideo	Spanish, Portunol/Brazilero
Uzbekistan	Republic, authoritarian rule	Tashkent	Uzbek, Russian, Kazakh, Tajik, other
Vanuatu	Republic	Port-Vila	English, French, Bislama (pidgin)
Venezuela	Federal republic	Caracas	Spanish, indigenous
Vietnam	Single-party communist state	Hanoi	Vietnamese, French, Chinese, English, Khmer
Yemen	Republic	San'aa	Arabic
Zambia	Republic	Lusaka	English, Tonga, Lozi, other indigenous
Zimbabwe	Parliamentary democracy	Harare	English, Shona, Sindebele, other

Source: *CIA Factbook,* 2005.

Table C
Defense Expenditures, Armed Forces, Refugees, and the Arms Trade

COUNTRY	MILITARY EXPENDITURES	NUMBERS IN ARMED FORCES	REFUGEE POPULATION	ARMS TRADE 2003 In $ millions based on 1990 price	
	(% of GDP)	2003	2003	Total Exports	Total Imports
Afghanistan	–	130,000	20	0	17
Albania	1.24	22,000	30	–	1
Algeria	3.32	308,700	169,030	–	513
Angola	4.89	130,000	13,380	0	0
Argentina	1.13	102,600	2,640	0	127
Armenia	2.74	45,600	239,290	–	0
Australia	1.85	53,600	56,260	30	485
Austria	0.78	34,600	16,110	2	55
Azerbaijan	1.94	81,500	330	–	0
Bahrain	–	21,400	–	–	–
Bangladesh	1.16	188,700	19,790	–	0
Belarus	1.23	182,900	640	60	0
Belgium	1.29	40,800	12,600	6	27
Belize	–	1,100	860	–	–
Benin	–	7,100	5,030	–	6
Bolivia	1.60	68,600	530	–	0
Bosnia and Herzegovina	9.50	18,800	22,520	0	0
Botswana	3.97	10,500	2,840	–	0
Brazil	1.49	673,200	3,190	0	87
Bulgaria	2.60	85,000	4,070	18	2
Burkina Faso	1.31	15,000	470	–	0
Burundi	6.20	56,000	40,970	–	0
Cambodia	2.30	192,000	80	0	0
Cameroon	1.54	32,100	58,580	–	0
Canada	1.17	61,600	133,090	556	94
Cape Verde	0.71	1,300	–	–	–
Central African Republic	1.13	3,600	44,750	–	0
Chad	1.50	34,800	146,400	0	0
Chile	3.48	114,100	470	0	156
China	2.34	3,750,000	299,350	404	2,548
Colombia	3.98	304,600	190	–	48
Congo, Democratic Republic	1.60	97,800	234,030	–	0
Congo, Republic	1.37	12,000	91,360	–	0
Costa Rica	–	17,000	13,510	–	0
Côte d'Ivoire	1.55	–	75,970	–	22
Croatia	2.07	30,800	4,390	0	0
Cuba	–	72,500	840	–	0
Cyprus	1.49	10,100	350	–	–
Czech Republic	2.07	62,600	1,520	48	111
Denmark	1.57	22,800	69,860	3	7
Djibouti	–	12,300	–	–	–
Dominican Republic	–	39,500	–	–	76
Ecuador	2.36	59,800	6,380	–	0
Egypt, Arab Republic	2.60	780,000	88,750	0	504
El Salvador	0.08	27,500	250	0	0

Table C *(Continued)*
Defense Expenditures, Armed Forces, Refugees, and the Arms Trade

COUNTRY	MILITARY EXPENDITURES	NUMBERS IN ARMED FORCES	REFUGEE POPULATION	ARMS TRADE 2003 In $ millions based on 1990 price	
	(% of GDP)	2003	2003	Total Exports	Total Imports
Equatorial Guinea	–	1,300	–	–	–
Eritrea	19.39	202,000	3,890	0	180
Estonia	1.85	8,100	10	0	16
Ethiopia	4.49	162,400	130,270	0	0
Fiji	–	3,500	–	–	–
Finland	1.20	30,100	10,840	10	125
France	2.58	360,400	130,840	1,753	120
Gabon	0.00	6,700	14,010	–	0
Gambia,The	0.30	800	7,470	–	0
Georgia	1.07	29,200	3,860	0	0
Germany	1.45	284,500	960,400	1,549	69
Ghana	0.66	7,000	43,950	–	0
Greece	4.14	181,600	2,770	0	1,957
Guatemala	0.48	50,400	720	–	0
Guinea	1.70	12,300	184,340	–	0
Guinea-Bissau	3.10	14,200	7550	–	0
Guyana	–	3,100	–	–	–
Haiti	–	5,300	–	–	–
Honduras	0.76	18,000	20	–	0
Hungary	1.84	47,400	7,020	0	0
Iceland	–	100	240	–	–
India	2.28	2,414,700	164,760	0	3,621
Indonesia	1.10	497,000	230	20	333
Iran, Islamic Republic	4.48	580,000	984,900	0	323
Iraq	–	132,000	134,190	0	0
Ireland	0.64	10,400	5,970	0	2
Israel	8.72	175,000	4,180	212	318
Italy	1.88	454,300	12,840	277	348
Jamaica	–	3,000	–	–	0
Japan	0.99	252,100	2,270	0	782
Jordan	8.53	110,500	1,200	0	258
Kazakhstan	0.94	100,300	15,830	0	62
Kenya	1.72	29,100	237,510	–	0
Korea, Democratic Republic	–	1,271,000	–	0	0
Korea, Republic	2.42	690,500	30	36	299
Kuwait	12.48	22,100	1,520	0	21
Kyrgyz Republic	1.38	15,900	5,590	76	9
Lao PDR	2.10	129,100	–	–	0
Latvia	1.75	8,100	20	0	29
Lebanon	4.34	85,100	2,520	0	0
Lesotho	3.10	2,000	–	–	0
Liberia	7.50	15,000	34,000	–	0
Libya	2.40	76,500	11,900	23	0
Liechtenstein	–	–	150	–	–
Lithuania	1.93	27,300	400	0	0

Table C *(Continued)*
Defense Expenditures, Armed Forces, Refugees, and the Arms Trade

COUNTRY	MILITARY EXPENDITURES	NUMBERS IN ARMED FORCES	REFUGEE POPULATION	ARMS TRADE 2003 In $ millions based on 1990 price	
	(% of GDP)	2003	2003	Total Exports	Total Imports
Luxembourg	0.87	1,500	1,200	–	–
Macedonia, FYR	2.78	20,400	190	–	0
Madagascar	1.20	21,600	–	–	–
Malawi	0.80	6,800	3,200	0	0
Malaysia	2.33	124,100	7,420	0	242
Mali	2.00	12,200	10,010	–	0
Malta	0.82	2,100	900	–	–
Mauritania	1.77	20,700	480	–	0
Mauritius	0.22	2,000	100	–	0
Mexico	.47	203,800	6,080	–	43
Moldova	0.40	10,300	21,020	0	0
Mongolia	2.30	15,800	6,080	–	0
Morocco	4.23	246,300	310	–	0
Mozambique	2.43	8,200	300	–	0
Myanmar	2.30	595,000	–	–	31
Namibia	2.76	15,000	19,800	–	5
Nepal	1.50	103,000	123,670	–	5
Netherlands	1.60	59,900	140,890	268	132
New Zealand	1.04	8,600	5,810	0	71
Nicaragua	0.87	14,000	300	0	0
Niger	1.10	10,700	330	–	0
Nigeria	1.02	160,500	9,170	0	51
Norway	1.99	26,600	46,110	150	0
Oman	12.20	46,100	–	0	14
Pakistan	4.07	909,000	1,124,300	0	–
Panama	1.20	13,000	1,450	–	0
Papua New Guinea	0.80	3,000	7,490	–	4
Paraguay	0.73	28,000	30	–	4
Peru	1.45	177,000	720	0	0
Philippines	1.01	150,000	110	–	8
Poland	1.86	184,400	1,840	89	420
Portugal	2.13	92,600	420	0	68
Qatar	–	12,400	50	–	–
Romania	2.36	177,000	2,010	22	46
Russian Federation	4.31	1,369,700	9,900	6,980	0
Rwanda	3.02	61,000	36,610	–	0
Saudi Arabia	8.74	215,000	240,840	0	487
Senegal	1.45	18,600	20,730	–	0
Serbia and Montenegro	4.18	109,200	291,400	0	0
Seychelles	1.65	800	–	–	–
Sierra Leone	2.09	13,000	61,190	–	0
Singapore	5.18	168,800	0	0	121
Slovak Republic	1.81	22,000	410	0	0
Slovenia	1.54	11,000	2,070	–	14
Somalia	–	–	370	–	0

Table C *(Continued)*
Defense Expenditures, Armed Forces, Refugees, and the Arms Trade

COUNTRY	MILITARY EXPENDITURES	NUMBERS IN ARMED FORCES	REFUGEE POPULATION	ARMS TRADE 2003 In $ millions based on 1990 price	
	(% of GDP)	2003	2003	Total Exports	Total Imports
South Africa	1.7	56,000	26,560	23	13
Spain	1.20	224,000	5,900	124	97
Sri Lanka	2.50	241,000	30	–	8
Sudan	2.20	115,000	138,160	–	0
Swaziland	1.70	–	690	–	0
Sweden	1.70	63,000	112,170	186	23
Switzerland	1.00	28,000	50,140	35	41
Syrian Arab Republic	6.90	427,000	3,680	0	15
Tajikistan	1.30	7,000	3,310	–	0
Tanzania	1.50	28,000	649,770	–	0
Thailand	1.30	427,000	119,050	5	163
Togo	1.60	9,000	12,400	–	0
Tunisia	1.64	47,000	100	–	0
Turkey	4.90	665,000	2,490	61	504
Turkmenistan	–	29,000	13,510	–	–
Uganda	2.50	62,000	230,900	–	19
Ukraine	2.90	403,000	2,880	234	0
United Arab Emirates	3.60	51,000	160	0	922
United Kingdom	2.40	213,000	276,520	525	555
United States	4.10	1,480,000	452,550	4,385	515
Uruguay	1.10	25,000	90	0	0
Uzbekistan	0.80	72,000	44,680	510	0
Venezuela, RB	1.30	105,000	60	0	0
Vietnam	–	524,000	15,360	–	7
Yemen, Rep.	7.00	137,000	61,880	–	30
Zambia	0.60	19,500	226,700	0	0
Zimbabwe	3.50	50,800	12,720	–	23
World	2.60	28,161,400	9,680,260	–	–

Sources: World Development Indicators 2003 (World Bank); *Refugees and Others of Concern to UNHCR*: 2002 Statistical Overview (UN High Commissioner on Refugees, 2003).

Note: Data for some countries are based on partial or uncertain data or rough estimates; see U.S. Department of State (2003).

Table D
Major Armed Conflicts, 1990-2005

COUNTRY	TYPE OF WAR	LOCATION OF WAR	ADVERSARIES (in interstate wars)	DATE WAR BEGAN	COMBAT STATUS (6/1/2005)
Afghanistan	interstate war	general	US, UK, NATO forces, Afghanistan	2002	ended by agreement 2003 although fighting continues against insurgents
Albania	civil war	southern regions		1996	order restored by UN peacekeepers 1997
Algeria	civil war	general		1992	continuing
Angola	civil war	general		1975	continuing
	regional civil war	Cabinda enclave		1978	continuing
Armenia	interstate war	Nagorno-Karabakh	Azerbaijan	1990	suspended by agreement 1994
Azerbaijan	interstate war	Nagorno-Karabakh	Armenia	1990	suspended by agreement 1994
Bangladesh	regional civil war	Chittagong		1973	suspended by agreement 1992
Bosnia-Herzegovina	civil and interstate war	general		1992	suspended by agreement 1995
Burundi	civil war	general		1988	continuing
Cambodia	civil war	general		1970	suspended by agreement 1992; fighting terminated by 1997-98
Cameroon	interstate war	Bakassi border region	Nigeria	1996	continuing
Chad	civil war	general		1965	continuing
Colombia	civil war	general		1986	continuing
Congo, Republic of the	civil war	general		1993	suspended by agreement 1994
	civil war	general		1997	continuing
Croatia	civil and interstate war	Slavonia/Krajina	Serbia-Montenegro	1991	suspended by agreement 1992
	regional civil war	Western Slavonia/ Krajina		1995	suspended by agreement 1995
Democratic Republic of the Congo (formerly Zaire)	civil war	general		1996	suspended by agreement 1997
	interstate war	general	Uganda, Rwanda, Burundi, Chad, Angola, Zimbabwe	1998	suspended by agreement 2002, although conflict is continuing
Djibouti	regional civil war	Afar		1991	suspended by agreement 2000
Ecuador	interstate war	border region	Peru	1995	suspended by agreement 1998
Egypt	civil war	general		1992	continuing
El Salvador	civil war	general		1979	suspended by agreement 1979
Eritrea	war of independence	Eritrea	Ethiopia	1962	suspended by agreement 1991
	interstate war	Hanish Islands	Yemen	1997	suspended by agreement 1998
	interstate war	border region with Ethiopia	Ethiopia	1998	suspended by agreement 2000
Ethiopia	against war of independence	Eritrea	Eritrea	1962	suspended by agreement 1991
	civil war	general		1974	suspended by agreement 1991
	interstate war	border region with Eritrea	Eritrea	1998	suspended by agreement 2000
France	interstate war	Kuwait/Iraq		1991	suspended by agreement 1991
Georgia	regional civil war	western region		1991	break in action 1993
	regional civil war	South Ossetia		1991	suspended by agreement 1996
	regional civil war	Abkhazia		1992	suspended by agreement 1994
Ghana	regional civil war	northern regions		1994	break in action 1995
Guatemala	civil war	general		1965	suspended by agreement 1996

Table D *(Continued)*
Major Armed Conflicts, 1990-2005

COUNTRY	TYPE OF WAR	LOCATION OF WAR	ADVERSARIES (in interstate wars)	DATE WAR BEGAN	COMBAT STATUS (6/1/2005)
Haiti	civil war	general		1991	suspended by US/UN intervention 1994
India	interstate war	Jammu-Kashmir	Pakistan	1982	continuing
	regional civil war	Jammu-Kashmir		1990	continuing
	regional civil war	Andhra Pradesh		1969	continuing
	regional civil war	Punjab		1981	break in action 1993
	regional civil war	Assam		1987	continuing
Indonesia	regional civil war	Irian Jaya/West Papua		1963	continuing
	regional civil war	East Timor		1975	suspended by agreement/UN peacekeeping mission, 1999; East Timor became ndependent state 2002
	regional civil war	Ambon		1999	continuing
	regional civil war	Borneo		1999	continuing
	regional civil war	Sumatra (Aceh)		1989	continuing
Iran	civil war	general		1978	break in action 1993
	regional civil war	northwestern Kurdish regions		1979	break in action 1995
Iraq	regional civil war	northern regions/Kur-distan		1974	continuing
	interstate war	Iraq/Kuwait	Kuwait, France, Saudi Arabia, Syria, United Kingdom, United States	1990/91	suspended by agreement 1991
	regional civil war	southern Shia regions		1991	continuing
	interstate war	central Iraq	United States and Great Britain	1998	continuing
	interstate war	Iraq	United States, Great Britain, others	2003	continuing interstate conflict ceased 2003; Iraqi elections restored state sovereignty 2004; conflict continues between Iraqi, US lead coalition forces throughout country
Israel	civil war	general, including occupied territories	Egypt, Syria, Lebanon, Jordan, Iraq and other Arab states including Israeli Palestinians	1948	continuing internal conflict beween Israeli state and Palestinian insurgents
Kurdistan	regional civil war	Turkish border region		1991	continuing
	civil war	general		1993	continuing
Kuwait	interstate war	Kuwait/Iraq	Iraq	1990	suspended by agreement 1991
Laos	civil war	general		1975	break in action 1990
	civil war	border region		2000	continuing
Lebanon	general, then regional civil war	southern zone, from 1990		1975	continuing
Liberia	civil war; possibly interstate	general, especially northern regions	Guinea	1989	continuing
Libya	civil war	general		1995	continuing
Mali	regional civil war	northern Tuareg regions		1990	suspended by agreement 1995
Mauritania	interstate war	border regions	Senegal	1989	suspended by agreement 1991
Mexico	regional civil war	Chiapas and other southern states		1994	continuing

Table D *(Continued)*
Major Armed Conflicts, 1990-2005

COUNTRY	TYPE OF WAR	LOCATION OF WAR	ADVERSARIES (in interstate wars)	DATE WAR BEGAN	COMBAT STATUS (6/1/2005)
Moldova	regional civil war	Trans-Dniestr		1991	suspended by agreement 1997
Morocco	against war of independence	western Sahara	Polisario Front (western Sahara)	1975	break in action 1991
Mozambique	civil war	general		1976	suspended by agreement 1992
Myanmar (Burma)	regional civil war	Kachin		1948	suspended by agreement 1994
	regional civil war	Shan		1948	continuing
	regional civil war	Karen		1949	continuing
	civil war	general		1991	break in action 1992
	regional civil war	Arakan		1992	suspended by agreement 1994
	regional civil war	Kaya		1992	continuing
Nicaragua	civil war	general		1970	suspended by agreement 1992
Niger	regional civil war	northern Tuareg regions		1991	continuing
	regional civil war	eastern region		1994	continuing
Nigeria	interstate war	Bakassi border region	Cameroon	1996	continuing
	regional civil war	Kaduna state		1997	continuing
	interstate war	Sierra Leone	Sierra Leone	1997	suspended by agreement 1999
Pakistan	interstate war	Kashmir	India	1982	continuing
	regional civil war	Karachi/Sind		1992	continuing
Papua New Guinea	regional civil war	Bougainville		1988	cease-fire and break in action 1998
Peru	civil war	general		1980	continuing
	interstate war	border region	Ecuador	1945	suspended by agreement 1998
Philippines	civil war	general		1969	continuing
	regional civil war	Mindanao		1974	suspended by agreement 2001
Russia	regional civil war	North Ossetia/ Ingushetia		1992	break in action 1992
	regional civil war	Moscow		1993	break in action 1993
	regional civil war	Chechnya		1994	ongoing
Rwanda	civil war	general		1990	break in action 1998
	interstate war	general	Dem. Republic of the Congo; Angola, Namibia, Chad, Zimbabwe	1998	suspended by agreement 1999, although conflict is continuing
Saudi Arabia	interstate war	Kuwait/Iraq	Iraq	1991	suspended by agreement 1991
Senegal	interstate war	border regions	Mauritania	1989	suspended by agreement 1991
	regional civil war	Casamance region		1984	suspended by agreement 2001
Serbia and Montenegro	interstate war	Slovenia	Slovenia	1991	suspended by agreement 1991
Sierra Leone	civil war	general		1991	suspended by agreement 1999; sporadic fighting continues
	interstate war	Sierra Leone	Nigeria	1997	
Slovenia	interstate war	Slovenia	Serbia-Montenegro	1991	suspended by agreement 1992
Somalia	civil war	general		1991	continuing
	regional civil war	Somaliland		1991	break in action 1995
South Africa	civil war	general		1948	suspended by agreement 1994
Spain	regional civil war	Basque region		1968	ongoing
Sri Lanka	regional civil war	Tamil areas/northeast		1977	continuing
Sudan	regional civil war	southern regions		1983	continuing
	regional civil war	Kassala/Darfur		2000	ongoing

Table D *(Continued)*
Major Armed Conflicts, 1990-2005

COUNTRY	TYPE OF WAR	LOCATION OF WAR	ADVERSARIES (in interstate wars)	DATE WAR BEGAN	COMBAT STATUS (6/1/2005)
Suriname	civil war	general		1986	suspended by agreement 1992
Syria	interstate war	Kuwait/Iraq	Iraq	1991	suspended by agreement 1991
Tajikistan	civil war	general		1992	continuing
Togo	civil war	general		1991	break in action 1991
Trinidad and Tobago	civil war	general		1990	break in action 1990
Turkey	regional civil war	southeastern Kurdish region/northern Iraq		1977	continuing
	regional civil war	western region		1991	break in action 1992
Uganda	regional civil war	northern region		early 1980's	continuing
	regional civil war	central region		1994	break in action 1995
	regional civil war	southeastern region		1995	break in action 1995
	interstate war	general	Dem. Republic of the Congo; Angola, Namibia, Chad, Zimbabwe	1998	suspended by agreement 1999
United Kingdom	regional civil war	northern Ireland		1969	suspended by agreement 1994
	interstate war	Kuwait/Iraq	Iraq	1991	suspended by agreement 1991
	interstate war	Iraq	Iraq	2003	continuing
United States	interstate war	Kuwait/Iraq	Iraq	1991	suspended by agreement 1991
	interstate war	Iraq	Iraq	2003	continuing
Venezuela	civil war	general		1992	break in action 1992
Yemen	civil war	general		1994	suspended by agreement 1994
	interstate war	Croatia	Croatia	1991	suspended by agreement 1992
	interstate war	Kosovo	NATO countries	1999	suspended by agreement 1999

Sources: The World Factbook 2003 (CIA); The Federation of American Scientists Military Analysis Network (2003).

Table E
World Countries: Basic Economic Indicators, 2003

COUNTRY	GROSS NATIONAL INCOME (GNI) 2003		PURCHASING POWER PARITY (GNI) 2003			AVERAGE ANNUAL % GROWTH IN GDP			STRUCTURE OF ECONOMIC OUTPUT (GDP) 2003 (value added in % of GDP)			
	Total ($US billions)	Per Capita ($US)	Total ($US billions)	Per Capita ($US)	Rank	1983	1993	2003	Agriculture	Industry	Manufacturing	Services
Afghanistan	–	–	–	–	–	–	–	–	–	–	–	–
Albania	6.29	1,740	14.92	4,710	130	1.10	9.60	6.00	24.72	19.15	9.79	56.13
Algeria	64.57	1,930	188.67	5,930	99	5.40	–2.10	6.80	10.25	55.06	7.00	34.70
Angola	10.73	740	25.79	1,910	171	4.20	–24.70	4.50	8.75	64.65	4.06	26.60
Antigua and Barbuda	0.72	9,160	0.76	9,730		4.68	5.39	3.20	–	–	–	–
Argentina	122.17	3,810	419.64	11,410	63	3.88	5.91	8.84	11.06	34.81	23.93	54.14
Armenia	2.90	950	11.58	3,790	145	–	–8.80	13.91	23.51	39.17	21.88	37.32
Aruba	–	–	0.00			–			–	–		
Australia	507.40	21,950	572.22	28,780	24	5.34	3.89	3.80	–	–	–	–
Austria	250.12	26,810	240.56	29,740	17	2.89	0.42	0.75	2.35	31.74	–	65.92
Azerbaijan	6.70	820	27.94	3,390	141	–	–23.10	11.20	14.34	54.51	23.26	31.15
Bahamas, The	–	–	–	–		3.60	1.90	–	–	–		–
Bahrain	–	–	–	–		6.38	12.87	–	–	–		–
Bangladesh	54.78	400	257.89	1,870	173	4.02	4.57	5.26	21.75	26.26	15.80	51.98
Barbados	2.52	9,260	4.08	15,060		0.18	0.99	1.32	–	–	–	–
Belarus	17.48	1,600	59.77	6,050	83	–	–7.60	6.78	9.78	30.08	23.03	60.14
Belgium	308.14	25,760	300.08	28,920	18	0.31	–0.96	1.11	1.32	26.48	–	72.19
Belize	0.90	3,370	1.73	6,320		–2.11	6.19	9.40	–	–	–	–
Benin	3.45	440	7.43	1,110	190	–4.35	3.52	4.81	35.68	14.37	9.03	49.95
Bermuda	–	–	–	–					–	–		
Bhutan	0.58	630	0.00			8.39	5.89	6.70	33,229.83	39.47	7.67	27.30
Bolivia	7.56	900	21.93	2,490	155	–4.04	4.27	2.45	14.85	30.09	14.85	55.06
Bosnia and Herzegovina	7.30	1,530	25.87	6,250	92	–	–	2.70	14.94	32.06	15.14	53.00
Botswana	7.24	3,530	14.42	8,370		13.15	1.92	5.41	2.36	45.18	4.28	52.46
Brazil	474.41	2,720	1,325.67	7,510	86	–3.41	4.90	–0.20	5.77	19.11	11.43	75.12
Bulgaria	19.36	2,130	58.97	7,540	89	3.43	–1.48	4.28	11.71	30.74	18.81	57.54
Burkina Faso	4.18	300	14.21	1,170	185	0.35	4.60	6.50	30.98	18.91	12.89	50.11
Burundi	0.58	90	4.51	630	203	3.72	–5.71	–1.20	49.00	19.00	–	32.00
Cambodia	4.06	300	26.75	2,000	168	–	–	5.16	34.47	29.65	22.17	35.88
Cameroon	11.75	630	32.05	1,990	174	6.87	–3.20	4.70	44.18	16.67	8.70	39.15
Canada	838.64	24,470	950.06	30,040	15	2.84	2.35	2.00	–	–	–	–
Cape Verde	0.78	1,440	2.41	5,130		10.80	7.06	5.00	6.83	19.72	7.96	73.45
Central African Republic	1.19	260	4.21	1,080	181	–8.13	0.34	–7.30	60.79	24.91	–	14.31
Chad	2.34	240	9.30	1,080	187	15.68	–15.71	11.30	45.58	13.49	11.52	40.93
Chile	69.14	4,360	154.73	9,810	76	–3.79	6.99	3.30	8.81	34.30	15.79	56.89
China	1,409.16	1,100	6,410.38	4,980	127	10.90	13.50	9.29	14.62	52.29	39.34	33.08
Colombia	75.20	1,810	285.70	6,410	88	1.58	2.37	3.95	12.32	29.43	14.11	58.26
Comoros	0.32	450	1.03	1,720		4.82	3.01	2.50	40.87	11.94	4.16	47.20
Congo, Democratic Republic	5.39	100	35.16	660	205	1.41	–13.47	5.60	–	–	–	–
Congo, Republic	2.68	650	2.73	730	203	5.85	–1.00	2.65	6.18	60.07	6.37	33.75
Costa Rica	16.58	4,300	36.60	9,140	74	2.86	7.41	6.49	8.76	28.74	21.20	62.50
Cote d'Ivoire	12.99	660	23.51	1,400	179	–3.90	–0.19	–3.78	26.17	18.62	10.73	55.22
Croatia	27.58	5,370	47.17	10,610	75	–	–8.03	4.27	8.35	30.10	19.28	61.54
Cuba	–	–	0.00	–	–	–	–	–	–	–	–	–
Cyprus	–	–	15.09	19,600		5.58	0.70	4.00	–	–	–	–
Czech Republic	85.55	7,150	159.13	15,600	55	–	0.06	3.11	3.48	39.37	27.24	57.14
Denmark	209.10	33,570	167.28	31,050	9	1.74	0.00	0.43	2.13	26.41	–	71.46
Djibouti	0.64	910	1.51	2,140		–	–3.90	3.53	–	–	–	–

Table E *(Continued)*
World Countries: Basic Economic Indicators, 2003

COUNTRY	GROSS NATIONAL INCOME (GNI) 2003		PURCHASING POWER PARITY (GNI) 2003			AVERAGE ANNUAL % GROWTH IN GDP			STRUCTURE OF ECONOMIC OUTPUT (GDP) 2003 (value added in % of GDP)			
	Total ($US billions)	Per Capita ($US)	Total ($US billions)	Per Capita ($US)	Rank	1983	1993	2003	Agriculture	Industry	Manufacturing	Services
Dominica	0.24	3,330	0.36	5,020		2.69	1.68	–0.63	–	–	–	–
Dominican Republic	15.30	2,130	55.14	6,310	90	4.59	3.02	–0.40	11.24	30.63	15.41	58.13
Ecuador	25.74	1,830	44.81	3,440	140	–2.53	0.30	2.66	7.68	28.71	10.66	63.61
Egypt, Arab Republic	82.26	1,390	266.31	3,940	131	7.40	2.88	3.20	16.14	34.04	18.94	49.82
El Salvador	15.29	2,340	32.09	4,910	107	1.54	7.37	1.83	8.53	32.11	23.71	59.36
Equatorial Guinea	–	–	–	–		–	6.30	14.70	6.78	88.92	–	4.30
Eritrea	0.90	190	4.47	1,020	189	–	13.45	3.00	13.87	24.69	11.29	61.44
Estonia	8.51	5,380	17.16	12,680	71	3.70	–5.91	5.14	4.49	28.48	18.00	67.03
Ethiopia	6.60	90	48.38	710	198	7.85	13.36	–3.69	41.83	10.73	–	47.44
Fiji	1.96	2,240	4.72	5,650		–3.96	2.61	4.80	–	–	–	–
Finland	160.95	27,060	143.13	27,460	28	2.80	–1.24	1.88	3.46	30.52	–	66.02
France	1,755.02	24,730	1,651.58	27,640	27	1.49	–0.89	0.47	2.71	24.47	–	72.82
Gabon	5.21	3,340	7.39	5,500	105	5.61	2.38	2.80	8.05	62.14	4.91	29.81
Gambia, The	0.37	270	2.47	1,740	160	10.88	3.01	6.70	30.14	14.64	5.38	55.22
Georgia	4.02	770	13.35	2,610	148	4.30	–29.30	11.09	20.49	25.45	18.76	54.06
Germany	2,390.69	25,270	2,279.12	27,610	21	1.55	–1.09	–0.10	1.14	29.45	–	69.41
Ghana	7.46	320	45.25	2,190	157	–4.56	4.85	5.20	35.80	24.87	8.48	39.33
Greece	171.71	13,230	219.52	19,900	47	–1.08	–1.60	4.28	6.87	23.83	–	69.30
Grenada	0.39	3,710	0.74	7,030		2.93	–2.64	5.77	–	–	–	–
Guatemala	24.38	1,910	50.33	4,090	120	–2.57	3.93	2.12	22.25	19.26	12.77	58.49
Guinea	3.60	430	16.44	2,080	164	–	3.96	1.20	24.57	36.44	4.00	39.00
Guinea-Bissau	0.23	140	1.01	680	193	–3.40	2.10	0.60	68.75	13.32	9.91	17.94
Guyana	0.70	900	3.06	3,980		–6.79	8.21	–0.65	–	–	–	–
Haiti	2.91	400	14.63	1,730	166	0.76	–2.44	0.40	–	–		–
Honduras	6.78	970	18.05	2,590	144	–0.92	6.23	3.01	13.48	30.72	20.24	55.80
Hong Kong, China	165.31	25,860	195.45	28,680		6.32	6.30	3.22	–	–	–	–
Hungary	78.51	6,350	140.17	13,840	59	0.72	–0.58	3.05	–	–	–	–
Iceland	10.29	30,910	8.83	30,570		–2.15	0.79	4.02	–	–	–	–
India	597.57	540	3,062.32	2,880	143	7.08	4.87	8.60	22.21	26.59	15.76	51.20
Indonesia	199.03	810	689.38	3,210	142	8.45	7.25	4.10	16.58	43.57	24.65	39.85
Iran, Islamic Republic	137.32	2,010	465.00	7,000	98	13.05	2.10	6.61	11.26	41.18	12.53	47.56
Iraq	–	–	0.00	–	–	–	–		–	–		
Ireland	125.89	27,010	123.44	30,910	14	–0.24	2.69	3.70	–	–		–
Israel	106.95	16,240	130.00	19,440	40	3.51	5.56	1.29	–	–	–	–
Italy	1,452.48	21,570	1,546.47	26,830	25	1.24	–0.88	0.26	2.65	27.80	19.71	69.55
Jamaica	7.53	2,980	10.02	3,790	133	2.27	1.97	2.26	5.15	29.81	12.56	65.04
Japan	4,374.44	34,180	3,628.85	28,450	20	1.61	0.25	2.66	–	–	–	–
Jordan	9.78	1,850	22.75	4,290	128	1.99	4.49	3.16	2.19	26.01	15.85	71.80
Kazakhstan	28.01	1,780	93.45	6,280	94	–	–9.20	9.20	7.84	38.29	15.55	53.88
Kenya	14.24	400	32.79	1,030	190	1.31	0.35	1.80	15.75	19.59	13.59	64.66
Kiribati	0.08	860	0.00			8.98	1.68	2.50	–	–	–	–
Korea, Dem. Republic	–	–	0.00	–	–	–	–		–	–	–	–
Korea, Republic	606.18	12,030	862.25	18,000	54	10.77	6.13	3.07	3.17	34.60	23.44	62.23
Kuwait	45.07	17,960	46.69	19,480	35	10.41	33.99	9.90	–	–	–	–
Kyrgyz Republic	1.85	340	8.56	1,690	147	–	–15.46	6.67	38.70	22.94	8.02	38.36
Lao PDR	2.08	340	9.78	1,730	175	–	5.91	5.00	48.57	25.94	19.18	25.50
Latvia	11.01	4,400	23.71	10,210	82	5.22	–4.98	7.46	4.52	24.43	14.52	71.04
Lebanon	18.12	4,040	21.77	4,840	119	–	7.00	2.70	12.23	20.05	9.43	67.72
Lesotho	1.38	610	5.56	3,100	139	–3.19	3.48	3.28	16.63	43.52	20.22	39.85
Liberia	0.38	110	–	–	196	–1.90	–32.98	–29.50	–	–	–	–

Table E *(Continued)*
World Countries: Basic Economic Indicators, 2003

COUNTRY	GROSS NATIONAL INCOME (GNI) 2003		PURCHASING POWER PARITY (GNI) 2003			AVERAGE ANNUAL % GROWTH IN GDP			STRUCTURE OF ECONOMIC OUTPUT (GDP) 2003 (value added in % of GDP)			
	Total ($US billions)	Per Capita ($US)	Total ($US billions)	Per Capita ($US)	Rank	1983	1993	2003	Agriculture	Industry	Manufacturing	Services
Libya	–	–	–	–	–	–	–		–	–	–	–
Liechtenstein	0.00	–	–	–					–	–	–	–
Lithuania	17.73	4,500	39.34	11,390	78	–	–16.23	8.96	7.27	33.77	21.35	58.96
Luxembourg	23.60	45,740	24.86	55,500		2.99	4.20	2.13	0.63	20.49	11.50	78.89
Macedonia, FYR	4.63	1,980	13.83	6,750	97	–	–7.47	3.23	12.22	30.44	19.05	57.35
Madagascar	5.39	290	13.47	800	197	0.90	2.10	9.79	29.20	15.37	13.71	55.43
Malawi	1.67	160	6.46	590	206	3.72	9.69	4.41	38.37	14.88	9.53	46.75
Malaysia	97.81	3,880	222.20	8,970	81	6.25	9.89	5.31	9.70	48.50	31.06	41.08
Maldives	0.68	2,350	–			–	–	9.19	–	–	–	–
Mali	4.16	290	11.13	960	200	4.80	–2.14	6.00	38.38	26.09	2.84	35.53
Malta	4.89	10,780	7.09	17,780		–0.61	4.48	–1.75	–	–	–	–
Marshall Islands	0.14	2,710	–			16.34	4.06	2.00	–	–	–	–
Mauritania	1.16	400	5.33	1,870	162	3.74	9.71	4.94	19.25	29.99	8.66	50.76
Mauritius	5.22	4,100	13.79	11,280	70	2.94	5.92	3.20	6.14	30.56	22.04	63.30
Mexico	613.27	6,230	918.65	8,980	80	–4.20	1.95	1.30	4.05	26.39	18.05	69.57
Micronesia, Fed. Sts.	0.26	2,070	–			–	8.14	2.40	–	–	–	–
Moldova	2.29	590	7.45	1,760	154	4.36	–1.20	6.29	22.54	24.70	18.39	52.76
Mongolia	1.25	480	4.51	1,820	170	5.83	–3.01	5.65	28.05	14.92	4.52	57.03
Morocco	43.00	1,310	118.57	3,940	132	–0.56	–1.01	5.24	16.82	29.59	16.62	53.59
Mozambique	4.11	210	19.95	1,060	188	–15.70	8.70	7.10	26.06	31.17	15.48	42.77
Myanmar	–	–	–	–	–	–	–	–	–	–	–	–
Namibia	4.60	1,930	13.41	6,660	85	–1.77	–2.01	3.74	10.83	25.56	12.27	63.61
Nepal	5.84	240	34.97	1,420	180	–2.98	3.50	3.09	40.57	21.59	8.14	37.85
Netherlands	497.39	26,230	463.31	28,560	13	1.76	0.65	–0.90	–	–	–	–
Netherlands Antilles	–	–	–						–	–	–	–
New Caledonia	–	–	–			–2.00	0.53	–	–	–	–	–
New Zealand	75.22	15,530	85.58	21,350	43	2.74	6.45	3.60	–	–	–	–
Nicaragua	3.98	740	17.40	3,180	158	4.61	–0.39	2.30	17.93	25.73	14.51	56.35
Niger	2.72	200	9.77	830	194	–4.75	1.45	5.32	39.86	16.76	6.56	43.38
Nigeria	49.96	350	122.56	900	199	–5.29	2.20	10.69	26.37	49.46	3.98	24.18
Norway	222.26	43,400	172.94	37,910	7	3.54	2.73	0.42	1.45	37.52	–	61.03
Oman	–	–	–	–	–	16.67	7.00	–	–	–	–	–
Pakistan	80.12	520	302.93	2,040	167	6.78	1.76	5.15	23.31	23.49	16.42	53.20
Palau	0.13	6,500	–			–	–12.30	1.50	–	–	–	–
Panama	12.07	4,060	19.15	6,420	104	–4.49	5.46	4.06	7.47	16.33	8.08	76.20
Papua New Guinea	2.74	500	12.40	2,250	149	3.22	18.20	2.70	25.67	39.15	8.99	35.19
Paraguay	6.04	1,110	26.46	4,690	106	–3.02	4.15	2.55	27.24	24.23	13.59	48.53
Peru	58.50	2,140	137.88	5,080	117	–11.80	4.76	3.76	10.32	29.29	16.16	60.39
Philippines	86.61	1,080	378.56	4,640	125	1.87	2.12	4.52	14.50	32.34	22.90	53.16
Poland	206.45	5,280	428.17	11,210	73	–	3.80	3.75	3.13	30.73	17.84	66.15
Portugal	144.52	11,800	184.97	17,710	46	–0.17	–2.04	–1.20	–	–	–	–
Qatar	–	–	–						–	–	–	–
Romania	55.87	2,260	155.22	7,140	101	6.10	1.51	4.90	11.86	36.09	29.96	52.05
Russian Federation	419.68	2,610	1,283.56	8,950	87	–	–8.67	7.35	5.16	34.17	–	60.67
Rwanda	1.66	220	10.80	1,290	183	5.98	–8.11	3.19	41.63	21.87	11.25	36.50
Samoa	0.26	1,440	1.03	5,780		0.43	4.10	3.54	–	–	–	–
San Marino	–	–	–			–	–	–	–			
Sao Tome and Principe	0.05	300	–			–	1.10	4.50	17.00	14.60	4.44	68.40
Saudi Arabia	214.75	9,240	297.97	13,230	54	–8.22	0.03	7.18	4.53	55.19	10.14	40.28
Senegal	6.40	540	16.63	1,620	176	2.18	–2.22	6.45	16.79	21.17	12.80	62.04

Table E *(Continued)*
World Countries: Basic Economic Indicators, 2003

COUNTRY	GROSS NATIONAL INCOME (GNI) 2003		PURCHASING POWER PARITY (GNI) 2003			AVERAGE ANNUAL % GROWTH IN GDP			STRUCTURE OF ECONOMIC OUTPUT (GDP) 2003 (value added in % of GDP)			
	Total ($US billions)	Per Capita ($US)	Total ($US billions)	Per Capita ($US)	Rank	1983	1993	2003	Agriculture	Industry	Manufacturing	Services
Serbia and Montenegro	20.49	1,910	–	–	–	–	–	3.00	–	–	–	–
Seychelles	0.68	7,490	–			–1.71	6.20	–5.07	3.26	35.06	22.58	61.69
Sierra Leone	0.76	150	2.81	530	208	–2.10	1.38	6.60	52.74	30.76	5.21	16.50
Singapore	90.22	21,230	102.76	24,180	32	8.52	12.26	1.09	0.11	34.93	27.91	64.96
Slovak Republic	32.40	4,940	72.46	13,440	60	–	–3.70	4.21	3.66	29.73	19.52	66.61
Slovenia	27.67	11,920	38.10	19,100	49	–	2.84	2.52	–	–	–	–
Solomon Islands	0.25	560	0.78	1,710		13.87	2.00	5.10	–	–	–	–
Somalia	–	–	–	–	–		–	–	–	–	–	–
South Africa	156.50	2,750	464.10	10,130	64	–3.44	1.23	1.85	3.81	30.99	18.92	65.20
Spain	829.64	17,040	910.41	22,150	39	1.77	–1.03	2.43	3.32	29.59	–	67.10
Sri Lanka	18.05	930	71.89	3,740	134	4.81	6.90	5.90	19.00	26.27	15.52	54.73
St. Kitts and Nevis	0.30	6,630	0.50	10,740		–0.97	6.66	0.00	2.99	28.32	9.34	68.69
St. Lucia	0.64	4,050	0.85	5,310		4.25	2.57	1.75	5.38	18.03	4.97	76.59
St. Vincent and Grenada	0.36	3,310	0.64	5,870		4.40	0.16	4.05	8.67	24.41	6.33	66.92
Sudan	16.35	460	58.88	1,750	169	2.06	4.57	6.00	–	–	–	–
Suriname	1.15	2,280	–	1,760		–4.08	–7.26	5.15	–	–	–	–
Swaziland	1.89	1,350	5.36	4,850	118	2.08	3.51	2.20	12.22	51.55	39.86	36.23
Sweden	301.12	28,910	239.19	26,710	29	1.88	–2.00	1.58	1.80	27.87	–	70.33
Switzerland	337.57	40,680	236.80	32,220	5	0.51	–0.23	–0.40	–	–	–	–
Syrian Arab Republic	20.63	1,160	59.64	3,430	136	1.43	5.18	2.50	23.49	28.55	25.47	47.96
Tajikistan	1.46	210	6.57	1,040	184	–	–16.40	10.20	23.40	20.25	–	56.35
Tanzania	10.24	300	22.19	620	207	–	1.21	7.10	45.03	16.36	7.25	38.61
Thailand	140.28	2,190	462.17	7,450	93	5.58	8.25	6.87	9.75	43.98	35.19	46.27
Timor-Leste	0.34	460	–			–	–	–2.97	–	–	–	–
Togo	1.70	310	7.97	1,640	172	–5.41	–15.10	2.70	40.79	22.16	9.28	37.05
Tonga	0.16	1,490	0.70	6,910		1.98	3.76	2.50	–	–	–	–
Trinidad and Tobago	10.15	7,790	13.64	10,390	77	–9.20	–1.45	13.18	1.17	48.81	7.06	50.02
Tunisia	23.94	2,240	67.76	6,850	96	4.68	2.19	5.57	12.05	28.13	17.81	59.82
Turkey	238.32	2,800	474.79	6,710	100	4.97	8.04	5.79	13.38	21.89	13.29	64.72
Turkmenistan	6.12	1,120	28.50	5,860	124	–	–10.00	16.90	–	–	–	–
Uganda	6.17	250	36.06	1,430	177	5.74	8.33	4.73	32.35	21.20	9.32	46.45
Ukraine	48.96	970	262.42	5,430	131	–	–14.23	9.40	14.06	40.30	24.99	45.64
United Arab Emirates	–	–	–	–	–	–5.39	–0.90	–	–	–	–	–
United Kingdom	1,830.59	28,320	1,642.62	27,690	26	3.54	2.33	2.22	0.97	26.59	–	72.44
United States	11,003.75	37,870	10,978.45	37,750	3	4.52	2.69	3.10	–	–	–	–
Uruguay	10.78	3,820	26.98	7,980	79	–10.27	2.66	2.50	12.80	27.30	18.90	59.90
Uzbekistan	9.83	420	44.12	1,720	152	–	–2.30	4.40	35.18	21.68	9.13	43.14
Vanuatu	0.28	1,180	0.61	2,900		5.58	6.00	2.00	–	–	–	–
Venezuela, RB	82.42	3,490	121.88	4,750	102	–3.76	0.28	–9.37	4.48	41.11	9.06	54.41
Vietnam	39.16	480	202.47	2,490	159	–	8.07	7.24	21.83	39.95	20.80	38.22
West Bank and Gaza	3.84	1,110	–	–	–	–	–	–1.67	6.18	12.00	9.84	81.81
Yemen, Rep.	10.04	520	15.80	820	202	–	4.07	3.80	14.98	39.98	5.20	45.05
Zambia	4.19	380	8.81	850	201	–1.97	6.80	5.10	22.80	27.01	12.04	50.19
Zimbabwe	–	–	–	–	156	1.59	1.05	–	–	–	–	–
World	36,361.71	5,510	51,400.58	8,190		2.64	1.72	2.83	–	–	–	–

Sources: *World Development Indicators, 2003* (World Bank)

Table F
World Countries: Population Growth, 1950-2025

COUNTRY	POPULATION (thousands)			AVERAGE ANNUAL POPULATION CHANGE (percent)		AVERAGE ANNUAL INCREMENT TO THE POPULATION (mid-year population, in thousands)		
	1950	2000[a]	2025[a]	1975-1980	2001-2015[a]	1985-1990	1995-2000	2005-2010
WORLD	**2,518,629.0**	**6,070,581.0**	**7,851,455.0**	**1.7**	**2.5**	**85,831.0**		
AFRICA								
Algeria	8,753.0	30,245.0	42,429.0	3.1	1.5	631.8	566.0	539.3
Angola	4,131.0	12,386.0	19,268.0	2.7	2.6	130.2	187.2	267.5
Benin	2,046.0	6,222.0	11,120.0	2.5	2.4	135.6	184.8	204.6
Botswana	419.0	1,725.0	1,614.0	3.5	0.5	43.3	21.5	-15.4
Burkina Faso	3,960.0	11,905.0	24,527.0	2.5	2.1	233.8	307.0	360.3
Burundi	2,456.0	6,267.0	12,328.0	2.3	1.7	95.2	123.0	166.5
Cameroon	4,466.0	15,127.0	20,831.0	2.8	1.7	326.5	370.9	374.8
Central African Republic	1,314.0	3,715.0	5,193.0	2.3	1.5	57.5	62.0	60.0
Chad	2,658.0	7,861.0	15,770.0	2.1	2.9	171.2	259.8	340.4
Congo Democratic Republic	12,184.0	48,571.0	95,448.0	3.0	2.6	1,146.2	1,234.4	1,907.8
Congo Republic	808.0	3,447.0	6,750.0	2.9	2.7	56.3	62.4	67.2
Côte d'Ivoire	2,775.0	15,827.0	22,140.0	3.9	1.6	411.1	350.2	394.2
Egypt	21,834.0	67,784.0	103,165.0	2.4	1.5	1,318.5	1,199.9	1,125.0
Equatorial Guinea	226.0	456.0	812.0	-0.7	–	8.7	11.1	13.6
Eritrea	1,140.0	3,712.0	7,261.0	2.6	2.3	35.4	134.9	153.2
Ethiopia	18,434.0	65,590.0	116,006.0	2.4	2.1	1,530.7	1,670.9	1,848.4
Gabon	469.0	1,258.0	1,915.0	3.1	2.2	11.6	13.9	8.6
Gambia	294.0	1,312.0	2,177.0	3.1	2.0	33.0	42.3	48.0
Ghana	4,900.0	19,593.0	30,618.0	1.9	1.6	435.4	380.2	285.2
Guinea	2,550.0	8,117.0	13,704.0	1.5	1.9	173.7	63.1	193.9
Guinea-Bissau	505.0	1,367.0	2,774.0	4.7	2.2	22.1	28.4	35.1
Kenya	6,265.0	30,549.0	39,917.0	3.8	1.4	723.5	604.9	206.8
Lesotho	734.0	1,785.0	1,608.0	2.5	0.8	41.5	39.6	11.5
Liberia	824.0	2,943.0	6,081.0	3.1	2.3	-2.9	236.3	107.3
Libya	1,029.0	5,237.0	7,785.0	4.4	1.9	92.8	92.2	136.3
Madagascar	4,230.0	15,970.0	30,249.0	2.5	2.5	308.1	433.2	590.5
Malawi	2,881.0	11,370.0	18,245.0	3.3	1.8	416.6	169.9	103.8
Mali	3,520.0	11,904.0	25,679.0	2.1	2.1	164.0	305.1	390.7
Mauritania	825.0	2,645.0	4,973.0	2.5	2.3	47.5	65.2	94.9
Morocco	8,953.0	29,108.0	40,721.0	2.3	1.4	565.7	535.1	515.0
Mozambique	6,442.0	17,861.0	25,350.0	2.8	1.6	78.7	359.2	75.1
Namibia	511.0	1,894.0	2,350.0	2.7	1.2	58.6	33.5	7.5
Niger	2,500.0	10,742.0	25,722.0	3.2	2.8	207.6	259.7	322.0
Nigeria	29,790.0	114,746.0	192,115.0	2.8	1.9	2,530.7	3,225.5	3,161.6
Rwanda	2,162.0	7,724.0	12,509.0	3.3	1.6	187.9	311.1	49.1
Senegal	2,500.0	9,393.0	15,663.0	2.8	2.0	191.6	278.9	336.2
Sierra Leone	1,944.0	4,415.0	7,593.0	2.0	1.9	106.3	143.8	162.5
Somalia	2,264.0	8,720.0	20,978.0	7.0	3.1	45.8	192.4	266.1
South Africa	13,683.0	44,000.0	42,962.0	2.2	0.4	943.4	383.4	-419.6
Sudan	9,190.0	31,437.0	47,536.0	3.1	2.0	634.6	902.5	1,059.5
Tanzania	7,886.0	34,837.0	53,435.0	3.1	1.7	799.7	832.1	970.9
Togo	1,329.0	4,562.0	7,551.0	2.7	1.9	122.0	160.9	115.3
Tunisia	3,530.0	9,586.0	12,843.0	2.6	1.3	168.9	124.3	104.8

Table F *(Continued)*
World Countries: Population Growth, 1950-2025

COUNTRY	POPULATION (thousands)			AVERAGE ANNUAL POPULATION CHANGE (percent)		AVERAGE ANNUAL INCREMENT TO THE POPULATION (mid-year population, in thousands)		
	1950	2000[a]	2025[a]	1975-1980	2001-2015[a]	1985-1990	1995-2000	2005-2010
Uganda	5,310.0	23,487.0	54,883.0	3.2	2.4	590.9	598.7	877.5
Zambia	2,440.0	10,419.0	14,401.0	3.4	1.2	210.9	174.1	190.9
Zimbabwe	2,744.0	12,650.0	12,857.0	3.0	0.6	308.9	78.1	-58.6
NORTH AND CENTRAL AMERICA						**353.9**		
Belize	69.0	240.0	356.0	1.7	–	5.0	6.3	7.2
Canada	13,737.0	30,769.0	36,128.0	1.2	0.6	369.8	331.8	289.5
Costa Rica	966.0	3,929.0	5,621.0	3.0	1.4	76.6	65.4	58.0
Cuba	5,850.0	11,202.0	11,479.0	0.9	0.3	93.2	48.4	37.3
Dominican Republic	2,353.0	8,353.0	10,955.0	2.4	1.3	141.1	136.4	147.1
El Salvador	1,951.0	6,209.0	8,418.0	2.1	1.6	87.1	110.8	117.7
Guatemala	2,969.0	11,428.0	19,456.0	2.5	2.4	255.9	317.6	366.3
Haiti	3,261.0	8,005.0	10,670.0	2.1	1.7	111.8	89.1	114.6
Honduras	1,380.0	6,457.0	10,115.0	3.4	2.1	117.2	151.1	134.9
Jamaica	1,403.0	2,580.0	3,263.0	1.2	1.1	18.3	16.8	23.9
Mexico	27,737.0	98,933.0	129,866.0	2.7	1.4	1,594.3	1,572.4	1,425.0
Nicaragua	1,134.0	5,073.0	8,318.0	3.1	2.1	91.0	107.6	100.9
Panama	860.0	2,950.0	4,290.0	2.5	1.3	44.8	39.8	32.5
Trinidad and Tobago	636.0	1,289.0	1,340.0	1.3	0.8	6.5	-4.9	-6.1
United States	157,813.0	285,003.0	358,030.0	0.9	0.8	2,296.3	2,503.8	2,429.2
SOUTH AMERICA								
Argentina	17,150.0	37,074.0	47,043.0	1.5	1.0	445.4	427.4	401.1
Bolivia	2,714.0	8,317.0	12,495.0	2.4	1.8	127.7	155.2	128.3
Brazil	53,975.0	171,796.0	216,372.0	2.4	1.1	2,756.3	1,975.6	1,285.4
Chile	6,082.0	15,224.0	19,651.0	1.5	1.0	212.2	189.7	148.3
Colombia	12,568.0	4,120.0	58,157.0	2.3	1.3	636.0	681.0	630.9
Ecuador	3,387.0	12,420.0	16,704.0	2.8	1.5	262.0	264.1	257.2
Guyana	423.0	759.0	724.0	0.7	–	-3.2	-3.5	4.1
Paraguay	1,488.0	5,470.0	9,173.0	3.2	2.1	113.5	141.6	162.8
Peru	7,632.0	25,952.0	35,622.0	2.7	1.3	472.9	491.4	432.4
Suriname	215.0	425.0	486.0	-0.5	–	3.9	3.3	1.4
Uruguay	2,239.0	3,342.0	3,875.0	0.6	0.6	19.4	23.7	26.7
Venezuela	5,094.0	24,277.0	31,189.0	3.4	1.5	465.5	397.3	351.7
ASIA								
Afghanistan	8,151.0	21,391.0	44,940.0	0.9	2.5	170.3	879.9	724.7
Armenia	1,354.0	3,112.0	2,866.0	1.8	0.3	-0.7	-13.8	7.6
Azerbaijan	2,896.0	8,157.0	10,222.0	1.6	0.7	103.6	23.6	61.8
Bangladesh	41,783.0	137,952.0	208,268.0	2.8	1.6	2,028.8	2,001.0	2,119.6
Bhutan	734.0	2,063.0	3,701.0	2.3	–	34.7	42.4	48.8
Cambodia	4,346.0	13,147.0	21,899.0	-1.8	1.5	313.1	271.6	314.6
China[b]	556,924.0	1,282,472.0	1,454,141.0	1.5	0.6	16,833.4	11,408.3	8,726.8
Georgia	3,527.0	5,262.0	4,429.0	0.7	-0.7	49.9	-53.5	-14.4
India	357,561.0	1,016,938.0	1,369,284.0	2.1	1.2	16,448.0	16,317.4	15,140.5
Indonesia	79,538.0	211,559.0	270,113.0	2.1	1.1	3,283.4	3,702.8	3,388.7
Iran	16,913.0	664,423.0	90,927.0	3.3	1.6	1,632.8	818.3	1,000.3

Table F *(Continued)*
World Countries: Population Growth, 1950-2025

COUNTRY	POPULATION (thousands)			AVERAGE ANNUAL POPULATION CHANGE (percent)		AVERAGE ANNUAL INCREMENT TO THE POPULATION (mid-year population, in thousands)		
	1950	2000[a]	2025[a]	1975-1980	2001-2015[a]	1985-1990	1995-2000	2005-2010
Iraq	5,158.0	23,224.0	41,707.0	3.3	1.9	488.2	623.7	719.5
Israel	1,258.0	6,042.0	8,598.0	2.3	1.5	87.4	107.5	73.6
Japan	83,625.0	127,034.0	123,444.0	0.9	-0.2	556.6	252.5	-30.4
Jordan	472.0	5,035.0	8,116.0	2.3	2.2	126.9	159.4	145.2
Kazakhstan	6,703.0	15,640.0	15,388.0	1.1	0.1	148.5	-42.0	85.9
Korea, North	10,815.0	22,268.0	24,665.0	1.6	0.6	307.4	27.2	4,294.4
Korea, South	18,859.0	46,835.0	50,165.0	1.6	0.4	943.5	459.2	322.0
Kuwait	152.0	2,247.0	3,930.0	6.2	2.1	81.8	70.6	90.4
Kyrgyzstan	1,740.0	4,921.0	6,484.0	1.9	1.1	76.8	30.0	80.9
Laos	1,755.0	5,279.0	8,635.0	1.2	2.2	110.7	130.3	155.3
Lebanon	1,443.0	3,478.0	4,554.0	-0.7	1.2	11.8	48.7	46.0
Malaysia	6,110.0	23,001.0	33,479.0	2.3	1.5	391.7	436.4	438.2
Mongolia	761.0	2,500.0	3,368.0	2.8	1.3	62.1	37.4	44.4
Myanmar (Burma)	17,832.0	47,544.0	59,760.0	2.1	1.0	452.8	317.4	162.3
Nepal	8,643.0	23,518.0	37,831.0	2.5	2.0	457.5	559.0	616.3
Oman	456.0	2,609.0	4,785.0	5.0	2.2	58.3	80.5	104.3
Pakistan	39,659.0	142,654.0	249,766.0	2.6	2.2	2,984.4	2,984.8	2,936.8
Philippines	1,996.0	75,711.0	108,589.0	2.3	1.6	1,450.5	1,658.1	1,666.1
Saudi Arabia	3,201.0	22,147.0	39,751.0	5.6	2.9	527.8	678.3	922.2
Singapore	1,022.0	4,016.0	4,905.0	1.3	1.1	56.1	134.2	169.1
Sri Lanka	7,483.0	18,595.0	21,464.0	1.7	1.1	234.4	186.9	153.5
Syria	3,495.0	16,560.0	26,979.0	3.1	2.1	391.1	399.2	431.5
Tajikistan	1,532.0	6,089.0	8,193.0	2.8	1.5	149.0	115.3	168.7
Thailand	19,626.0	60,925.0	73,869.0	2.4	0.6	755.5	598.8	468.5
Turkey	21,484.0	68,281.0	88,995.0	2.1	1.1	1,083.1	895.5	732.4
Turkmenistan	1,211.0	4,643.0	6,549.0	2.5	1.1	85.4	83.3	95.8
United Arab Emirates	70.0	2,820.0	3,944.0	14.0	1.8	76.1	38.6	40.0
Uzbekistan	6,314.0	24,913.0	33,774.0	2.6	1.3	473.1	381.7	485.8
Vietnam	27,369.0	78,137.0	104,649.0	2.2	1.2	1,321.7	1,178.3	1,123.4
Yemen	4,316.0	18,017.0	43,204.0	3.2	3.0	436.3	524.0	782.1
EUROPE								
Albania	1,215.0	3,113.0	3,629.0	1.9	1.0	60.3	50.7	34.4
Austria	6,935.0	8,102.0	7,979.0	-0.1	-0.1	32.1	17.8	11.3
Belarus	7,745.0	10,034.0	8,950.0	0.6	-0.5	46.7	-7.5	-1.4
Belgium	8,639.0	10,251.0	10,516.0	0.1	0.0	22.2	20.9	5.3
Bosnia-Herzegovina	2,661.0	3,977.0	4,183.0	0.9	0.5	29.7	96.0	15.5
Bulgaria	7,251.0	8,099.0	6,609.0	0.3	-0.7	-9.9	-95.1	-74.3
Croatia	3,850.0	4,446.0	4,088.0	0.5	-0.3	10.1	-34.6	11.2
Czech Republic	8,925.0	10,269.0	9,806.0	0.6	-0.2	-0.1	-10.6	-14.7
Denmark	4,271.0	5,322.0	5,469.0	0.2	0.1	5.5	20.8	12.0
Estonia	1,101.0	1,369.0	1,017.0	0.6	-0.5	7.0	-10.5	-4.9
Finland	4,009.0	5,177.0	5,289.0	0.3	0.1	16.9	12.3	4.8
France	41,829.0	59,296.0	64,165.0	0.4	0.3	312.8	236.0	142.8
Germany	68,376.0	82,282.0	81,959.0	-0.1	-0.2	339.1	229.8	152.4
Greece	7,566.0	10,903.0	10,707.0	1.3	0.1	44.5	22.4	11.0
Hungary	9,338.0	10,012.0	8,865.0	0.3	-0.6	-55.4	-31.4	-31.1

Table F *(Continued)*
World Countries: Population Growth, 1950-2025

COUNTRY	POPULATION (thousands)			AVERAGE ANNUAL POPULATION CHANGE (percent)		AVERAGE ANNUAL INCREMENT TO THE POPULATION (mid-year population, in thousands)		
	1950	2000[a]	2025[a]	1975-1980	2001-2015[a]	1985-1990	1995-2000	2005-2010
Iceland	143.0	282.0	325.0	0.9	-	2.7	1.8	1.1
Ireland	2,969.0	3,819.0	4,668.0	1.4	0.8	-6.4	37.2	31.9
Italy	47,104.0	57,536.0	52,939.0	0.4	-0.4	4.0	74.2	-67.5
Latvia	1,949.0	2,373.0	1,857.0	0.4	-0.7	12.3	-23.5	-12.9
Lithuania	2,567.0	3,501.0	3,035.0	0.7	-0.2	22.2	-10.4	-3.6
Macedonia	1,230.0	2,024.0	2,199.0	1.4	0.4	6.9	11.0	7.2
Moldova	2,341.0	4,283.0	4,096.0	0.9	-0.2	49.9	-5.8	16.0
Netherlands	10,114.0	15,898.0	17,123.0	0.7	0.4	92.0	86.6	62.5
Norway	3,265.0	4,473.0	4,859.0	0.4	0.4	17.9	24.4	18.2
Poland	24,824.0	38,671.0	37,337.0	0.9	0.0	178.7	8.5	11.2
Portugal	8,405.0	10,016.0	9,834.0	1.4	-0.1	5.1	15.9	9.6
Romania	16,311.0	22,480.0	20,806.0	0.9	-0.3	69.0	-56.3	-49.8
Russian Federation	102,192.0	145,612.0	124,428.0	0.6	-0.5	820.8	-422.7	-281.7
Serbia-Montenegro	7,131.0	10,555.0	10,230.0	0.9	0.1	21.0	-5.2	-4.6
Slovak Republic	3,463.0	5,391.0	5,397.0	1.0	0.0	23.6	9.3	6.1
Slovenia	1,473.0	1,990.0	1,859.0	1.0	-0.2	4.6	3.6	1.2
Spain	28,009.0	40,752.0	40,369.0	1.1	0.0	163.2	49.1	-2.8
Sweden	7,014.0	8,856.0	9,055.0	0.3	0.0	40.5	9.5	0.5
Switzerland	4,694.0	7,173.0	6,801.0	-0.1	-0.1	54.8	19.2	7.9
Ukraine	37,298.0	49,688.0	40,775.0	0.4	-0.7	142.7	-432.6	-264.4
United Kingdom	49,816.0	58,689.0	63,275.0	0.0	0.0	189.0	178.9	94.6
OCEANIA								
Australia	8,219.0	19,153.0	23,205.0	0.9	0.7	246.8	209.7	167.0
Fiji	289.0	814.0	965.0	1.9	–	7.8	11.3	12.8
New Zealand	1,908.0	3,784.0	4,379.0	0.2	0.5	12.3	50.8	38.5
Papua New Guinea	1,798.0	5,334.0	8,443.0	2.5	1.9	89.3	116.4	125.1
Solomon Islands	90.0	437.0	783.0	3.5	–	11.0	13.8	14.3

a Data include projections based on 1990 base year population data
b Includes Hong Kong and Macao

Source: United Nations Population Division and International Labour Organisation. *World Resources 2000-2001;* (World Resources Institute), U.S. Bureau of the Census International Data Base (2000).

Table G
World Countries: Basic Demographic Data, 1990-2003

COUNTRY	CRUDE BIRTH RATE (births per 1,000 population)		LIFE EXPECTANCY AT BIRTH (years)		LIFE EXPECTANCY OF FEMALES AND MALES				TOTAL FERTILITY RATE		PERCENTAGE OF POPULATION IN SPECIFIC AGE GROUPS					
					Females		Males				1990			2003		
	1990	2003	1990	2003	1990	2003	1990	2003	1993	2003	<15	15-65	>65	<15	15-65	>65
Afghanistan	48.58	–	41.52	–	41.64	–	41.40	–	6.90	–	44.08	52.99	2.93	–	–	–
Albania	24.9	17.05	72.28	74.33	75.40	76.66	69.30	72.12	3.03	2.23	32.75	61.89	5.35	27.35	65.37	7.28
Algeria	31.12	21.64	67.37	70.88	68.80	72.32	66.00	69.50	4.49	2.72	42.00	54.44	3.56	33.92	62.04	4.05
Angola	51	50.28	45.46	46.68	47.10	48.32	43.90	45.12	7.20	7.00	46.88	50.19	2.93	47.63	49.44	2.93
Antigua and Barbuda	20.1	17.96	73.75	75.36	76.35	78.29	71.28	72.57	1.78	1.70	–	–	–	20.67	71.15	8.19
Argentina	20.96	18.24	71.64	74.46	75.26	78.02	68.20	71.06	2.90	2.43	30.63	60.44	8.94	27.03	63.18	9.79
Armenia	22.5	9.36	71.72	74.98	75.20	78.69	68.40	71.44	2.62	1.15	30.37	64.02	5.61	20.52	69.32	10.16
Australia	15.4	12.50	77.00	79.76	80.12	82.74	74.02	76.92	1.91	1.75	21.90	66.96	11.14	19.98	67.53	12.49
Austria	11.6	9.50	75.53	79.06	78.89	81.99	72.33	76.26	1.45	1.39	17.40	67.63	14.97	16.17	67.80	16.03
Azerbaijan	26.4	16.26	70.80	–	74.80	–	67.00	–	2.74	2.07	33.12	62.06	4.82	27.00	65.51	7.49
Bahamas, The	19.6	18.33	69.24	69.79	73.30	74.19	65.38	65.59	2.12	2.12	32.42	62.87	4.71	28.37	66.04	5.58
Bahrain	27.46	20.67	71.43	73.36	73.62	75.99	69.32	70.86	3.76	2.30	31.62	65.93	2.45	27.16	70.31	2.52
Bangladesh	32.68	27.54	54.76	62.40	54.68	63.06	54.84	61.78	4.12	2.88	44.33	52.54	3.13	35.47	61.18	3.35
Barbados	15.7	13.91	74.94	74.76	77.50	77.31	72.50	72.33	1.74	1.75	24.90	63.43	11.67	20.63	69.20	10.17
Belarus	13.9	9.41	70.84	68.17	75.60	74.19	66.30	62.45	1.91	1.25	23.06	66.29	10.65	16.83	69.13	14.04
Belgium	12.4	10.70	75.97	78.30	79.40	81.42	72.70	75.32	1.62	1.61	18.15	66.77	15.07	16.96	66.29	16.75
Belize	35.24	26.95	72.55	71.25	73.82	72.83	71.34	69.75	4.49	3.15	44.14	51.60	4.26	37.66	57.93	4.41
Benin	45.46	37.92	51.87	52.96	53.82	54.76	50.02	51.24	6.62	5.20	48.33	48.47	3.20	44.97	52.37	2.67
Bhutan	–	34.60	–	63.54	–	64.82	–	62.31	–	5.09	42.20	53.97	3.83	42.60	53.32	4.08
Bolivia	36.06	28.76	58.31	64.14	60.04	66.26	56.66	62.12	4.85	3.75	41.23	55.14	3.63	38.35	57.28	4.37
Bosnia and Herzegovina	14.2	11.96	71.44	74.01	74.10	77.16	68.90	71.00	1.70	1.30	23.86	69.94	6.20	17.22	71.86	10.92
Botswana	38.16	29.30	56.76	38.02	58.60	37.82	55.00	38.22	5.07	3.74	45.70	51.95	2.35	41.52	56.18	2.31
Brazil	23.56	19.04	65.60	68.72	69.42	72.77	61.96	64.86	2.74	2.14	34.75	60.93	4.32	27.52	67.10	5.38
Brunei	25.82	18.36	74.16	76.79	76.38	79.26	72.04	74.43	3.20	2.47	34.11	62.77	3.12	30.52	66.48	2.99
Bulgaria	12.1	8.40	71.64	72.08	75.15	75.63	68.30	68.71	1.81	1.23	20.42	66.59	12.99	14.38	69.20	16.41
Burkina Faso	47.38	42.64	45.39	42.79	46.50	43.40	44.34	42.20	7.02	6.19	47.47	49.80	2.72	46.88	50.40	2.71
Burundi	46.4	38.06	43.59	41.58	45.20	41.78	42.06	41.38	6.80	5.67	45.60	51.19	3.21	45.32	52.14	2.53
Cambodia	40.76	29.01	50.31	54.02	51.80	55.51	48.90	52.60	5.56	3.90	39.42	57.64	2.94	40.99	55.81	3.20
Cameroon	41.12	35.24	54.18	48.00	55.74	48.93	52.70	47.11	6.00	4.60	44.75	51.63	3.63	41.13	55.17	3.71
Canada	15	10.50	77.38	79.34	80.65	82.57	74.26	76.28	1.83	1.52	20.74	68.05	11.21	18.15	69.01	12.84
Cape Verde	36.4	29.78	65.29	69.18	68.00	72.10	62.70	66.41	5.50	3.46	43.28	52.03	4.69	41.60	54.79	3.61
Central African Republic	40.6	35.40	47.59	41.83	49.70	42.40	45.58	41.28	5.46	4.60	43.25	52.70	4.04	41.84	54.66	3.50
Chad	46.6	44.77	46.16	48.34	47.82	49.86	44.58	46.90	7.06	6.19	45.45	51.01	3.55	48.01	49.18	2.81
Chile	23.5	16.74	73.70	76.44	76.80	79.58	70.74	73.45	2.58	2.15	30.06	63.81	6.13	26.92	65.66	7.42
China	21.06	14.54	68.87	70.80	70.47	72.56	67.36	69.13	2.10	1.88	27.69	66.75	5.56	23.65	69.05	7.30
Colombia	27.32	22.20	68.27	71.93	72.48	74.99	64.26	69.01	3.07	2.48	36.04	59.67	4.28	31.77	63.47	4.76
Comoros	40.28	31.81	55.97	61.56	57.40	62.94	54.60	60.26	5.80	4.05	46.97	50.37	2.66	41.95	55.42	2.63
Congo, Dem. Rep.	47.9	44.97	51.55	45.21	53.38	45.86	49.80	44.58	6.70	6.70	47.26	49.88	2.86	47.89	49.55	2.56
Congo Republic	44.48	44.17	51.23	51.67	53.88	53.76	48.70	49.68	6.29	6.29	45.51	51.15	3.33	46.89	50.05	3.06
Costa Rica	27.4	17.33	76.80	78.64	79.00	81.10	74.70	76.30	3.20	2.30	36.46	59.34	4.20	29.81	64.38	5.80
Côte d'Ivoire	41.58	36.54	49.80	45.11	51.12	45.55	48.54	44.68	6.18	4.54	47.51	49.84	2.65	41.49	55.91	2.60
Croatia	11.7	9.85	72.17	73.99	75.93	78.29	68.59	69.90	1.63	1.45	20.51	68.14	11.35	16.16	68.08	15.76
Cuba	17.6	12.57	75.03	76.86	76.94	78.93	73.22	74.89	1.69	1.62	23.05	68.54	8.41	20.34	69.24	10.42
Cyprus	18.3	13.28	76.54	78.16	78.82	80.58	74.36	75.85	2.42	1.90	25.99	63.13	10.88	21.81	66.24	11.95
Czech Republic	12.7	8.90	71.38	75.17	75.42	78.50	67.54	72.00	1.89	1.18	21.45	66.05	12.50	15.48	70.65	13.88
Denmark	12.3	12.00	74.81	77.14	77.73	79.50	72.02	74.90	1.67	1.76	17.04	67.36	15.60	18.56	66.53	14.91
Djibouti	40.36	35.76	47.77	42.96	49.44	42.87	46.18	43.05	5.98	5.20	42.99	54.28	2.73	42.79	54.20	3.01
Dominica	23	17.77	73.16	76.70	75.00	78.93	71.40	74.58	2.70	1.90	–	–	–	25.02	66.60	8.38
Dominican Rep.	28.24	22.41	65.93	67.13	68.00	70.13	63.96	64.28	3.38	2.60	37.04	59.45	3.50	31.95	63.51	4.54

Table G *(Continued)*
World Countries: Basic Demographic Data, 1990-2003

COUNTRY	CRUDE BIRTH RATE (births per 1,000 population)		LIFE EXPECTANCY AT BIRTH (years)		LIFE EXPECTANCY OF FEMALES AND MALES				TOTAL FERTILITY RATE		PERCENTAGE OF POPULATION IN SPECIFIC AGE GROUPS					
					Females		Males				1990			2003		
	1990	2003	1990	2003	1990	2003	1990	2003	1993	2003	<15	15-65	>65	<15	15-65	>65
Ecuador	29.34	22.46	68.12	71.05	70.64	73.68	65.72	68.54	3.71	2.69	38.93	56.94	4.13	32.72	62.41	4.87
Egypt, Arab Rep.	31.14	23.70	62.80	69.13	64.28	70.78	61.40	67.56	3.97	3.08	39.89	56.29	3.82	33.51	62.18	4.31
El Salvador	30.04	24.82	65.62	70.42	69.86	73.50	61.58	67.48	3.85	2.83	40.78	55.00	4.23	34.67	60.35	4.98
Equatorial Guinea	43.62	40.08	47.16	52.17	48.79	54.18	45.60	50.26	5.89	5.42	42.32	53.69	3.98	43.58	52.75	3.67
Eritrea	–	37.28	48.94	51.10	50.76	52.32	47.20	49.94	6.50	4.79	44.20	53.13	2.66	44.33	53.02	2.65
Estonia	14.2	9.60	69.48	71.18	74.60	77.27	64.60	65.38	2.04	1.35	22.18	66.11	11.71	16.14	68.63	15.22
Ethiopia	50.52	39.96	45.00	42.03	46.54	42.89	43.54	41.21	6.91	5.60	45.46	51.60	2.94	45.41	51.77	2.82
Fiji	24.96	21.93	99.73	69.66	68.88	71.44	64.68	67.96	3.09	2.60	38.21	58.62	3.17	32.29	63.92	3.78
Finland	13.1	10.80	74.81	78.32	78.88	81.71	70.94	75.09	1.78	1.76	19.30	67.32	13.38	17.62	67.04	15.34
France	13.4	12.70	76.75	79.26	80.93	82.90	72.76	75.80	1.78	1.89	20.24	65.78	13.98	18.60	65.27	16.13
French Polynesia	26.24	19.77	69.59	73.94	72.30	77.14	67.00	70.90	3.25	2.50	34.69	61.74	3.57	30.77	63.70	5.53
Gabon	36.24	34.79	51.88	52.97	53.60	54.35	50.24	51.65	5.09	4.05	36.98	57.02	5.99	40.43	54.04	5.53
Gambia, The	44.7	36.34	49.28	53.38	51.40	55.00	47.26	51.84	5.90	4.80	42.02	55.16	2.82	40.53	56.22	3.25
Georgia	17	8.16	72.31	73.47	76.10	77.72	68.70	69.41	2.21	1.10	24.60	66.07	9.33	18.39	67.30	14.31
Germany	11.4	8.60	75.21	78.33	78.58	81.30	72.00	75.50	1.45	1.34	16.09	68.96	14.95	14.87	67.82	17.31
Ghana	39.2	30.89	57.16	54.38	58.90	54.91	55.50	53.87	5.50	4.40	45.34	51.75	2.91	41.89	53.68	4.43
Greece	10.1	9.30	76.94	77.99	79.50	80.70	74.50	75.40	1.40	1.27	19.25	67.07	13.69	14.65	66.66	18.68
Grenada	–	24.44	–	73.15	–	76.48	–	69.97	–	3.00	–	–	–	34.20	58.79	7.01
Guam	28.7	23.23	74.35	77.95	77.91	80.39	70.95	75.62	3.34	3.75	30.83	66.19	2.99	34.16	59.10	6.74
Guatemala	39.32	33.27	61.42	66.10	64.18	69.13	58.80	63.20	5.33	4.29	45.95	50.88	3.16	42.49	54.05	3.46
Guinea	46.4	37.46	43.71	46.16	44.20	46.67	43.25	45.67	5.90	5.02	46.73	50.70	2.57	43.64	53.81	2.55
Guinea-Bissau	49.76	48.56	42.35	45.54	43.98	47.03	40.80	44.11	7.10	6.59	41.73	54.16	4.11	44.45	52.12	3.44
Guyana	24.82	21.82	63.66	62.18	67.06	66.78	60.42	57.79	2.61	2.31	33.49	62.74	3.77	29.89	65.29	4.82
Haiti	35.78	32.26	53.10	51.87	54.98	54.09	51.30	49.75	5.42	4.20	44.34	51.85	3.81	39.00	57.54	3.46
Honduras	38.02	29.47	64.92	66.10	67.36	69.16	62.60	63.20	5.16	4.00	45.21	51.84	2.95	40.76	55.91	3.33
Hong Kong, China	12.32	7.07	77.58	80.12	80.40	82.70	74.90	77.67	1.27	0.96	21.49	70.05	8.46	15.78	72.64	11.57
Hungary	12.1	9.50	69.32	72.58	73.70	76.81	65.14	68.56	1.84	1.30	20.24	66.41	13.36	16.33	68.99	14.68
Iceland	18.7	14.10	77.94	79.88	80.30	82.23	75.70	77.64	2.31	1.99	24.89	64.52	10.59	22.46	65.66	11.88
India	30.2	24.38	59.13	63.42	59.22	64.24	59.04	62.64	3.80	2.86	36.42	59.23	4.35	32.40	62.53	5.07
Indonesia	25.42	20.92	61.71	66.89	63.50	68.86	60.00	65.02	3.06	2.41	35.69	60.45	3.86	29.73	65.39	4.88
Iran, Islamic Rep.	30.8	17.65	64.65	69.45	65.42	70.51	63.92	68.44	4.68	2.00	45.47	50.97	3.56	29.49	65.79	4.72
Iraq	38.44	29.08	61.28	63.11	62.66	64.36	59.96	61.92	5.88	4.05	44.22	52.91	2.87	39.39	57.55	3.05
Ireland	15.1	15.50	74.58	77.68	77.40	80.34	71.90	75.14	2.12	1.98	27.35	61.32	11.33	21.31	67.52	11.17
Israel	22.2	20.15	76.09	78.76	78.00	80.80	74.28	76.82	2.82	–	31.28	59.64	9.08	27.36	62.96	9.68
Italy	10	9.40	76.86	79.83	80.26	82.90	73.62	76.90	1.26	1.29	15.87	68.81	15.33	14.02	66.98	19.00
Jamaica	25.2	19.30	73.24	75.84	75.16	77.96	71.42	73.82	2.94	2.30	35.15	57.46	7.39	29.66	63.46	6.88
Japan	10	9.20	78.84	81.68	81.91	85.30	75.91	78.24	1.54	1.33	18.40	69.62	11.99	14.20	67.25	18.55
Jordan	35.8	28.03	68.48	72.12	70.34	73.75	66.70	70.57	5.40	3.50	46.83	49.99	3.18	37.44	59.39	3.17
Kazakhstan	21.7	15.30	68.34	61.32	73.10	66.57	63.80	56.33	2.72	1.80	31.53	62.60	5.87	24.51	67.43	8.06
Kenya	38.5	33.76	57.11	45.41	58.84	45.76	55.46	45.08	5.64	4.80	49.13	47.88	2.98	42.11	55.22	2.67
Kiribati	32.18	27.46	56.80	63.13	59.10	66.40	54.60	60.02	4.04	3.60	–	–	–	38.32	59.53	2.15
Korea, Dem. Rep.	20.72	16.64	65.52	62.82	68.08	64.56	63.08	61.17	2.39	2.07	26.92	68.83	4.26	25.68	67.68	6.64
Korea, Republic	16.28	12.14	70.28	74.15	73.88	77.73	66.86	70.74	1.77	1.45	25.84	69.16	5.00	20.71	71.73	7.56
Kuwait	24.8	20.04	74.88	77.02	76.84	79.13	73.01	75.02	3.44	2.52	36.60	62.19	1.21	24.79	73.34	1.87
Kyrgyz Republic	29.3	19.09	68.30	65.03	72.60	69.46	64.20	60.81	3.69	2.40	37.57	57.45	4.98	31.68	62.19	6.14
Lao PDR	45.16	35.42	49.72	54.74	51.00	56.02	48.50	53.51	6.00	4.80	43.64	53.38	2.98	41.80	54.70	3.50
Latvia	14.2	8.80	69.27	70.69	74.60	76.14	64.20	65.50	2.02	1.29	21.43	66.57	12.00	15.12	69.34	15.55
Lebanon	37.72	18.89	67.90	70.88	69.90	72.73	66.00	69.12	3.22	2.22	34.85	59.99	5.17	30.28	63.82	5.89
Lesotho	36.88	32.95	57.58	37.20	58.98	38.28	56.24	36.17	5.08	4.30	41.36	54.64	4.00	41.42	53.44	5.13
Liberia	47.6	42.88	45.06	47.05	46.60	48.17	43.60	45.99	6.80	5.79	46.67	50.42	2.91	44.13	53.11	2.76

Table G *(Continued)*
World Countries: Basic Demographic Data, 1990-2003

COUNTRY	CRUDE BIRTH RATE (births per 1,000 population)		LIFE EXPECTANCY AT BIRTH (years)		LIFE EXPECTANCY OF FEMALES AND MALES				TOTAL FERTILITY RATE		PERCENTAGE OF POPULATION IN SPECIFIC AGE GROUPS					
					Females		Males				1990			2003		
	1990	2003	1990	2003	1990	2003	1990	2003	1993	2003	<15	15-65	>65	<15	15-65	>65
Libya	30.04	26.82	68.50	72.65	70.38	75.30	66.70	70.12	4.72	3.32	45.03	52.48	2.49	32.53	63.77	3.70
Lithuania	15.3	8.80	71.16	71.86	76.17	77.70	66.39	66.30	2.03	1.25	22.61	66.43	10.97	17.65	68.16	14.18
Luxembourg	12.9	11.50	75.19	78.32	78.80	81.69	71.76	75.11	1.62	1.63	17.45	69.41	13.13	18.95	66.13	14.92
Macedonia, FYR	18.8	13.66	71.62	73.60	73.56	76.11	69.78	71.22	2.06	–	26.10	66.46	7.44	21.50	67.84	10.66
Madagascar	46.02	38.33	52.76	55.69	54.30	57.23	51.30	54.22	6.22	5.19	41.96	54.86	3.18	44.05	52.92	3.03
Malawi	50.88	44.36	44.61	37.55	45.16	38.00	44.08	37.12	7.00	5.99	47.27	50.05	2.68	44.87	51.76	3.37
Malaysia	28.9	21.30	70.51	72.99	72.73	75.50	68.40	70.60	3.77	2.85	36.50	59.80	3.71	33.02	62.56	4.41
Maldives	39.22	28.87	61.71	69.47	61.40	70.97	62.00	68.04	5.70	4.00	46.74	50.02	3.23	39.18	56.83	3.99
Mali	50.7	47.53	44.99	40.62	46.00	41.64	44.02	39.65	–	6.40	47.19	49.75	3.06	47.15	50.06	2.78
Malta	15.2	10.00	75.50	78.53	77.85	81.13	73.26	76.05	2.05	1.41	23.59	65.95	10.45	19.42	67.97	12.61
Mauritania	44.06	34.26	49.06	51.01	50.70	52.65	47.50	49.45	6.02	4.60	44.84	51.90	3.26	42.57	54.30	3.13
Mauritius	21.3	16.38	69.40	72.32	73.40	75.84	65.60	68.96	2.32	1.97	29.76	65.03	5.21	25.00	68.64	6.36
Mexico	28	19.30	70.79	73.63	73.90	76.54	67.82	70.87	3.31	2.21	38.55	57.49	3.96	32.33	62.51	5.16
Micronesia, Fed. Sts.	34.6	25.06	63.45	68.88	65.50	71.06	61.50	66.81	4.76	3.45	–	–	–	36.41	59.82	3.77
Moldova	17.7	11.21	68.32	67.02	71.80	70.76	65.00	63.46	2.39	1.40	27.88	63.79	8.33	20.35	68.57	11.08
Mongolia	30.92	22.26	62.66	65.92	64.00	67.56	61.38	64.36	4.03	2.40	41.67	54.32	4.01	31.70	64.22	4.08
Morocco	31.16	–	63.48	68.59	65.25	70.66	61.79	66.62	4.01	2.66	38.87	57.37	3.76	32.94	62.69	4.37
Mozambique	45.18	40.06	43.44	40.72	44.84	41.59	42.10	39.89	6.34	5.00	44.29	52.46	3.25	42.27	54.10	3.63
Myanmar	30.5	23.05	54.70	57.34	56.92	59.98	52.58	54.82	3.76	2.80	35.62	60.33	4.05	31.94	63.55	4.51
Namibia	38.7	35.37	57.52	40.33	58.94	40.04	56.16	40.62	5.39	4.80	42.84	53.52	3.64	41.85	54.33	3.82
Nepal	38.54	31.04	53.58	60.16	52.94	59.90	54.18	60.40	5.26	4.07	43.13	53.32	3.55	40.09	56.09	3.82
Netherlands	13.2	12.40	76.88	78.49	80.11	80.90	73.80	76.20	1.62	1.75	18.24	68.92	12.84	18.34	67.69	13.96
Netherlands Antilles	18.9	15.44	74.46	76.30	77.56	79.33	71.50	73.42	2.29	2.09	27.13	65.96	6.91	23.61	67.74	8.65
New Caledonia	23.92	20.02	70.96	73.70	74.70	77.66	67.40	69.94	2.93	2.50	32.74	63.09	4.16	29.03	64.35	6.61
New Zealand	17.5	13.80	75.38	79.15	78.40	81.40	72.50	77.00	2.18	1.94	23.42	65.45	11.13	21.92	66.41	11.68
Nicaragua	38.96	29.07	64.49	98.83	67.42	71.17	61.70	66.59	4.80	3.44	46.26	50.97	2.77	40.93	56.01	3.06
Niger	55.5	48.23	42.09	46.38	42.40	46.69	41.80	46.08	7.64	7.10	47.70	49.83	2.47	48.86	48.85	2.29
Nigeria	43.82	42.57	49.05	44.91	50.68	45.39	47.50	44.45	6.51	5.65	45.48	51.73	2.79	44.05	53.31	2.64
Norway	14.4	12.00	76.54	79.05	79.80	81.61	73.43	76.60	1.93	1.80	18.94	64.74	16.32	19.74	65.39	14.88
Oman	43.72	26.03	69.00	74.24	71.00	75.78	67.10	72.77	7.38	3.96	46.39	51.20	2.41	41.30	55.92	2.78
Pakistan	41.4	32.29	59.10	64.09	60.00	65.43	58.24	62.83	5.84	4.50	42.95	54.14	2.91	40.10	56.55	3.35
Panama	25.88	20.31	72.44	75.02	74.68	77.44	70.30	72.72	3.01	2.42	35.32	59.72	4.97	29.98	64.20	5.81
Papua New Guinea	33.48	32.59	55.13	57.23	55.90	58.10	54.40	56.40	5.55	4.30	40.36	57.21	2.42	40.87	56.64	2.49
Pataguay	35.1	29.56	68.14	70.95	70.44	73.26	65.94	68.75	4.60	3.84	42.00	54.33	3.67	38.41	58.05	3.54
Peru	28.8	22.11	65.80	70.01	68.24	72.32	63.48	67.81	3.68	2.68	38.27	57.76	3.97	32.98	62.15	4.86
Philippines	32.44	25.46	65.63	69.92	67.78	71.84	63.58	68.09	4.12	3.24	40.36	56.15	3.48	36.00	60.04	3.96
Poland	14.3	9.20	70.89	74.60	75.50	78.90	66.50	70.50	2.04	1.24	25.11	64.83	10.05	17.64	69.88	12.48
Portugal	11.8	10.80	73.66	76.18	77.30	79.51	70.20	73.00	1.43	1.42	19.98	66.41	13.61	17.25	67.59	15.16
Puerto Rico	18.5	15.07	74.79	76.71	79.10	81.53	70.68	72.13	2.20	1.90	27.22	63.14	9.64	23.37	66.33	10.31
Qatar	24.66	13.69	72.22	75.03	73.72	75.42	70.80	74.66	4.34	2.48	27.85	71.33	0.82	24.21	72.73	3.06
Romania	13.7	9.60	69.74	70.09	73.05	74.03	66.59	66.33	1.84	1.27	23.57	66.04	10.40	16.65	69.46	13.90
Russian Federation	13.4	9.82	68.90	65.70	74.26	72.05	63.80	59.66	1.89	1.28	22.96	67.01	10.03	16.34	70.44	13.22
Rwanda	42.04	43.29	40.19	39.79	41.52	40.19	38.92	39.41	7.15	5.69	47.57	49.93	2.51	45.71	51.28	3.01
Samoa	33.2	28.52	66.29	69.54	67.92	72.59	64.74	66.65	4.76	4.00	40.13	56.11	3.76	35.47	59.56	4.98
Sao Tome and Principe	40.44	30.59	62.15	66.03	64.20	68.84	60.20	63.35	5.07	4.35	–	–	–	38.65	54.94	6.41
Saudi Arabia	36.28	30.99	69.00	73.33	70.48	75.13	67.60	71.61	6.56	5.30	41.99	55.45	2.56	40.23	56.89	2.88
Senegal	43.04	34.47	49.54	52.31	51.68	54.09	47.50	50.61	6.20	4.90	45.45	51.67	2.88	43.62	53.69	2.69

Table G *(Continued)*
World Countries: Basic Demographic Data, 1990-2003

COUNTRY	CRUDE BIRTH RATE (births per 1,000 population)		LIFE EXPECTANCY AT BIRTH (years)		LIFE EXPECTANCY OF FEMALES AND MALES				TOTAL FERTILITY RATE		PERCENTAGE OF POPULATION IN SPECIFIC AGE GROUPS					
					Females		Males				1990			2003		
	1990	2003	1990	2003	1990	2003	1990	2003	1993	2003	<15	15-65	>65	<15	15-65	>65
Serbia and Montenegro	14.9	11.00	71.64	72.78	74.30	75.28	69.10	70.40	2.08	1.74	23.40	67.05	9.55	19.60	66.40	14.01
Seychelles	23.7	19.20	70.30	73.18	73.52	77.31	67.22	69.25	2.82	2.01	–	–	–	27.95	65.14	6.92
Sierra Leone	49.1	43.74	35.20	37.38	36.64	38.71	33.82	36.11	6.50	5.59	43.87	53.05	3.08	43.60	53.81	2.59
Singapore	18.4	11.37	74.34	–	76.92	–	71.88	–	1.87	1.37	21.47	73.00	5.54	20.72	71.67	7.61
Slovak Republic	15.2	9.60	70.93	73.39	75.44	77.51	66.64	69.47	2.09	1.17	25.27	64.44	10.29	18.16	70.40	11.44
Slovenia	11.2	8.60	73.25	76.09	77.30	79.99	69.40	72.37	1.46	1.22	19.09	69.23	11.68	14.99	70.37	14.64
Solomon Islands	39.64	38.42	64.48	69.51	65.40	70.98	63.60	68.10	5.87	5.26	45.93	51.58	2.49	44.69	52.73	2.58
Somalia	51.94	49.99	41.55	47.40	42.70	49.16	40.46	45.73	7.25	6.95	47.81	49.45	2.74	47.80	49.79	2.41
South Africa	31.88	24.90	61.93	45.70	65.00	46.49	59.00	44.95	3.32	2.80	37.40	59.30	3.30	31.98	63.63	4.39
Spain	10.3	10.40	76.84	79.56	80.51	83.50	73.34	75.80	1.33	1.26	19.38	66.80	13.82	14.98	67.96	17.06
Sri Lanka	20.8	18.84	70.24	74.04	72.90	76.24	67.70	71.94	2.53	1.98	32.65	62.15	5.21	25.10	68.20	6.70
St. Kitts and Nevis	22.03	17.19	67.15	71.52	69.20	74.19	65.20	68.98	2.68	2.11	–	–	–	26.10	62.14	11.76
St. Lucia	27.77	17.10	70.97	74.00	73.33	75.91	68.73	72.18	3.28	2.10	–	–	–	30.59	64.04	5.37
St. Vincent and the Grenadines	21.2	17.90	70.46	72.93	73.26	75.90	67.79	70.10	2.57	2.10	–	–	–	25.73	67.14	7.14
Sudan	38.3	32.80	52.17	58.63	53.60	60.04	50.80	57.29	5.42	4.40	43.31	53.92	2.77	39.52	56.90	3.58
Suriname	29.04	21.21	68.66	70.37	71.22	73.39	66.22	67.50	2.64	2.40	36.07	59.69	4.24	28.67	65.78	5.55
Swaziland	41	34.63	56.64	42.53	59.00	42.59	54.40	42.47	5.30	4.20	45.84	51.38	2.78	42.07	55.09	2.84
Sweden	14.5	11.00	77.54	80.11	80.40	82.43	74.81	77.91	2.31	1.71	17.92	64.31	17.77	17.47	64.99	17.54
Switzerland	12.5	9.70	77.24	80.49	80.71	83.14	73.94	77.97	1.59	1.41	16.89	68.75	14.36	16.59	67.77	15.64
Syrian Arab Rep.	37	28.77	66.41	70.49	68.52	72.92	64.40	68.18	5.34	3.40	47.80	49.48	2.72	38.19	58.66	3.15
Tajikistan	38.8	22.49	69.29	66.34	71.90	69.36	66.80	6.46	5.05	2.87	43.19	53.00	3.81	36.54	58.82	4.64
Tanzania	43.76	37.97	50.05	42.67	51.56	43.12	48.62	42.24	6.25	5.02	46.47	51.02	2.51	44.71	52.85	2.44
Thailand	20.9	15.05	68.51	69.33	71.00	71.66	66.14	67.11	2.27	1.80	31.92	63.75	4.33	22.86	70.54	6.60
Timor-Leste	–	46.26	–	62.21	–	64.16	–	60.36	–	7.58	41.60	56.64	1.75	49.53	48.81	1.66
Togo	42.72	35.43	50.48	49.70	52.08	50.76	48.96	48.70	6.60	4.90	45.50	51.29	3.21	43.23	53.59	3.18
Tonga	30.12	22.79	68.75	71.48	70.80	73.65	66.80	69.41	4.16	3.40	–	–	–	33.77	57.80	8.43
Trinidad and Tobago	21.92	15.53	71.11	72.34	73.52	74.40	68.82	70.39	2.36	1.75	33.50	60.33	6.17	23.73	69.82	6.45
Tunisia	25.2	16.63	70.31	73.16	72.10	75.28	68.60	71.14	3.50	2.00	37.61	58.18	4.21	27.48	66.46	6.06
Turkey	24.8	20.90	65.69	68.64	67.96	71.00	63.52	66.40	3.00	2.43	34.99	60.73	4.28	28.25	65.82	5.92
Turkmenistan	34.2	22.33	66.22	64.47	69.70	68.28	62.90	60.84	4.17	2.70	40.49	55.75	3.76	33.81	61.71	4.48
Uganda	50.32	43.98	46.75	43.20	47.12	43.62	46.40	42.80	6.98	5.99	48.44	49.08	2.48	49.77	48.39	1.85
Ukraine	12.7	8.70	70.14	68.29	74.90	73.81	65.60	63.03	1.85	1.20	21.43	66.44	12.13	15.99	68.86	15.15
United Arab Emirates	23.76	17.29	73.53	75.40	74.70	76.85	72.42	74.02	4.12	3.00	30.31	68.24	1.44	24.84	72.04	3.12
United Kingdom	13.9	11.60	75.88	77.63	78.80	80.13	73.10	75.25	1.83	1.64	19.14	65.13	15.73	18.23	65.73	16.05
United States	16.7	13.88	75.21	77.41	78.80	80.01	71.80	74.92	2.08	2.01	21.91	65.70	12.39	20.98	66.62	12.40
Uruguay	18	15.72	72.62	75.38	76.46	79.09	68.96	71.86	2.51	2.19	26.04	62.43	11.53	24.36	63.03	12.62
Uzbekistan	33.7	20.19	69.17	66.68	72.40	70.01	66.10	63.51	4.07	2.35	40.91	55.08	4.00	33.29	61.86	4.85
Vanuatu	37.26	31.71	64.45	68.66	66.10	70.25	62.88	67.13	5.54	4.26	44.31	51.01	4.68	40.85	56.23	2.92
Venezuela, RB	28.56	22.60	71.25	73.87	74.22	76.85	68.42	71.03	3.43	2.69	38.16	58.19	3.65	32.17	63.10	4.73
Vietnam	28.78	18.41	64.78	69.91	66.84	72.45	62.82	67.50	3.62	1.87	38.84	56.37	4.79	30.62	64.12	5.26
Virgin Islands (U.S.)	21.4	16.89	74.09	78.33	77.40	79.86	70.93	76.88	2.59	2.20	–	–	–	25.77	65.12	9.11
West Bank and Gaza	–	34.40	–	72.96	–	75.42	–	70.61	–	4.89	–	–	–	45.10	51.79	3.11
World	–	20.80	65.25	66.76	67.31	68.80	63.28	64.87	3.12	2.61	32.36	61.47	6.18	28.88	64.01	7.12
Yemen, Republic	47.2	40.87	52.18	57.71	52.60	58.40	51.78	57.05	7.53	6.00	48.83	48.66	2.51	45.25	52.13	2.62
Zambia	45.72	38.21	49.15	36.48	50.06	36.85	48.28	36.13	6.32	5.04	49.20	48.37	2.43	46.83	50.48	2.69
Zimbabwe	37.46	28.86	56.16	38.53	58.04	38.00	54.36	39.03	4.78	3.65	44.28	52.93	2.79	43.44	53.46	3.10

Sources: United Nations Population Division; *World Resources 2000-2001* (World Resources Institute); *World Development Indicators*, 2003 (World Bank).

Table H
World Countries: Mortality, Health, and Nutrition, 1990–2004

COUNTRY	MORTALITY						HEALTH			NUTRITION	
	Infant & Child Mortality		Adult Mortality								
	Infant Mortality (per 1,000 live births)	Children (under 5) Mortality (per 1,000 live births)	Male (per 1,000 live births)		Female (per 1,000 live births)		Health Expenditure per Capita ($US)	Physicians per 1,000 People	Hospital Beds per 1,000 People	Prevalence of Under-nourishment (% of population)	Prevalence of Child Malnutrition (% of children under five)
	2002–2003	2002–2003	1990	2000-2003	1990	2000-2003	2002	2000–2003	1998–2002	1998–2002	1998–2002
Afghanistan	–	–	486	–	476	–	14	0.19	–	–	49.30
Albania	18	21	–	209	–	95	52	1.39	3.26	6	13.60
Algeria	35	41	193	155	156	119	77	1.00	2.10	5	6.00
Angola	154	260	514	492	420	386	38	–	–	40	30.50
Antiqua and Barbuda	11	12	142	183	85	133	470	0.76	3.88	–	–
Argentina	17	20	188	184	95	92	238	–	3.29	2.5	5.40
Armenia	30	33	216	223	119	106	42	3.53	4.25	34	2.60
Australia	–	–	125	100	68	54	1,995	2.49	7.90	–	0.00
Austria	4.5	–	154	124	74	60	1,969	3.30	8.60	15	–
Azerbaijan	75	91	216	261	96	150	27	3.54	8.51	30	6.80
Bahamas, The	11	14	186	267	70	161	1,127	1.06	3.94	–	–
Bahrain	12	15	201	120	147	93	517	1.60	2.90	–	8.70
Bangladesh	46	69	322	262	308	252	11	0.23	–	2.5	52.20
Barbados	11	13	140	180	82	115	669	1.28	7.56	–	–
Belarus	13	17	254	381	98	133	93	4.50	12.58	15	–
Belgium	4	5	141	128	75	67	2,159	3.90	7.30	–	–
Belize	33	39	194	200	123	124	176	1.05	2.13	–	–
Benin	91	154	447	384	369	328	20	0.06	–	21	22.90
Bhutan	70	85	–	268	–	222	12	0.16	–	–	18.70
Bolivia	53	66	307	264	250	219	63	0.73	1.67	8	7.60
Bosnia and Herzegovina	14	17	186	200	109	93	130	1.34	3.22	32	4.10
Botswana	82	112	–	703	–	669	171	–	–	9	12.50
Brazil	33	35	193	259	135	136	206	2.06	3.11	11	5.70
Brunei	5	6	149	144	98	94	430	1.01	–	–	–
Bulgaria	12.3	–	217	239	97	103	143	3.38	7.20	19	–
Burkina Faso	107	207	429	559	338	507	11	0.04	1.42	68	37.70
Burundi	114	190	460	648	379	603	3	0.05	–	33	45.10
Cambodia	97	140	392	373	319	264	32	0.16	–	25	45.20
Cameroon	95	166	430	488	361	440	31	–	–	43	22.20
Canada	–	–	127	101	70	57	2,222	2.10	3.90	–	–
Cape Verde	26	35	245	210	218	121	69	–	–	–	–
Central African Republic	115	180	485	620	381	573	11	0.04	–	34	23.20
Chad	117	200	487	449	397	361	14	0.03	–	4	28.00
Chile	8	9	165	151	92	67	246	1.09	2.67	11	0.80
China	30	37	160	161	135	110	63	1.64	2.45	13	10.00
Columbia	18	21	222	238	127	115	151	1.35	1.46	71	6.70
Comoros	54	73	365	381	307	325	10	–	–	–	25.40
Congo, Dem. Rep.	129	205	–	571	–	493	4	–	–	37	31.00
Congo, Rep.	81	108	370	475	273	406	18	0.25	–	4	–
Costa Rica	8	10	134	131	88	78	383	0.90	1.68	14	5.10
Côte d'Ivoire	117	192	352	553	294	494	44	–	–	7	21.20
Croatia	6	7	207	154	96	117	369	2.37	6.00	3	0.60
Cuba	–	–	125	143	83	94	197	5.91	5.13	2.5	3.90
Cyprus	4	5	118	116	72	59	882	2.98	–	–	–
Czech Republic	3.9	–	231	173	95	76	504	3.50	8.80	25	–
Denmark	4.4	–	155	126	101	79	2,835	3.66	4.50	–	–
Djibouti	97	138	472	590	387	541	54	0.19	–	–	18.20
Dominica	12	14	154	183	113	105	205	0.47	2.65	–	–
Dominican Republic	29	35	157	234	109	146	154	1.88	1.50	4	5.30
Ecuador	24	27	205	199	141	120	91	1.48	1.55	3	14.30
Egypt, Arab Rep.	33	39	230	210	190	147	59	2.12	2.10	11	8.6

Table H *(Continued)*
World Countries: Mortality, Health, and Nutrition, 1990–2004

COUNTRY	MORTALITY						HEALTH			NUTRITION	
	Infant & Child Mortality		Adult Mortality								
	Infant Mortality (per 1,000 live births)	Children (under 5) Mortality (per 1,000 live births)	Male (per 1,000 live births)		Female (per 1,000 live births)		Health Expenditure per Capita ($US)	Physicians per 1,000 People	Hospital Beds per 1,000 People	Prevalence of Undernourishment (% of population)	Prevalence of Child Malnutrition (% of children under five)
	2002–2003	2002–2003	1990	2000-2003	1990	2000-2003	2002	2000–2003	1998–2002	1998–2002	1998–2002
El Salvador	32	36	283	250	165	148	178	1.27	1.65	73	10.3
Equatorial Guinea	97	146	488	339	400	280	83	0.21	–	–	–
Eritrea	45	85	433	493	347	441	8	–	–	5	39.60
Estonia	8	9	286	316	106	114	263	3.16	6.72	46	–
Ethiopia	112	169	448	594	358	535	5	0.03	–	6	47.20
Fiji	16	20	220	240	161	180	94	0.36	–	–	–
Finland	3.1	–	183	144	70	63	1,852	3.10	7.50	–	–
France	–	–	168	139	69	61	2,348	3.30	8.20	–	–
Gabon	60	91	402	380	332	330	159	0.29	–	27	11.90
Gambia, The	90	123	530	373	432	320	18	–	–	27	17.20
Georgia	41	45	195	250	90	133	25	3.91	4.30	13	3.10
Germany	4.2	–	160	125	78	62	2,631	3.30	9.10	–	–
Ghana	59	95	334	379	270	326	17	0.09	–	24	22.1
Greece	4	5	117	114	67	47	1,198	4.40	4.90	–	–
Grenada	18	23	–	202	–	159	285	0.65	5.27	–	–
Guam	–	–	183	–	–	–	–	–	–	–	–
Guatemala	35	47	–	286	–	182	93	1.10	0.98	26	22.70
Guinea	104	160	529	432	495	366	22	0.09	–	9	23.20
Guinea-Bissau	126	204	544	495	533	427	9	–	–	47	25.00
Guyana	52	69	263	299	172	209	53	0.48	3.87	–	13.60
Haiti	76	118	353	524	291	373	29	0.25	0.71	22	17.20
Honduras	32	41	202	221	141	157	60	0.65	1.06	2.5	16.60
Hong Kong, China	–	–	122	–	64	–	–	1.32	–	–	–
Hungary	7.7	7.3	305	295	133	123	496	3.10	8.20	21	–
Iceland	3	4	109	85	63	51	2,916	3.60	14.60	–	–
India	63	87	236	250	241	191	30	0.51	–	6	46.70
Indonesia	31	41	275	–	219	–	26	0.16	–	4	27.30
Iran, Islamic Rep.	33	39	170	170	174	139	104	1.05	1.60	10	10.90
Iraq	102	125	193	258	154	208	11	0.54	1.45	–	15.90
Ireland	5.1	–	134	108	78	62	2,255	2.40	9.70	–	–
Israel	5	6	120	99	72	56	1,496	3.91	6.16	–	–
Italy	4.3		131	100	61	50	1,737	4.40	4.90	–	–
Jamaica	17	20	155	169	97	127	180	0.85	2.12	7	3.80
Japan	–	–	109	98	54	44	2,476	2.00	16.50	–	–
Jordan	23	28	205	199	152	144	165	2.05	1.80	13	4.40
Kazakhstan	63	73	306	366	136	201	56	3.61	7.02	33	4.20
Kenya	79	123	357	578	287	529	19	0.13	–	36	19.9
Kiribati	49	66	–	269	–	208	49	0.30	–	–	–
Korea, Dem. Rep.	42	55	223	238	116	192	0.3	2.97	–	2.5	27.90
Korea, Rep.	5	5	239	186	117	71	577	1.40	6.10	5	–
Kuwait	8	9	130	100	80	68	547	1.53	2.76	6	1.70
Kyrgyz Republic	59	68	291	335	143	299	14	2.60	5.50	22	5.80
Lao PDR	82	91	464	355	389	299	10	–	–	4	40.00
Latvia	10	12	310	321	118	117	203	2.91	8.20	3	–
Lebanon	27	31	210	192	150	136	568	3.25	2.70	12	3.00
Lesotho	79	110	–	667	–	630	25	0.05	–	46	17.90
Liberia	157	235	–	448	–	385	4	–	–	2.5	26.50
Libya	13	16	234	210	185	157	121	–	4.30	2.5	4.70
Liechtenstein	10	11	–	–	–	–	–	–	–	–	–
Lithuania	8	811	288	293	107	103	255	4.03	9.22	11	–
Luxembourg	4.9	–	165	135	85	64	2,951	2.60	8.00	–	–

Table H *(Continued)*
World Countries: Mortality, Health, and Nutrition, 1990–2004

COUNTRY	MORTALITY						HEALTH			NUTRITION	
	Infant & Child Mortality		Adult Mortality								
	Infant Mortality (per 1,000 live births)	Children (under 5) Mortality (per 1,000 live births)	Male (per 1,000 live births)		Female (per 1,000 live births)		Health Expenditure per Capita ($US)	Physicians per 1,000 People	Hospital Beds per 1,000 People	Prevalence of Under-nourishment (% of population)	Prevalence of Child Malnutrition (% of children under five)
	2002–2003	2002–2003	1990	2000-2003	1990	2000-2003	2002	2000–2003	1998–2002	1998–2002	1998–2002
Macedonia, FYR	10	11	147	160	100	89	124	2.19	4.83	37	5.90
Madagascar	78	126	434	385	377	322	5	0.14	0.42	33	33.10
Malawi	112	178	479	701	436	653	14	0.01	1.34	2.5	25.40
Malaysia	7	7	198	202	125	113	149	0.70	2.01	29	19.00
Maldives	55	72	208	228	284	226	96	0.78	–	–	30.20
Mali	122	220	433	518	351	446	12	–	0.24	10	33.20
Malta	5	6	134	111	81	46	957	2.93	4.96	–	–
Marshall Islands	53	61	–	302	–	230	210	0.47	–	–	–
Mauritania	77	107	441	357	365	302	14	0.14	–	6	31.80
Mauritius	16	18	240	228	134	109	113	0.85	–	5	14.90
Mexico	23	28	187	180	117	101	379	1.50	1.10	11	7.50
Micronesia, Fed. Sts.	19	23	316	243	256	188	143	0.60	–	–	–
Moldova	26	32	269	325	146	165	27	2.69	5.89	28	–
Monaco	4	4	–	–	–	–	3,656	5.86	19.57	–	–
Mongolia	56	68	251	280	211	199	27	2.67	–	7	12.70
Morocco	36	39	234	174	184	113	55	0.48	0.98	47	9.00
Mozambique	101	147	418	674	321	612	11	0.02	–	6	26.10
Myanmar	76	107	–	343	–	245	315	0.30	–	22	28.20
Namibia	48	65	373	695	318	661	99	–	–	17	24.00
Nepal	61	82	350	314	376	314	12	0.05	0.17	27	48.30
Netherlands	4.8	5.7	117	97	67	65	2,298	3.10	10.80	–	–
Netherlands Antilles	–	–	145	–	88	–	–	–	6.15	–	–
New Zealand	5	6	143	104	93	67	1,255	2.10	6.20	–	–
Nicaragua	30	38	220	225	147	161	60	1.64	1.48	34	9.60
Niger	154	262	515	473	413	308	7	0.03	0.12	9	40.10
Nigeria	98	198	476	443	401	393	19	0.27	–	20	28.7
Norway	3.4	–	132	107	68	61	4,033	3.56	14.60	–	–
Oman	10	12	217	187	157	135	246	1.26	2.20	–	17.80
Pakistan	74	98	232	221	230	198	13	0.66	–	26	35.00
Palau	23	28	–	–	–	–	439	1.09	–	–	–
Panama	18	24	146	145	94	93	355	1.68	2.21	14	8.10
Papua New Guinea	69	93	425	359	386	329	22	0.05	–	–	–
Paraguay	25	29	–	173	–	129	82	1.17	1.34	13	–
Peru	26	34	228	190	173	139	93	0.98	1.47	22	7.10
Philippines	27	36	273	249	208	142	28	1.16	–	2.5	31.80
Poland	6	7	264	217	102	86	303	2.30	4.90	2.5	–
Portugal	4	5	164	164	82	66	1,092	3.20	4.00	–	–
Qatar	11	15	194	173	135	121	935	2.21	1.65	–	5.50
Romania	18	20	237	260	114	117	128	1.89	7.49	4	3.20
Russian Federation	16	21	316	428	116	156	150	4.17	10.83	37	5.50
Rwanda	118	203	493	667	409	599	11	0.02	–	3	24.30
Samoa	19	24	262	242	202	151	88	0.33	–	–	1.90
San Marino	4	5	–	–	–	–	2,475	–	–	–	–
Sao Tome and Principe	75	118	–	269	–	226	36	–	–	–	12.90
Saudi Arabia	22	26	192	181	158	116	345	1.40	2.30	24	–
Senegal	78	137	488	355	399	303	27	0.10	0.40	11	22.70
Serbia-Montenegro	12	14	168	180	101	100	120	2.02	5.31	50	1.90
Seychelles	11	15	221	268	113	122	425	–	–	–	–
Sierra Leone	166	284	601	587	492	531	6	–	–	5	27.20
Singapore	–	–	138	114	80	61	898	1.40	–	–	3.40
Slovak Republic	7	8	247	216	100	83	265	3.60	7.80	2.5	–

Table H *(Continued)*
World Countries: Mortality, Health, and Nutrition, 1990–2004

COUNTRY	MORTALITY						HEALTH			NUTRITION	
	Infant & Child Mortality		Adult Mortality								
	Infant Mortality (per 1,000 live births)	Children (under 5) Mortality (per 1,000 live births)	Male (per 1,000 live births)		Female (per 1,000 live births)		Health Expenditure per Capita ($US)	Physicians per 1,000 People	Hospital Beds per 1,000 People	Prevalence of Under-nourishment (% of population)	Prevalence of Child Malnutrition (% of children under five)
	2002–2003	2002–2003	1990	2000-2003	1990	2000-2003	2002	2000–2003	1998–2002	1998–2002	1998–2002
Slovenia	4	4	211	170	91	76	922	2.19	5.16	22	–
Solomon Islands	19	22	–	221	–	154	29	0.14	–	–	21.10
Somalia	133	225	–	516	–	452	–	–	–	–	25.80
South Africa	53	66	–	567	–	502	206	0.69	–	–	11.50
Spain	4	4	146	121	60	49	1,192	2.90	4.10	–	–
Sri Lanka	13	15	182	244	122	124	32	0.43	–	27	32.90
St. Kits and Nevis	19	22	227	243	165	148	467	1.07	6.36	–	–
St. Lucia	16	18	205	–	144	–	229	0.56	3.38	–	–
St. Vincent and the Grenadines	23	27	202	246	119	165	180	0.46	1.85	–	19.50
Sudan	63	93	464	341	398	291	19	0.16	–	11	40.70
Suriname	30	39	216	230	137	138	188	0.45	3.74	19	13.20
Swaziland	105	153	260	627	196	587	66	0.18	–	4	10.30
Sweden	2.8	–	115	–	66	–	2,489	3.05	3.60	–	–
Switzerland	4.3	–	127	99	62	54	4,219	3.60	17.90	–	–
Syrian Arab Republic	16	18	237	170	177	132	58	1.40	1.40	61	6.90
Tajikistan	76	95	168	293	106	204	6	2.18	6.40	44	–
Tanzania	104	165	444	569	373	520	13	0.02	–	20	29.40
Thailand	23	26	207	245	123	150	90	0.24	1.99	26	17.60
Timor-Leste	87	124	–	–	–	–	47	–	–	–	42.60
Togo	78	140	389	460	321	406	91	0.06	–	12	25.10
Tonga	15	19	260	226	194	159	91	0.34	–	–	–
Trinidad and Tobago	17	20	182	209	148	133	264	0.79	5.11	2.5	5.90
Tunisia	19	24	190	169	174	99	126	0.66	1.70	3	4.00
Turkey	33	39	–	218	–	120	172	1.30	2.60	9	8.30
Turkmenistan	79	102	250	–	135	–	79	3.17	7.11	19	12.00
Uganda	81	140	526	617	461	567	18	0.05	–	3	22.90
Ukraine	15	20	268	365	105	135	40	2.97	8.74	2.5	3.20
United Arab Emirates	7	8	130	143	101	93	802	2.02	2.64	4	7.00
United Kingdom	5.3	–	126	108	76	67	2,031	2.10	4.10	–	–
United States	–	–	173	147	91	84	5,274	5.49	3.60	–	–
Uruguay	12	14	178	185	90	89	361	3.65	4.39	26	–
Uzbekistan	57	69	207	282	109	176	21	2.89	5.34	17	7.90
Vanuatu	31	38	288	240	241	185	44	–	–	–	12.10
Venezuela, RB	18	21	186	178	112	99	184	1.94	1.47	19	4.40
Vietnam	19	23	215	203	153	139	23	0.53	1.67	16.49	33.80
Yemen, Rep.	82	113	363	278	336	226	23	0.22	0.60	49	46.10
Zambia	102	182	434	725	377	687	20	0.07	–	44	28.10
Zimbabwe	78	126	305	650	270	612	118	0.06	–	–	13.00
World	56.80	85.54	219.00	234.55	168.29	165.69	523.75	1.65	–	36	–

Sources: World Development Indicators, 2003 (World Bank); *World Resources 2000–2001* (World Resources Institute)

Table I
World Countries: Education and Literacy, 2000–2002

	Pupil-teacher ratio, primary	Public spending on education (% of GDP)	Primary school enrollment (% net)	Secondary school enrollment (% net)	Tertiary school enrollment (% net)	Literacy rate, adult total (% of people ages 15 and above)	Literacy rate, adult male (% of males ages 15 and above)	Literacy rate, adult female (% of females ages 15 and above)	Literacy rate, youth total (% of people ages 15-24)	Literacy rate, youth male (% of males ages 15-24)	Literacy rate, youth female (% of females ages 15-24)
	2000-02	2000-02	2000-02	2000-02	2000-02	2000-02	2000-02	2000-02	2000-02	2000-02	2000-02
Afghanistan	61		–		1						
Albania	22		97	74	15	99	99	98	99	99	99
Algeria	28		95	67	21	69	78	60	90	94	86
Andorra	12										
Angola	–	2.80	–		1						
Antiqua and Barbuda	–	3.84									
Argentina	18	4.60	–	81	56	97	97	97	99	98	99
Armenia	12	3.17	94	84	28	99	100	99	100	100	100
Aruba	18	5.10	98	78	29						
Australia	–	4.89	97	88	74						
Austria	13	5.79	90	89	48						
Azerbaijan	15	3.15	80	76	16						
Bahamas, The	17		86	76							
Bahrain	16		90	87	33	88	92	84	99	98	99
Bangladesh	56	2.38	85	44	6	41	50	31	50	58	41
Barbados	16	7.61	100	90	–	100	100	100	100	100	100
Belarus	16	6.00	94	85	62	100	100	100	100	100	100
Belgium	12	–	100	95	60						
Belize	21	5.24	99	68	2	77	77	77	84	84	85
Benin	62	3.26	–	20	–	40	55	26	56	73	38
Bermuda	9		100	86	–						
Bhutan	39	5.15									
Bolivia	24	6.31	95	71	39	87	93	81	97	99	96
Bosnia-Herzegovina						95	–	91	100	100	100
Botswana	27	2.15	81	54	5	79	76	82	89	85	93
Brazil	23	4.30	97	72	18	86	–	87	94	–	96
Brunei	13	–	–	–	13	–	–	–	–	93	–
Bulgaria	17	3.53	90	87	38	99	99	98	100	100	100
Burkina Faso	45		36	9	1	–	–	–	–	–	–
Burundi	50	3.95	57	9	2	50	58	44	66	67	65
Cambodia	56	1.83	93	18	3	69	81	59	80	85	76
Cameroon	57	3.81	–		5	68	77	60	90	92	88
Canada	17	5.25	–	98	58						
Cape Verde	28	7.94	99	58	5	76	85	68	89	92	86
Central African Republic						49	65	33	59	70	47
Chad	68	–	63	10	–	46	55	38	70	76	64
Chile	33	4.22	86	79	42	96	96	96	99	99	99
China	20	–	95		13	91	95	87	99	99	99
Colombia	27	5.20	87	54	24	92	92	92	97	97	98
Comoros	37	3.87	–		2	56	63	49	59	66	52
Congo, Dem. Rep.	–		–		–						
Congo, Republic	65	3.21	54		4	83	89	77	98	98	97
Costa Rica	23	5.08	90	50	20	96	96	96	98	98	99
Côte d'Ivoire	42	4.56	61	21	–	–	–	–	60	70	52
Croatia	18	4.49	89	87	39	98	99	97	100	100	100
Cuba	11	8.96	93	86	34	97	97	97	100	100	100
Cyprus	19	5.83	96	92	26	97	99	95	100	100	100
Czech Republic	17	4.16	88	89	34						
Denmark	10	8.50	100	93	63						
Dijbouti	34	–	34	17	1						
Dominica	19	–	81	91	34		84	84			
Dominican Republic	39	2.27	92	36		84			92	91	92
Ecuador	24	1.99	100	50	–	91	92	90	96	96	96
Egypt, Arab Rep.	22		90	81	–	–	–	–	–	–	–
El Salvador	26	2.87	90	49	17	80	82	77	89	90	88
Equatorial Guinea	43	0.61	85	37	–						
Eritrea	47	4.07	45	22	2						
Estonia	14	5.48	96	87	64	100	100	–	100	100	100
Ethiopia	65	4.59	47	15	2	42	49	34	57	63	52
Fiji	28	5.63	100	76	–	–	–	–	–	–	–
Finland	16	6.24	100	94	86						
France	19	5.65	100	93	54						
Gabon	49	3.93	–		–						
Gambia, The	38	2.79	79	33							
Georgia	14	2.24	89	61	38						
Germany	14	4.57	–	88	–						
Ghana	31	–	63	33	3	74	82	66	92	94	90
Greece	13	3.91	97	85	68	97	99	96	100	100	100
Grenada	19	5.12	–								

Table I *(Continued)*
World Countries: Education and Literacy, 2000–2002

	Pupil-teacher ratio, primary	Public spending on education (% of GDP)	Primary school enrollment (% net)	Secondary school enrollment (% net)	Tertiary school enrollment (% net)	Literacy rate, adult total (% of people ages 15 and above)	Literacy rate, adult male (% of males ages 15 and above)	Literacy rate, adult female (% of females ages 15 and above)	Literacy rate, youth total (% of people ages 15-24)	Literacy rate, youth male (% of males ages 15-24)	Literacy rate, youth female (% of females ages 15-24)
	2000-02	2000-02	2000-02	2000-02	2000-02	2000-02	2000-02	2000-02	2000-02	2000-02	2000-02
Guatemala	30		87	30	9	70	77	62	80	86	74
Guinea	45	1.85	65	21	–						
Guinea-Bissau	–	–	–	–	–						
Guyana	26	8.41	–	77	6						
Haiti						52	54	50	66	66	67
Honduras	34	–	87		15	80	80	80	89	87	91
Hong Kong, China	20	4.11	98	72	26						
Hungary	10	5.15	91	92	44	99	99	99	100	100	100
Iceland	–	6.17	100	85	55						
India	41	–	83		11	61	68	45	73	80	65
Indonesia	21	1.32	92	–	15	88	92	83	98	99	98
Iran, Islamic Rep.	24	4.93	87		21	77	84	70	94	96	92
Iraq	19		–	–	14						
Ireland	19	4.35	95	82	50						
Israel	12	7.34	100	89	58	95	97	93	100	100	99
Italy	11	4.98	99	91	53	99	99	98	100	100	100
Jamaica	34	6.09	95	75	17	88	84	91	94	91	98
Japan	20	3.56	100	100	49						
Jordan			91	81	31	91	96	86	99	99	100
Kazakhstan	19	3.03	91	87	45	99	100	99	100	100	100
Kenya	34	6.98	66	25	–	84	90	79	96	96	95
Kiribati	–										
Korea, Rep.	31	4.31	100	87	85						
Kuwait	13		83	–	–	83	85	81	93	92	94
Kyrgyz Republic	24	3.09	89		42						
Lao PDR	31	2.77	85	35	5	66	77	55	79	86	73
Latvia	14	5.53	88	–	69	100	100	100	100	100	100
Lebanon	17	2.68	91		44						
Lesotho	47	10.36	86	22	3	81	74	90	91	83	99
Liberia	–		–	–	–	56	72	39	71	86	55
Libya	–	–	–		58	82	92	71	97	100	94
Lithuania	16	5.92	94	93	64	100	100	100	100	100	100
Luxembourg	12	–	96	80	12						
Macedonia, FYR	21	–	92	81	27						
Madagascar	52	2.85	79	–	2						
Malawi	62	5.98	–	29	–	62	76	49	72	82	63
Malaysia	20	7.89	95	69	27	89	92	85	97	97	97
Makdives	20	–	92	51		97	97	97	99	99	99
Mali	57	–	44	–	2	–	27	12	24	32	17
Malta	19	4.66	97	82	24	93	92	93	99	98	100
Marshall Islands	17	11.21	–								
Mauritania	41	–	68	16	3	41	51	31	50	57	42
Mauritius	25	4.68	90	71	15	84	88	81	95	94	95
Mexico	27	5.16	99	60	21	91	93	89	97	97	96
Micronesia, Fed. Sts.		7.00			–						
Moldova	19	4.92	79	69	30	99	100	99	100	100	100
Monaco	16										
Mongolia	31	9/01	79	77	37	98	98	98	98	97	98
Morocco	28	6.49	90	36	11	51	63	38	70	77	61
Mozambique	67	–	55	12	–	46	62	31	63	77	49
Myanmar (Burma)	33	–	84	35	12	85	89	81	91	92	91
Namibia	22	7.22	78	44	7	83	84	83	92	91	94
Nepal	36	3.39	–		5	44	62	26	63	78	46
Netherlands		4.99	99	90	57	97	97	97	98	98	98
Netherlands Antilles	20		88	63	14						
New Zealand	18	6.66	100	93	74						
Nicaragua	35	3.12	85	39	18	77	77	77	86	84	89
Niger	42	2.33	38	6	1	17	25	9	24	34	15
Nigeria	42		–		8	67	74	59	89	91	87
Norway		7.00	100	95	74						
Oman	21	4.64	72	69	7	74	82	65	99	100	97
Pakistan	40	–	–		3	–	–	–	–	–	–
Palau	–	11.15	–		–						
Panama	24	4.46	100	63	43	92	93	92	97	97	97
Papua New Guinea	35	–	69	24	–						
Paraguay	–	4.76	92	50	19	92	93	90	96	96	96
Peru	29	2.95	100	69	32	85	91	80	97	98	96
Philippines	35	3.20	93	56	31	93	93	93	95	94	96
Poland	11	5.56	98	–	60		95	91			
Portugal	11	5.90	100	85	53	93			100	100	100

Table I *(Continued)*
World Countries: Education and Literacy, 2000–2002

	Pupil-teacher ratio, primary	Public spending on education (% of GDP)	Primary school enrollment (% net)	Secondary school enrollment (% net)	Tertiary school enrollment (% net)	Literacy rate, adult total (% of people ages 15 and above)	Literacy rate, adult male (% of males ages 15 and above)	Literacy rate, adult female (% of females ages 15 and above)	Literacy rate, youth total (% of people ages 15-24)	Literacy rate, youth male (% of males ages 15-24)	Literacy rate, youth female (% of females ages 15-24)
	2000-02	2000-02	2000-02	2000-02	2000-02	2000-02	2000-02	2000-02	2000-02	2000-02	2000-02
Qatar	12	–	94	82	22	–	–	–	–	–	–
Romania	17	3.28	88	80	30	97	98	96	98	98	98
Russian Federation	17	3.11	–		70	100	100	99	100	100	100
Rwanda	60	–	87	–	3	69	75	63	85	86	84
Samoa	25	4.79	95	61	7	99	99	98	99	99	100
San Marino	–										
Sao Tome and Principe	33		97	29	1						
Saudi Arabia	12	–	54	53	25	78	84	69	94	95	92
Senegal	49	–	58		–	39	49	30	53	61	44
Serbia and Montenegro	20	–	–	–	–						
Seychelles	14	5.18	–	100	–	–	–	–	–	–	–
Sierra Leone	37	–	–		2						
Singapore						93	97	89	100	99	100
Slovak Republic	19	4.03	87	87	32	100	100	100	100	100	100
Slovenia	12			93	66	100	100	100	100	100	100
Solomon Islands		–	–								
South Africa	35	5.29	89	66	15	86	87	85	92	92	92
Spain	14	4.41	100	94	59	98	99	97	100	100	100
Sri Lanka	23	–	–		–	92	95	90	97	97	97
St. Kitts and Nevis	17	3.22	100		1						
St. Lucia	22	7.71	99	76							
St. Vincent and Grenadines	18	9.98	90	58							
Sudan	29		–		–	60	71	49	79	84	74
Suriname	19		97	63	12						
Swaziland	31	7.06	75	32	5	81	82	80	91	90	92
Sweden	11	7.31	100	99	76						
Switzerland	14	5.52	99	87	44						
Syrian Arab Rep.	24		98	43	–	83	91	74	95	97	93
Tajikistan	22	2.80	–	79	16	99	100	99	100	100	100
Tanzania	53	–	69		1	77	85	69	92	94	89
Thailand	19	–	86		37	93	95	91	98	98	98
Timor-Leste	51				12						
Togo	35	2.63	91	–	–	60	74	45	77	88	67
Tonga	21	4.88	–	–	4	–	–	–	–	–	–
Trinidad and Tobago	19	4.31	91	70	9	98	99	98	100	100	100
Tunisia	22	–	97	68	23	73	83	63	94	98	91
Turkey		3.65	88	–	25	87	94	79	96	98	93
Turkmenistan	–						–	–	–	–	–
Uganda	53	–	–	14	3	69	79	59	80	86	74
Ukraine	19	5.43	84	85	62	100	100	100	100	100	100
United Arab Emirates	15	1.59	83	71	35	77	76	81	91	88	95
United Kingdom	17	–	100	96	64						
United States	–	–	93	85	81						
Uruguay	21	3.15	90	72	37	98	97	98	99	99	99
Uzbekistan			–		16	99	100	99	100	100	100
Vanuata	29	11.03	93	28	4						
Venezuela, RB			91	59	40	93	94	93	98	98	99
Vietnam	25		94	65	12	–	–	–	–	–	–
World	24	4.38	87		26	79	80	73	–	90	–
Yemen, Rep.	–	9.55	72	–	–	49	69	29	68	84	51
Zambia	43	1.99	68	23	–	80	86	74	89	91	87
Zimbabwe	39	4.70	80	38	4	90	94	86	98	99	96

Source: World Development Indicators 2003 (World Bank).

Table J
World Countries: Agricultural Operations, 1999–2004

COUNTRY			AGRICULTURAL INPUTS				AGRICULTURAL OUTPUTS AND PRODUCTIVITY			
			Agricultural Machinery							
	Arable Land (hectares per person)	Irrigated Land (hectares)	Land under cereal production (hectares)	Fertilizer consumption (100 grams per hectare of arable land)	Tractors per agricultural worker	Tractors per hectares of arable land	Crop production index (1999–2001 = 100)	Food Production index (1999–2001 = 100)	Livestock production index (1999–2001 = 100)	Agricultural value added per worker (constant 2000 US$)
	2002	2002	2003	2002	2002	2002	2004	2004	2004	2001-2003
Afghanistan	0.28	2,386,000	–	26.30	0.00014	0.01	–	–	–	277
Albania	0.18	340,000	160,100	611.71	0.01061	1.37	100.9	105.9	108.6	1,393
Algeria	0.24	560,000	2,901,520	127.85	0.03653	1.27	128	118	104.6	2,113
Angola	0.23	75,000	1,281,277	–	0.00240	0.34	119.2	112.9	100	161
Antigua and Barbuda	0.10	–	35	–	0.03000	3.00	112.9	107.8	104.9	2,855
Argentina	0.92	1,561,000	10,595,710	219.44	0.20522	0.89	104.9	101.4	94.4	9,627
Armenia	0.16	280,000	197,420	227.94	0.09402	3.70	118.7	115.2	108.3	2,809
Australia	2.46	2,545,000	18,282,500	471.96	0.71267	0.65	92.1	93.3	93.2	22,847
Austria	0.17	4,000	782,194	1,497.48	1.87500	23.72	103.8	100.4	99.3	25,117
Azerbaijan	0.22	1,455,000	765,396	99.23	0.03110	1.69	125.5	121.3	113.7	1,076
Bahamas, The	0.03	1,000	160	1,000.00	0.02400	1.50	111.8	104.9	96.3	–
Bahrain	0.00	4,000	–	500.00	0.00500	0.75	96	110.2	113.7	–
Bangladesh	0.06	4,597,000	11,618,840	1,775.28	0.00014	0.07	105	104.6	102.6	313
Barbados	0.06	1,000	100	506.88	0.09750	3.66	91.2	94.1	95.3	18,798
Belarus	0.56	131,000	2,104,000	1,334.29	0.09585	1.11	129.6	115.1	107.7	2,766
Belgium	–	–	300,210	–	–	–	106.3	101.5	99.7	41,876
Belize	0.26	3,000	18,929	671.43	0.04259	1.64	94.7	101.3	143.6	4,713
Benin	0.39	12,000	1,050,547	187.61	0.00012	0.01	134.1	138	115.6	606
Bermuda	0.02	–	–	1,000.00	0.04500	4.50	–	–	–	–
Bhutan	0.17	40,000	63,870	–	–	–	92.5	94.5	99.1	186
Bolivia	0.34	132,000	744,063	47.38	0.00385	0.21	119.9	110.7	109.2	755
Bosnia and Herzegovina	0.24	3,000	315,076	326.98	0.34118	2.91	102.1	100.5	88.5	5,246
Botswana	0.22	1,000	180,550	124.32	0.01700	1.62	75.6	102.4	107.7	407
Brazil	0.34	2,920,000	19,818,930	1,302.48	0.06360	1.37	126.7	123.7	123.6	3,227
Brunei	0.03	1,000	240	–	0.07200	0.80	111.5	132.9	136.4	–
Bulgaria	0.43	592,000	1,604,403	494.61	1.12528	0.96	106	97.7	95.9	6,826
Burkina Faso	0.37	25,000	3,604,000	3.84	0.00037	0.05	140.5	123.6	108.8	164
Burundi	0.14	74,000	211,700	25.75	0.00005	0.02	104.2	104.4	100.2	101
Cambodia	0.28	270,000	2,090,000	–	0.00055	0.07	115.1	113.3	101.8	300
Cameroon	0.38	33,000	828,300	58.56	0.00013	0.01	103.3	103	102.4	1,215
Canada	1.46	785,000	18,194,300	571.39	1.97466	1.60	102.8	102.2	103.5	36,212
Cape Verde	0.09	3,000	50,000	52.38	0.00040	0.04	90.3	94.5	102.1	1,666
Central African Republic	0.51	–	190,500	3.11	0.00002	0.00	102.8	108.2	114	425
Chad	0.43	20,000	1,927,000	48.61	0.00006	0.00	112.6	111.8	107.6	257
Chile	0.13	1,900,000	685,700	2,295.66	0.05482	2.72	107	106.7	107.7	6,341
China	0.11	54,937,000	77,111,240	2,776.59	0.00181	0.65	115.8	117.5	120.2	349
Columbia	0.05	900,000	1,165,534	3,015.70	0.00568	0.92	110.8	106.4	103.4	2,788
Comoros	0.14	–	15,700	37.50	–	–	102.6	103.4	109.1	386
Congo, Dem. Rep.	0.13	11,000	2,014,780	15.69	0.00018	0.04	97.3	97.7	97.1	197
Congo Rep.	0.05	1,000	10,300	12.42	0.00122	0.37	104.9	108.1	120.2	347
Costa Rica	0.06	108,000	52,167	6,736.09	0.02147	3.11	91.8	94	97.1	4,472
Côte d'Ivoire	0.19	73,000	1,377,000	351.61	0.00122	0.12	86.4	90.4	113.2	802
Croatia	0.33	5,000	689,455	1,176.47	0.02755	0.29	99.8	102.6	105.4	9,302
Cuba	0.24	870,000	337,417	456.90	0.10317	2.92	116.4	111.7	96	–
Cyprus	0.09	40,000	51,400	2,139.72	0.55323	23.82	96.3	105.3	110.4	–
Czech Republic	0.30	24,000	1,555,736	1,201.84	0.21367	3.08	109.8	104.1	95.3	5,280
Denmark	0.42	447,000	1,488,586	1,304.93	1.20588	5.40	101	99.7	100.6	36,420
Djibouti	0.00	1,000	8	–	0.00002	0.60	100.4	107.1	108.5	70

Table J *(Continued)*
World Countries: Agricultural Operations, 1999–2004

COUNTRY			AGRICULTURAL INPUTS				AGRICULTURAL OUTPUTS AND PRODUCTIVITY			
			Agricultural Machinery							
	Arable Land (hectares per person)	Irrigated Land (hectares)	Land under cereal production (hectares)	Fertilizer consumption (100 grams per hectare of arable land)	Tractors per agricultural worker	Tractors per hectares of arable land	Crop production index (1999–2001 = 100)	Food Production index (1999–2001 = 100)	Livestock production index (1999–2001 = 100)	Agricultural value added per worker (constant 2000 US$)
	2002	2002	2003	2002	2002	2002	2004	2004	2004	2001-2003
Dominica	0.07	–	135	1,086.00	0.01125	1.80	98	97.3	100	4,659
Dominican Republic	0.13	275,000	132,100	818.46	0.00321	0.17	107.8	103.2	100.4	4,142
Ecuador	0.13	865,000	891,715	1,416.80	0.01179	0.91	99.7	112.5	126.8	1,491
Egypt, Arab Rep.	0.04	3,400,000	2,773,000	4,375.18	0.01058	3.09	105.1	108.8	119.2	1,996
El Salvador	0.10	45,000	327,341	838.38	0.00440	0.52	89.8	100.4	104.9	1,628
Equatorial Guinea	0.27	–	–	–	0.00123	0.13	93.8	93.4	101.9	654
Eritrea	0.12	21,000	349,716	73.52	0.00031	0.09	70.7	84.1	95.9	57
Estonia	0.45	4,000	263,135	439.85	0.67232	8.54	110.5	108.9	100.9	3,440
Ethiopia	0.15	190,000	6,720,350	151.00	0.00012	0.03	104.8	105.9	106.9	109
Fiji	0.24	3,000	7,020	614.50	0.05303	3.50	90.3	92.5	97.7	1,966
Finland	0.42	64,000	1,191,600	1,331.82	1.49231	8.82	109.2	106.7	104.9	32,031
France	0.31	2,600,000	8,949,510	2,150.79	1.54523	6.85	104.3	100	97.3	39,038
French Polynesia	0.01	1,000	–	4,346.67	0.00853	9.67	109	105.5	93.8	–
Gabon	0.25	15,000	19,500	9.23	0.00728	0.46	101.5	101.1	100.7	1,805
Gambia, The	0.18	2,000	146,000	32.00	0.00008	0.02	65.6	69.2	104.5	220
Georgia	0.15	469,000	348,556	355.44	0.04390	2.74	87	99.7	109	1,503
Germany	0.14	485,000	6,866,977	2,200.26	1.02362	8.01	105.5	102.8	101	22,911
Ghana	0.21	11,000	1,461,530	74.22	0.00063	0.09	121.9	121.6	111.2	346
Greece	0.25	1,431,000	1,282,500	1,490.62	0.33187	9.20	95.4	96.7	96.3	9,144
Grenada	0.02	–	300	–	0.00150	0.60	99.7	96.5	101.4	3,645
Guam	0.03	–	15	–	0.00400	1.60	108	107	98.8	–
Guatemala	0.11	130,000	666,450	1,369.12	0.00215	0.32	103.2	104.4	93.3	2,247
Guinea	0.12	95,000	778,000	35.56	0.00016	0.06	104.7	108.9	117.6	231
Guinea-Bissau	0.21	17,000	133,500	80.00	0.00004	0.01	102.1	102.2	103.1	252
Guyana	0.63	150,000	133,200	372.48	0.06600	0.76	100.1	105.8	145.3	3,645
Haiti	0.09	75,000	452,500	178.59	0.00006	0.02	100.6	102.3	108.3	460
Honduras	0.16	80,000	405,490	470.30	0.00679	0.50	120.1	108.5	101.5	1,223
Hungary	0.45	230,000	2,788,450	1,086.52	0.23996	2.46	119.4	110.3	99.3	3,990
Iceland	0.02	–	–	25,554.29	0.83846	155.71	88.4	104.3	105.1	52,472
India	0.15	57,198,000	98,487,200	995.58	0.00564	0.94	102.1	104.1	111.2	406
Indonesia	0.10	4,815,000	14,832,050	1,459.51	0.00189	0.46	113.1	114.8	125.8	547
Iran, Islamic Rep.	0.23	7,500,000	8,738,000	859.88	0.03718	1.58	116.6	111.9	103.1	2,480
Iraq	0.24	3,525,000	–	1,110.96	0.09507	1.03	–	–	–	–
Ireland	0.29	–	300,000	5,236.40	0.97484	13.83	102.1	98.4	98.2	–
Israel	0.05	194,000	81,300	2,405.33	0.36029	7.25	94.2	103.2	117.2	–
Italy	0.14	2,750,000	4,138,664	1,728.78	1.36066	20.03	95.6	95.5	98.1	21,437
Jamaica	0.07	25,000	821	1,287.36	0.01171	1.77	97.4	98.8	101.8	1,957
Japan	0.03	2,607,000	1,985,665	2,906.29	0.82641	45.90	95.8	97.9	99.6	26,417
Jordan	0.06	75,000	56,745	1,135.59	0.03005	1.96	127.4	118.5	98.8	996
Kazakhstan	1.45	2,350,000	13,720,400	30.14	0.03807	0.23	98	98.8	112	1,436
Kenya	0.15	90,000	1,898,400	310.34	0.00105	0.28	96	101.1	108.7	148
Kiribati	0.02	–	–	–	0.00180	0.90	105.1	107.3	128.6	705
Korea, Dem. Rep.	0.11	1,460,000	1,303,996	1,064.80	0.01933	2.57	110	109.3	113.6	–
Korea, Rep.	0.04	1,138,000	1,107,322	4,096.80	0.09563	12.25	90.5	92.5	98.7	9,792
Kuwait	0.01	13,000	1,660	807.69	0.00657	0.71	109.6	124.6	120.6	
Kyrgyz Republic	0.27	1,072,000	588,825	205.20	0.04552	1.89	94.9	81	69.4	961
Lao PDR	0.17	175,000	858,000	76.29	0.00051	0.12	121.9	119.8	102	460
Latvia	0.78	20,000	439,100	273.08	0.39167	3.08	123.1	118.5	108.2	2,513
Lebanon	0.04	104,000	57,750	2,318.82	0.19302	4.88	95.6	99.8	117.7	45,298

Table J *(Continued)*
World Countries: Agricultural Operations, 1999–2004

COUNTRY			AGRICULTURAL INPUTS				AGRICULTURAL OUTPUTS AND PRODUCTIVITY			
			Agricultural Machinery							
	Arable Land (hectares per person)	Irrigated Land (hectares)	Land under cereal production (hectares)	Fertilizer consumption (100 grams per hectare of arable land)	Tractors per agricultural worker	Tractors per hectares of arable land	Crop production index (1999–2001 = 100)	Food Production index (1999–2001 = 100)	Livestock production index (1999–2001 = 100)	Agricultural value added per worker (constant 2000 US$)
	2002	2002	2003	2002	2002	2002	2004	2004	2004	2001-2003
Lesotho	0.19	1,000	264,540	342.42	0.00717	0.61	111.2	105.3	100	499
Liberia	0.12	3,000	120,000	–	0.00039	0.09	96.7	97.4	114.9	–
Libya	0.33	470,000	342,600	341.05	0.39356	2.19	94.8	100.9	100.5	–
Lithuania	0.84	7,000	862,100	662.12	0.51150	3.49	112.4	108.4	97.8	4,424
Luxembourg	–	–	28,864	–	–	–	100.8	94.2	94	–
Macedonia, FYR	0.28	55,000	194,797	393.99	0.49541	9.54	92.9	95.1	103.2	3,096
Madagascar	0.18	1,090,000	1,421,140	30.93	0.00060	0.12	102.8	102.5	101.3	173
Malawi	0.21	30,000	1,704,000	839.17	0.00031	0.06	91.8	95.6	101.8	128
Malaysia	0.07	365,000	699,000	6,833.33	0.02412	2.41	117.1	116.9	116.7	4,851
Maldives	0.01	–	5	–	–	–	112.6	112.6	–	–
Mali	0.41	138,000	3,037,900	90.13	0.00055	0.06	108.1	106.6	118.4	247
Malta	0.02	2,000	2,840	777.78	0.25000	5.56	76.6	100.8	101.8	–
Mauritania	0.18	49,000	163,975	59.43	0.00058	0.08	99.5	107.3	108.4	271
Mauritius	0.08	22,000	55	2,500.00	0.00627	0.37	103.4	105.8	113.4	4,846
Mexico	0.25	6,320,000	10,802,000	690.28	0.03818	1.31	105.6	107.8	108.6	2,866
Moldova	0.43	300,000	850,079	54.99	0.08798	2.22	109.8	108.9	110.5	706
Mongolia	0.49	84,000	223,100	37.13	0.01634	0.42	105	92.3	92.1	698
Morocco	0.28	1,345,000	5,558,800	475.23	0.01147	0.58	133.8	122.6	100.6	1,711
Mozambique	0.23	107,000	2,101,355	59.29	0.00073	0.14	105.9	104.2	104	146
Myanmar	0.20	1,996,000	7,334,000	134.14	0.00050	0.09	116.6	116.6	122.8	–
Namibia	0.41	7,000	236,900	3.68	0.01013	0.39	106.6	95.7	91.3	1,036
Nepal	0.13	1,135,000	3,360,298	278.24	0.00042	0.14	112.4	108.8	106.8	208
Netherlands	0.06	565,000	219,700	3,668.12	0.63889	16.32	98.7	93	90.9	38,339
Netherlands Antilles	0.04	–	–	–	–	0.25	–	–	–	–
New Caledonia	0.02	10,000	1,076	1,800.00	0.04621	38.82	104.6	101.7	99.9	–
New Zealand	0.38	285,000	139,087	5,685.98	0.45238	5.07	105.2	115.7	115.5	29,045
Nicaragua	0.36	94,000	497,414	279.46	0.00742	0.15	117.9	127.5	134.1	1,988
Niger	0.39	66,000	8,044,200	11.08	0.00003	0.00	122.5	118.3	103.5	174
Nigeria	0.23	233,000	23,017,000	55.03	0.00198	0.10	104.6	105.1	108.8	871
Norway	0.19	127,000	327,000	2,112.51	1.30000	14.93	103	99.9	98.3	38,043
Oman	0.01	62,000	2,490	3,219.21	0.00042	0.39	88.9	91.1	96.6	–
Pakistan	0.15	17,800,000	12,474,000	1,381.38	0.01249	1.49	105.2	109.3	112.5	695
Panama	0.19	35,000	135,950	524.07	0.03240	1.48	106.2	107.8	106.5	3,605
Papua New Guinea	0.04	–	2,900	536.36	0.00060	0.53	102.2	107.8	112.7	443
Paraguay	0.55	67,000	681,500	507.18	0.02257	0.55	118.8	110.9	100.7	2,544
Peru	0.14	1,195,000	1,138,077	740.56	0.00437	0.36	94.9	90.4	81.9	1,770
Philippines	0.07	1,550,000	6,579,000	1,268.46	0.00091	0.20	109.5	113.5	123.2	1,040
Poland	0.36	100,000	8,163,257	1,085.84	0.32810	9.80	95.8	107.2	106.2	1,397
Portugal	0.19	650,000	449,718	1,040.20	0.27750	8.49	99.7	101	100.7	5,677
Puerto Rico	0.01	40,000	260	–	–	–	99.9	100.2	100.2	–
Qatar	0.03	13,000	1,500	–	0.02050	0.46	98.3	143.9	159.3	–
Romania	0.43	3,077,000	5,108,896	347.01	0.11474	1.80	132.6	125.4	119.1	3,621
Russian Federation	0.86	4,600,000	36,645,500	119.39	0.08311	0.52	116.9	114.3	107.7	2,323
Rwanda	0.14	6,000	307,264	137.09	0.00001	0.01	113.8	112.7	105.5	234
Samoa	0.34	–	–	583.33	0.00448	0.16	103.7	101.4	94.8	1,645
Sao Tome and Principe	0.05	10,000	1,000	–	0.00278	1.79	101.8	102	102.8	226
Saudi Arabia	0.16	1,620,000	666,100	1,059.17	0.01460	0.28	99.8	108.2	104.1	14,618
Senegal	0.25	71,000	1,331,210	136.14	0.00022	0.03	81.9	85.3	101.1	265
Serbia and Montenegro	0.42	29,000	3,284,644	–	–	–	126.6	113.5	94.5	–

Table J *(Continued)*
World Countries: Agricultural Operations, 1999–2004

COUNTRY			AGRICULTURAL INPUTS				AGRICULTURAL OUTPUTS AND PRODUCTIVITY			
			Agricultural Machinery							
	Arable Land (hectares per person)	Irrigated Land (hectares)	Land under cereal production (hectares)	Fertilizer consumption (100 grams per hectare of arable land)	Tractors per agricultural worker	Tractors per hectares of arable land	Crop production index (1999–2001 = 100)	Food Production index (1999–2001 = 100)	Livestock production index (1999–2001 = 100)	Agricultural value added per worker (constant 2000 US$)
	2002	2002	2003	2002	2002	2002	2004	2004	2004	2001-2003
Seychelles	0.01	–	–	170.00	0.00133	4.00	96.8	98.9	99.5	554
Sierra Leone	0.10	30,000	231,900	5.61	0.00008	0.02	109.4	110	112.3	295
Singapore	0.00	–	–	24,180.00	0.02167	6.50	100	66.6	70.7	32,073
Slovenia	0.08	3,000	97,904	4,159.94	–	–	94.2	105.7	111.9	30,713
Solomon Islands	0.04	–	1,300	–	0.00005	0.05	107.2	107.6	112.8	–
Somalia	0.11	200,000	–	4.78	0.00060	0.16	–	–	–	–
South Africa	0.33	1,498,000	4,467,400	654.17	0.04355	0.49	98.5	104.4	109.8	2,251
Spain	0.34	3,780,000	6,592,191	1,572.06	0.77545	6.89	103	103.7	111	15,656
Sri Lanka	0.05	638,000	946,430	3,102.82	0.00270	1.15	94.4	95.2	105.9	745
St. Kitts and Nevis	0.15	–	–	2,428.57	0.03875	2.21	99.6	89.6	100	2,123
St. Lucia	0.03	3,000	–	3,357.50	0.00973	3.65	116.4	89.3	82.4	1,738
St.. Vincent and the Grenadines	0.06	1,000	200	3,047.14	0.00667	1.14	106	101.2	87	2,477
Sudan	0.50	1,950,000	9,885,800	42.81	0.00153	0.07	124.4	114.1	107.1	720
Suriname	0.13	51,000	51,350	982.46	0.04290	2.33	105.2	105.6	105.2	3,002
Swaziland	0.16	70,000	61,250	393.26	0.03264	2.22	93.9	102.2	115.2	1,189
Sweden	0.30	115,000	1,153,890	1,000.37	1.17021	6.16	108.6	101	97.2	31,960
Switzerland	0.06	25,000	154,248	2,274.82	0.73684	27.38	90.1	99.3	101.7	–
Syrian Arab Republic	0.27	1,333,000	3,109,601	702.82	0.06631	2.26	116.6	119.8	107.9	2,768
Tajikistan	0.15	719,000	394,065	300.00	0.02437	2.15	133.6	121.4	122	454
Tanzania	0.11	170,000	2,827,939	17.87	0.00052	0.19	104.7	103.8	109.1	290
Thailand	0.26	4,957,000	11,545,300	1,071.79	0.01081	1.39	105.3	99.8	89.3	620
Timor-Leste	0.08	–	70,400	–	0.00036	0.16	106.3	106	104.6	269
Togo	0.53	18,000	813,000	67.95	0.00007	0.00	111.9	105.3	109.6	405
Tonga	0.17	–	–	–	0.01250	0.88	103	102.2	99.9	3,341
Trinidad and Tobago	0.06	4,000	2,200	434.40	0.05510	3.60	80.2	117.1	142.7	2,135
Tunisia	0.28	381,000	1,236,500	368.10	0.03664	1.27	94.5	96.5	98.7	2,639
Turkey	0.37	5,215,000	13,906,950	672.05	0.06601	3.74	104.7	105.2	106.8	1,766
Turkmenistan	0.39	1,800,000	963,000	528.65	0.07082	2.70	122.6	109.2	96.8	1,352
Uganda	0.21	9,000	1,408,000	18.25	0.00050	0.09	107.6	108.9	117.1	231
Ukraine	0.67	2,262,000	10,684,400	180.99	0.11852	1.24	125.6	116.1	107.8	1,400
United Arab Emirates	0.02	76,000	35	4,666.67	0.00535	0.51	59.1	53.1	119.1	–
United Kingdom	0.10	170,000	3,060,000	3,130.54	0.98425	8.69	99.3	98.1	97.2	26,471
United States	0.61	22,500,000	57,888,330	1,096.48	1.65176	2.73	110.5	106.8	102	47,566
Uruguay	0.39	181,000	533,000	991.76	0.17460	2.54	122	106.6	98.4	7,363
Uzbekistan	0.18	4,281,000	1,778,200	1,601.92	0.05655	3.79	106.3	105.2	104.7	1,601
Vanuatu	0.15	–	1,300	–	0.00234	0.25	96.4	96.2	95.9	1,099
Venezuela, RB	0.10	575,000	856,265	1,229.51	0.06218	2.01	93.4	99.6	103.4	6,071
Vietnam	0.08	3,000,000	8,359,100	2,948.06	0.00577	2.43	118.3	118.7	118.9	296
Yemen, Rep.	0.08	500,000	534,702	75.42	0.00223	0.42	100	106.5	113	524
Zambia	0.51	46,000	849,900	123.89	0.00195	0.11	106.5	106.8	98.9	210
Zimbabwe	0.25	117,000	1,673,096	341.61	0.00668	0.75	80.2	87.8	101.7	267

Source: World Development Indicators 2003 (World Bank).

Table K
Land Use and Deforestation, 1990–2002

COUNTRY	LAND AREA	RURAL POPULATION DENSITY	LAND USE						FOREST AREA	AVERAGE ANNUAL DEFORESTATION
	(sq. km.)	(people per sq. km.)	Arable Land (% of land area)		Permanent Cropland (% of land area)		Other Land Use (% of land area)		(% of total land area)	Decline in % of forest area
	2002	2002	1990	2002	1990	2002	1990	2002	2000	1990-2000
Afghanistan	652,090	262	12.1	12.1	0.22	0.22	87.65	87.65	2.07	0.80
Albania	27,400	307	21.1	21.1	4.56	4.42	74.31	74.49	36.17	0.80
Algeria	2,381,740	171	3.0	3.2	0.23	0.25	96.79	96.53	0.90	-1.30
Angola	1,246,700	282	2.3	2.4	0.40	0.24	97.27	97.35	55.95	0.20
Antiqua and Barbuda	440	598	18.2	18.2	4.55	4.55	77.27	77.27	20.45	–
Argentina	2,736,690	12	10.6	12.3	0.44	0.48	89.00	87.21	12.66	0.60
Armenia	28,200	202	–	17.6	–	2.30	–	80.14	12.45	0.80
Australia	7,682,300	3	6.2	6.3	0.02	0.04	93.74	93.67	20.12	0.00
Austria	82,730	188	17.2	16.8	0.95	0.86	81.81	82.33	46.97	-0.20
Azerbaijan	82,600	220	–	21.6	–	2.74	–	75.68	13.24	-1.30
Bahamas, The	10,010	428	0.8	0.8	0.20	0.40	99.00	98.80	84.12	–
Bahrain	710	2525	2.9	2.8	2.90	5.63	94.20	91.55	–	–
Bangladesh	130,170	1249	70.2	61.6	2.30	3.15	27.50	35.25	10.25	-1.30
Barbados	430	823	37.2	37.2	2.33	2.33	60.47	60.47	4.65	
Belarus	207,480	54	–	27.0	–	0.60	–	72.38	45.32	-3.20
Belgium	30,230	–	–	–	–	–	–	–	–	0.2
Belize	22,800	196	2.3	3.1	1.10	1.40	96.62	95.53	59.12	–
Benin	110,620	144	14.6	23.1	0.95	2.40	84.45	74.55	23.96	2.30
Bermuda	50	0	20.0	20.0	–	–	–	–	–	–
Bhutan	47,000	542	2.4	3.1	0.40	0.43	97.19	96.49	64.17	–
Bolivia	1,084,380	109	1.9	2.7	0.14	0.19	97.92	97.14	48.94	0.30
Bosnia-Herzegovina	51,200	231	–	19.5	–	1.88	–	78.65	44.39	0.00
Botswana	566,730	232	0.7	0.7	0.01	0.01	99.26	99.34	21.93	0.90
Brazil	8,459,420	53	6.0	7.0	0.80	0.90	93.21	92.13	64.30	0.40
Brunei	5,270	1042	0.6	1.7	0.76	0.76	98.67	97.53	83.87	–
Bulgaria	110,630	76	34.9	30.3	2.71	2.06	62.43	67.61	33.35	-0.60
Burkina Faso	273,600	225	12.9	15.9	0.20	0.19	86.93	83.92	25.91	0.20
Burundi	25,680	648	36.2	38.4	14.02	14.21	49.77	47.39	3.66	9.00
Cambodia	176,520	292	20.9	21.0	0.57	0.61	78.50	78.43	52.88	0.60
Cameroon	465,400	131	12.8	12.8	2.64	2.58	84.59	84.62	51.26	0.90
Canada	9,220,970	14	5.0	5.0	0.01	0.01	95.02	95.02	26.52	0.00
Cape Verde	4,030	387	10.2	10.4	0.50	0.74	89.33	88.83	21.09	–
Central African Republic	622,980	114	3.1	3.1	0.14	0.15	96.78	96.75	36.77	0.10
Chad	1,259,200	175	2.6	2.9	0.02	0.02	97.38	97.12	10.08	0.60
Chile	748,800	108	3.7	2.6	0.33	0.43	95.93	96.92	20.75	0.10
China	9,327,420	559	13.3	15.3	0.83	1.22	85.91	83.49	17.53	-1.20
Colombia	1,038,700	459	3.2	2.2	1.63	1.50	95.19	96.29	47.75	0.40
Comoros	2,230	480	35.0	35.9	15.70	23.32	49.33	40.81	3.59	–
Congo, Dem. Rep.	2,267,050		2.9	3.0	0.52	0.49	96.53	96.56	59.64	0.40
Congo, Rep.	341,500	642	0.5	0.6	0.12	0.15	99.43	99.30	64.60	0.10
Costa Rica	51,060	700	5.1	4.4	4.90	5.88	90.01	89.72	38.54	0.80
Côte d'Ivoire	318,000	296	7.6	9.7	11.01	11.95	81.35	78.30	22.38	0.80
Croatia	55,920	126	–	26.1	–	2.25	–	71.06	31.88	-0.10
Cuba	109,820	103	29.6	24.3	7.38	10.20	63.03	65.51	21.38	-1.30
Cyprus	9,240	313	11.5	7.8	5.52	4.44	83.01	87.77	18.61	

Table K *(Continued)*
Land Use and Deforestation, 1990–2002

COUNTRY	LAND AREA	RURAL POPULATION DENSITY	LAND USE						FOREST AREA	AVERAGE ANNUAL DEFORESTATION
	(sq. km.)	(people per sq. km.)	Arable Land (% of land area)		Permanent Cropland (% of land area)		Other Land Use (% of land area)		(% of total land area)	Decline in % of forest area
	2002	2002	1990	2002	1990	2002	1990	2002	2000	1990-2000
Czech Republic	77,280	84	–	39.7	–	3.05	–	57.23	34.06	0.00
Denmark	42,430	35	60.4	53.6	0.24	0.19	39.35	46.17	10.72	-0.20
Djibouti	23,180	10830	0.0	0.0	–	–	–	–	0.26	–
Dominica	750	402	6.7	6.7	14.67	20.00	78.67	73.33	61.33	–
Dominican Republic	48,380	263	21.7	22.7	9.30	10.33	69.00	67.01	28.44	0.00
Ecuador	276,840	286	5.8	5.9	4.77	4.93	89.43	89.22	38.13	1.20
Egypt, Arab Rep.	995,450	1309	2.3	2.9	0.37	0.50	97.34	96.58	0.07	-3.40
El Salvador	20,720	365	26.5	31.9	12.55	12.07	60.91	56.08	5.84	4.60
Equatorial Guinea	28,050	184	4.6	4.6	3.57	3.57	91.80	91.80	62.46	–
Eritrea	101,00	691	–	5.0	–	0.03	–	95.02	15.69	0.30
Estonia	42,390	68	–	14.5	–	0.40	–	85.11	48.60	-0.60
Ethiopia	1,000,000	567	–	9.9	–	0.74	–	89.33	4.59	0.80
Fiji	18,270	202	8.8	10.9	4.38	4.65	86.86	84.40	44.61	–
Finland	304,590	97	7.4	7.2	0.02	0.03	92.53	92.75	72.01	0.00
France	550,100	78	32.7	33.5	2.17	2.06	65.12	64.40	27.89	-0.40
French Polynesia	3,660	3781	0.5	0.8	5.74	6.01	93.72	93.17	28.69	–
Gabon	257,670	69	1.1	1.3	0.63	0.66	98.23	98.08	84.71	0.00
Gambia, The	10,000	378	18.2	25.0	0.50	0.50	81.30	74.50	48.10	-1.00
Georgia	69,490	280	–	11.5	–	3.81	–	84.69	43.00	0.00
Germany	348,950	85	34.3	33.8	1.27	0.59	64.42	65.62	30.78	0.00
Ghana	227,540	307	11.9	18.4	6.59	9.45	81.54	72.18	27.84	1.70
Greece	128,900	160	22.5	21.1	8.29	8.76	69.22	70.16	27.92	-0.90
Grenada	340	3157	5.9	5.9	29.41	29.41	64.71	64.71	14.71	–
Guam	550	1915	9.1	9.1	14.55	16.36	76.36	74.55	38.18	–
Guatemala	108,430	526	12.0	12.5	4.47	5.03	83.54	82.43	26.28	1.70
Guinea	245,720	616	3.0	3.7	2.03	2.60	95.00	93.73	28.20	0.50
Guinea-Bissau	28,120	322	10.7	10.7	4.16	8.82	85.17	80.51	77.77	0.90
Guyana	196,850	100	2.4	2.4	0.11	0.15	97.45	97.41	85.75	
Haiti	27,560	669	28.3	28.3	11.61	11.61	60.09	60.09	3.19	5.70
Honduras	111,890	289	13.1	9.5	3.20	3.22	83.73	87.24	48.11	1.00
Hong Kong, China	1,042	–	–	–	–	–	–	–	–	–
Hungary	92,100	77	54.7	50.1	2.53	2.06	42.73	47.84	19.98	-0.40
Iceland	100,250	297	0.1	0.1	–	–	–	–	0.31	–
India	2,973,190	466	54.9	54.4	2.12	2.83	43.01	42.78	21.56	-0.10
Indonesia	1,811,570	588	11.2	11.3	6.47	7.29	82.35	81.40	57.95	-0.10
Iran, Islamic Rep.	1,636,200	151	9.3	9.2	0.81	1.26	89.92	89.56	4.46	0.00
Iraq	437,370	137	12.1	13.1	0.66	0.78	87.22	86.08	1.83	0.00
Ireland	68,890	142	15.1	16.3	0.04	0.03	84.85	83.70	9.57	-3.00
Israel	21,710	157	15.8	15.6	4.05	3.96	80.15	80.47	6.08	-4.00
Italy	294,110	228	30.6	28.2	10.06	9.44	59.29	62.38	34.01	-0.30
Jamaica	10,830	647	11..0	16.1	9.23	10.16	79.78	73.78	30.01	1.50
Japan	364,500	603	13.1	12.1	1.30	0.94	85.62	86.94	66.07	0.00
Jordan	88,930	369	3.3	3.3	1.01	1.18	95.73	95.50	0.97	0.00
Kazakhstan	2,699,700	30	–	8.0	–	0.05	–	91.97	4.50	-2.20
Kenya	569,140	441	7.4	8.1	0.88	0.99	91.74	90.93	30.04	0.50
Kiribati	730	2885	2.7	2.7	50.68	50.68	46.58	46.58	38.36	–

Table K *(Continued)*
Land Use and Deforestation, 1990–2002

COUNTRY	LAND AREA	RURAL POPULATION DENSITY	LAND USE						FOREST AREA	AVERAGE ANNUAL DEFORESTATION
	(sq. km.)	(people per sq. km.)	Arable Land (% of land area)		Permanent Cropland (% of land area)		Other Land Use (% of land area)		(% of total land area)	Decline in % of forest area
	2002	2002	1990	2002	1990	2002	1990	2002	2000	1990-2000
Korea, Dem. Rep.	120,410	352	19.0	20.8	1.49	1.66	79.50	77.58	68.18	0.00
Korea, Rep.	98,730	481	19.8	17.1	1.58	1.95	78.64	80.99	63.28	0.10
Kuwait	17,820	689	0.2	0.7	0.06	0.11	99.72	99.16	0.28	-5.20
Kyrgyz Republic	191,800	244	–	7.0	–	0.34	–	92.64	5.23	-2.80
Lao PDR	230,800	480	3.5	4.0	0.26	0.35	96.27	95.66	54.42	0.40
Latvia	62,050	50	–	29.5	–	0.47	–	70.01	47.11	-0.40
Lebanon	10,230	253	17.9	16.6	11.93	13.98	70.19	69.40	3.52	0.90
Lesotho	30,350	379	10.4	10.9	0.13	0.13	89.42	89.00	0.46	0.30
Liberia	96,320	467	4.2	3.9	2.23	2.28	93.62	93.77	36.14	2.00
Libya	1,759,540	35	1.0	1.0	0.20	0.19	98.78	98.78	0.20	-1.40
Lichtenstein	160	–	25.0	25.0	–	–	–	–	43.75	–
Lithuania	62.680	37	–	46.7	–	0.94	–	52.31	31.81	-0.20
Luxembourg	2,586	–	–	–	–	–	–	–	–	–
Macedonia, FYR	25,430	146	–	22.3	–	1.81	–	75.93	35.63	-0.20
Madagascar	581,540	386	4.7	5.1	1.04	1.03	94.28	93.90	20.17	0.90
Malawi	94,080	395	19.3	24.4	1.22	1.49	79.49	74.06	27.23	2.40
Malaysia	328,550	557	5.2	5.5	15.97	17.61	78.85	76.91	58.72	1.20
Maldives	300	5126	13.3	13.3	13.33	26.67	73.33	60.00	3.33	–
Mali	1,220,190	167	1.7	3.8	0.03	0.03	98.28	96.15	10.81	0.70
Malta	320	379	37.5	28.1	3.13	3.13	59.38	68.75	0.00	–
Marshall Islands	181	–	–	–	–	–	–	–	–	–
Mauritania	1,025,220	226	0.4	0.5	0.01	0.01	99.60	99.51	0.31	2.70
Mauritius	2,030	702	49.3	49.3	2.96	2.96	47.78	47.78	7.88	0.60
Mexico	1,908,690	102	12.6	13.0	1.00	1.31	86.43	85.70	28.92	1.10
Micronesia, Fed. Sts.	702	–	–	–	–	–	–	–	–	–
Moldova	32,880	134	–	56.1	–	9.12	–	34.82	9.88	-0.20
Mongolia	1,566,500	88	0.9	0.8	0.00	0.00	99.12	99.23	6.80	0.50
Morocco	446,300	153	19.5	18.8	1.65	1.99	78.84	79.20	6.78	0.00
Mozambique	784,090	288	4.4	5.4	0.29	0.30	95.31	94.34	39.03	0.20
Myanmar	657,550	353	14.5	15.0	0.76	1.14	84.69	83.86	52.34	1.40
Namibia	823,290	166	0.8	1.0	0.00	0.00	99.20	99.00	9.77	0.90
Nepal	143,000	659	16.0	22.4	0.46	0.66	83.55	76.97	27.27	1.80
Netherlands	33,880	182	25.9	27.0	0.89	0.97	73.17	71.99	11.07	-0.30
Netherlands Antilles	800	832	10.0	10.0	–	–	–	–	1.25	–
New Caledonia	18,280	929	0.5	0.3	0.33	0.22	99.18	99.51	20.35	–
New Zealand	267,990	37	9.4	5.6	5.05	6.99	85.58	87.42	29.65	-0.50
Nicaragua	121,400	120	10.7	15.9	1.61	1.94	87.69	82.20	27.00	3.00
Niger	1,266,700	200	2.8	3.5	0.01	0.01	97.15	96.45	1.05	3.70
Nigeria	910,770	239	32.4	33.2	2.78	3.07	64.78	63.77	14.84	2.60
Norway	306,250	129	2.8	2.8	–	–	–	–	28.96	-0.40
Oman	309,500	1534	0.1	0.1	0.15	0.14	99.74	99.74	0.00	0.00
Pakistan	770,880	447	26.6	27.8	0.59	0.87	72.84	71.31	3.06	1.10
Palau	460	–	–	8.7	–	4.35	–	86.96	76.09	–
Panama	74,430	231	6.7	7.4	2.08	1.98	91.21	90.66	38.64	1.60
Papua New Guinea	452,860	2007	0.4	0.5	1.28	1.44	98.30	98.08	67.57	0.40
Paraguay	397,300	78	5.3	7.6	0.22	0.24	94.47	92.16	58.83	0.50

Table K *(Continued)*
Land Use and Deforestation, 1990–2002

COUNTRY	LAND AREA	RURAL POPULATION DENSITY	LAND USE						FOREST AREA	AVERAGE ANNUAL DEFORESTATION
	(sq. km.)	(people per sq. km.)	Arable Land (% of land area)		Permanent Cropland (% of land area)		Other Land Use (% of land area)		(% of total land area)	Decline in % of forest area
	2002	2002	1990	2002	1990	2002	1990	2002	2000	1990-2000
Peru	1,280,000	192	2.7	2.9	0.33	0.48	96.94	96.63	50.95	0.40
Philippines	298,170	559	18.4	19.1	14.76	16.77	66.86	64.11	19.42	1.40
Poland	306,290	102	47.3	45.5	1.13	0.99	51.60	53.55	29.72	-0.10
Portugal	91,500	173	25.6	21.7	8.54	7.81	65.85	70.44	40.07	-1.70
Puerto Rico	8,870	2662	7.3	3.9	5.64	5.52	87.03	90.53	25.82	0.20
Qatar	11,000	234	0.9	1.6	0.09	0.27	99.00	98.09	0.09	–
Romania	229,870	103	41.0	40.9	2.57	2.18	56.41	56.94	27.99	-0.20
Russian Federation	16,888,500	32	–	7.3	–	0.11	–	92.58	50.41	0.00
Rwanda	24,670	684	35.7	45.2	12.36	10.90	51.97	43.86	12.44	3.90
Samoa	2,830	227	19.4	21.2	23.67	24.38	56.89	54.42	37.10	–
Soa Tome and Principe	960	1138	2.1	7.3	40.63	48.96	57.29	43.75	28.13	–
Saudi Arabia	2,149,690	79	1.6	1.7	0.04	0.09	98.38	98.24	0.70	0.00
Senegal	192,530	208	12.1	12.8	0.13	0.24	87.79	86.98	32.23	0.70
Serbia-Montenegro	102,000	116	27.5	–	2.81	–	69.70	–	11.36	0.00
Seychelles	450	2860	2.2	2.2	11.11	13.33	86.67	84.44	66.67	–
Sierra Leone	71,620	607	6.8	7.5	0.75	0.91	92.46	91.62	14.73	2.90
Singapore	670	–	1.5	1.5	1.49	1.49	97.01	97.01	2.99	0.00
Slovak Republic	48,800	–	–	–	–	–	–	–	–	-0.30
Slovenia	20,120	603	–	8.3	–	1.49	–	90.16	55.02	-0.20
Solomon Islands	27,990	1950	0.6	0.6	1.86	2.04	97.53	97.32	90.60	–
Somalia	627,340	638	1.6	1.7	0.03	0.04	98.34	98.29	11.98	1.00
South Africa	1,214,470	128	11.1	12.1	0.71	0.79	88.23	87.06	7.34	0.10
Spain	499,440	65	30.7	27.5	9.68	9.97	59.61	62.53	28.77	-0.60
Sri Lanka	64,630	1588	13.5	14.2	15.86	15.47	70.60	70.35	30.02	1.80
St. Kitts and Nevis	360	437	22.2	19.4	5.56	2.78	72.22	77.78	11.11	–
St. Lucia	610	2455	8.2	6.6	21.31	22.95	70.49	70.49	14.75	–
St. Vincent and the Grendadines	390	670	12.8	17.9	17.95	17.95	69.23	64.10	15.38	–
Sudan	2,376,000	125	5.5	6.8	0.10	0.18	94.43	92.99	25.94	1.40
Suriname	156,000	188	0.4	0.4	0.07	0.06	99.56	99.57	90.47	–
Swaziland	17,200	446	10.5	10.3	0.70	0.70	88.84	88.95	30.35	-1.20
Sweden	411,620	56	6.9	6.5	0.01	0.01	93.08	93.48	65.92	0.00
Switzerland	39,550	579	9.9	10.3	0.53	0.61	89.58	89.05	30.32	-0.40
Syrian Arab Republic	183,780	177	26.6	25.0	4.03	4.51	69.39	70.50	2.51	0.00
Tajikistan	140,600	488	–	6.6	–	0.90	–	92.48	2.84	-0.50
Tanzania	883,590	578	4.0	4.5	1.02	1.24	95.02	94.23	43.92	0.20
Thailand	510,890	310	34.2	31.1	6.09	6.85	59.67	62.09	28.89	0.70
Timor-Leste	14,870	1098	4.7	4.7	3.90	4.51	91.39	90.79	34.10	–
Togo	54,390	124	38.6	46.1	1.65	2.21	59.74	51.65	9.38	3.40
Tonga	720	397	23.6	23.6	43.06	43.06	33.33	33.33	5.56	–
Trinidad and Tobago	5,130	437	14.4	14.6	8.97	9.16	76.61	76.22	50.49	0.80
Tunisia	155,360	117	18.7	17.8	12.50	13.76	68.78	68.41	3.28	-0.20
Turkey	769,630	90	32.0	33.7	3.94	3.36	64.04	62.94	13.29	-0.20
Turkmenistan	469,930	142	–	3.9	–	0.14	–	95.92	7.99	0.00
Uganda	197,100	410	25.4	25.9	9.39	10.65	65.25	63.47	21.26	2.00
Ukraine	579,350	48	–	56.2	–	1.58	–	42.25	16.54	-0.30

Table K *(Continued)*
Land Use and Deforestation, 1990–2002

COUNTRY	LAND AREA	RURAL POPULATION DENSITY	LAND USE						FOREST AREA	AVERAGE ANNUAL DEFORESTATION
	(sq. km.)	(people per sq. km.)	Arable Land (% of land area)		Permanent Cropland (% of land area)		Other Land Use (% of land area)		(% of total land area)	Decline in % of forest area
	2002	2002	1990	2002	1990	2002	1990	2002	2000	1990-2000
United Arab Emirates	83,600	621	0.4	0.9	0.24	2.28	99.34	96.82	3.84	-2.00
United Kingdom	240,880	107	27.5	23.9	0.27	0.21	72.24	75.91	11.60	-0.80
United States	9,158,960	37	20.3	19.2	0.22	0.22	79.50	80.56	24.67	-0.20
Uruguay	175,020	20	7.2	7.4	0.26	0.23	92.54	92.34	7.38	-5.00
Uzbekistan	414,240	357	–	10.8	–	0.83	–	88.35	4.75	-0.20
Vanuatu	12,190	531	2.5	2.5	7.38	7.38	90.16	90.16	36.67	–
Venezuela, RB	882,050	130	3.2	2.8	0.88	0.92	95.91	96.32	56.13	0.40
Vietnam	325,490	901	16.4	20.6	3.21	6.74	80.39	72.67	30.17	-0.50
Virgin Islands (U.S.)	340	1461	11.8	11.8	2.94	2.94	85.29	85.29	41.18	–
Yemen, Rep.	527,970	903	2.9	2.9	0.20	0.25	96.92	96.84	0.85	1.80
Zambia	743,390	117	7.1	7.1	0.03	0.04	92.91	92.89	42.03	2.40
Zimbabwe	386,850	255	7.5	8.3	0.31	0.34	92.22	91.34	49.22	1.50

Source: World Development Indicators 2003 (World Bank).

Table L
World Countries: Energy Production and Use, 1990–2002

COUNTRY	COMMERCIAL ENERGY PRODUCTION		COMMERCIAL ENERGY USE			COMMERCIAL ENERGY USE PER CAPITA			NET ENERGY IMPORTS[a]	
	Thousand Metric Tons (Kilotons) of oil equivalent		Thousand Metric Tons (Kilotons) of oil equivalent		Average Annual % Growth	Kilogram of oil equivalent		Average Annual % Growth	% of Commercial Energy Use	
	1990	2002	1990	2002	1980-2000	1990	2002	1980-2000	1990	2002
Albania	2,449	771	2,662	1,944	-5.6	812.33	617.09	-6.3	8.00	60.34
Algeria	104,507	150,292	23,874	30,845	3.5	954.12	984.82	1	-337.74	-387.25
Angola	28,652	51,548	6,280	8,815	2.8	672.38	671.81	-0.3	-356.24	-484.78
Argentina	48,456	81,692	46,110	56,297	2.2	1,428.00	1,543.23	0.8	-5.09	-45.11
Armenia	–	738	–	1,938	–	–	631.69	–	–	61.92
Australia	157,712	255,192	87,536	112,712	2.4	5,129.53	5,732.25	1	-80.17	-126.41
Austria	8,080	9,926	25,260	30,443	1.6	3,269.61	3,774.24	1.1	68.01	67.39
Azerbaijan	–	19,753	–	11,728	–	–	1,435.14	–	–	-68.43
Bahrain	13,437	15,270	4,829	6,865	–	9,600.40	9,837.41	–	-178.26	-122.43
Bangladesh	10,747	16,747	12,815	21,004	4.1	116.47	154.80	1.9	16.14	20.27
Belarus	–	3,589	–	24,771	1.8	–	2,495.82	–	–	85.51
Belgium	12,490	13,164	48,685	56,887	–	4,884.42	5,505.37	1.6	74.35	76.86
Benin	1,774	1,546	1,678	2,231	2.3	356.26	340.50	-0.8	-5.72	30.70
Bolivia	4,923	8,152	2,774	4,310	3.8	415.95	498.54	1.5	-77.47	-89.14
Bosnia and Herzegovina	–	3,318	–	4,324	–	–	1,051.64	–	–	23.27
Brazil	97,616	161,737	133,531	190,664	2.7	902.50	1,092.72	1	26.90	15.17
Brunei	15,300	20,115	1,459	2,156	–	5,677.04	6,148.99	–	-948.66	-832.98
Bulgaria	9,613	10,512	28,820	19,019	-2.8	3,305.80	2,416.95	-2.2	66.64	44.73
Cameroon	12,090	12,004	5,031	6,569	2.5	431.44	416.57	-0.2	-140.31	-82.74
Canada	273,680	385,412	209,089	250,035	1.6	7,523.62	7,972.55	0.4	-30.89	-54.14
Chile	7,640	8,783	13,629	24,708	5.7	1,040.46	1,584.96	4	43.94	64.45
China	902,689	1,220,812	879,923	1,228,574	3.7	775.14	959.52	2.4	-2.59	0.63
Colombia	48,479	72,275	25,048	27,397	2.5	716.27	625.02	0.5	-93.54	-163.81
Congo, Dem. Rep.	12.019	16,134	11,903	15,402	2.7	318.52	298.61	-0.6	-0.97	-4.75
Congo, Rep.	9,005	13,199	1,056	923	-1.1	423.42	252.42	-4	-752.75	-1330.01
Costa Rica	1,032	1,762	2,025	3,564	4.2	664.15	904.17	1.5	49.04	50.56
Côte d'Ivoire	3,382	6,528	4,408	6,555	3.3	373.56	396.96	-0.1	23.28	0.41
Croatia	–	3,706	–	8,222	–	–	1,851.80	–	–	54.93
Cuba	6,271	6,480	16,524	14,197	-1.6	1,555.20	1,261.84	-2.4	62.05	54.36
Cyprus	6	45	1,536	2,467	–	2,255.51	3,224.98	–	99.61	98.18
Czech Republic	38,474	30,668	47,379	41,725	-1.2	4,571.94	4,090.29	-1.2	18.80	26.50
Denmark	9,735	28,754	17,581	19,749	0.6	3,420.43	3,674.71	0.4	44.63	-45.60
Dominican Republic	1,031	1,513	4,139	8,167	4.0	586.43	948.23	2.1	75.09	81.47
Ecuador	16,474	22,209	6,128	9,048	2.1	597	67	-0.3	-	
Egypt, Arab Rep.	54,869	59,766	31,895	52,393	4.6	608.20	789.39	2.3	-72.03	-14.07
El Salvador	1,722	2,367	2,535	4,299	2.2	496.09	669.92	0.6	32.07	44.94
Estonia	–	3,160	–	4,514	–	–	3,324.01	–	–	30.00
Ethiopia	14,158	18,445	15,151	19,934	2.6	296.03	296.56	-0.1	6.55	7.47
Finland	12,081	16,089	29,171	35,622	1.7	5,850.58	6,851.70	1.3	58.59	54.83
France	111,439	134,379	227,276	265,881	1.9	4,005.92	4,469.72	1.5	50.97	49.46
Gabon	14,630	12,690	1,242	1,590	-0.2	1,303.25	1,208.74	-3.1	-1077.94	-698.11
Georgia	–	1,325	–	2,559	–	–	494.30	–	–	48.22
Germany	186,159	134,771	356,221	346,352	-0.2	4,484.55	4,197.80	-0.5	47.74	61.09
Ghana	4,392	5,974	5,337	8,344	3.6	349.35	411.07	0.5	17.71	28.40
Greece	9,200	10,232	22,181	29,025	3.0	2,182.95	2,637.44	2.5	58.52	64.75
Guatemala	3,390	5,408	4,478	7,384	3.5	511.83	615.75	0.9	24.30	26.76
Haiti	1,253	1,515	1,585	2,081	0.4	244.86	251.13	-1.6	20.95	27.20
Honduras	1,694	1,618	2,416	3,426	2.8	496.30	504.08	-0.2	29.88	52.77
Hong Kong, China	43	48	10,662	16,377	–	1,869.05	2,413.00	–	99.60	99.71
Hungary	14,325	10,834	28,553	24,449	-1.0	2,754.75	2,505.07	-0.7	49.83	57.43

Table L *(Continued)*
World Countries: Energy Production and Use, 1990–2002

COUNTRY	COMMERCIAL ENERGY PRODUCTION		COMMERCIAL ENERGY USE			COMMERCIAL ENERGY USE PER CAPITA			NET ENERGY IMPORTS[a]	
	Thousand Metric Tons (Kilotons) of oil equivalent		Thousand Metric Tons (Kilotons) of oil equivalent		Average Annual % Growth	Kilogram of oil equivalent		Average Annual % Growth	% of Commercial Energy Use	
	1990	2002	1990	2002	1980-2000	1990	2002	1980-2000	1990	2002
Iceland	1,400	2,462	2,172	3,404	–	8,524.33	11,819.44	–	35.54	27.67
India	334,056	438,797	365.377	538,305	3.8	430.10	513.34	1.8	8.57	18.49
Indonesia	161,308	240,908	94,836	156,086	4.8	532.09	736.89	3.1	-70.09	-54.34
Iran, Islamic Rep.	179,738	240,522	68,775	133,960	5.6	1,264.25	2,043.94	3.1	-161.34	-79.55
Iraq	106,715	105,414	20,841	28,996	4.3	1,152.84	1,199.47	1.3	-412.04	-263.55
Ireland	3,467	1,499	10,575	15,303	2.7	3,016.43	3,893.89	2.3	67.22	90.20
Israel	433	722	12,112	20,954	5.2	2,599.14	3,191.29	2.6	96.43	96.56
Italy	25,548	26,590	152,553	172,720	1.3	2,689.63	2,993.93	1.2	83.25	84.61
Jamaica	485	463	2,943	3,914	3.5	1,231.38	1,493.30	2.6	83.52	88.17
Japan	76,129	98,133	445,916	516,927	2.6	3,609.57	4,057.54	2.2	82.93	81.02
Jordan	162	261	3,499	5,359	4.9	1,103.79	1,036.29	0.5	95.37	95.13
Kazakhstan	–	95,780	–	46,455	–	–	3,123.03	–	–	-106.18
Kenya	10,272	12,877	12,479	15,324	2.2	534.34	488.89	-0.7	17.69	15.97
Korea, Dem. Rep.	28,725	18.358	32,874	19,537	1.9	1,647.32	868.74	0.5	12.62	6.03
Korea, Rep.	21,908	36,206	92,650	203,498	9.1	2,161.24	4,271.58	8	76.35	82.21
Kuwait	50,401	105,991	7,579	22,189	1.0	3,566.59	9,503.11	0.2	-565.01	-377.67
Kyrgyz Republic	–	1,204	–	2,536	–	–	506.80	–	–	52.52
Latvia	–	1,870	–	4,266	–	–	1,824.64	–	–	56.17
Lebanon	143	192	2,309	5,369	4.8	635.21	1,208.90	2.8	93.81	96.42
Libya	73,173	69,519	11,541	18,704	3.6	2,680.21	3,433.04	1	-534.03	-271.68
Lithuania	–	4,915	–	8,589	–	–	2,475.93	–	–	42.78
Luxembourg	31	56	3,571	4,041	–	9,350.62	9,111.61	–	99.13	98.61
Malaysia	48,727	80,243	22,455	51,753	7.7	1,233.66	2,129.35	4.9	-117.00	-55.05
Malta	–	–	774	892	–	2,150.00	2,246.85	–	–	–
Mexico	194,482	229,888	124,057	157,308	2.1	1,490.60	1,560.31	0.2	-56.77	-46.14
Moldova	–	66	–	2,993	–	–	703.41	–	–	97.79
Morocco	773	589	6,725	10,753	4.3	279.71	362.78	2.2	88.51	94.52
Mozambique	6,846	8,041	7,203	8,045	-0.8	509.01	436.32	-2.6	4.96	0.05
Myanmar	10,651	15,825	10,683	12,578	1.3	263.74	257.82	-0.4	0.30	-25.81
Namibia	–	301	–	1,188	–	–	598.59	–	–	74.66
Nepal	5,501	7,618	5,806	8,515	2.7	320.03	352.96	0.4	5.25	10.53
Netherlands	60,316	59,924	66,491	77,923	1.4	4,446.96	4,826.75	0.8	9.29	23.10
Netherlands Antilles	–	–	1,493	1,482	–	7,878.63	6,782.48	–	–	–
New Zealand	12,153	14,876	13,914	18,013	3.8	4,035.38	4,572.87	2.7	12.66	17.42
Nicaragua	1,495	1,659	2,118	2,908	2.7	553.87	544.37	0	29.41	42.95
Nigeria	150,453	192,660	70,905	95,675	2.6	737.04	718.34	-0.4	-112.19	-101.37
Norway	120,304	232,221	21,492	26,515	1.8	5,067.08	5,842.88	1.3	-459.76	-775.81
Oman	38,312	62,516	4,562	10,825	11.0	2,803.93	4,265.17	6.6	-739.81	-477.52
Pakistan	34,360	49,677	43,424	65,806	4.8	402.17	454.14	2.2	20.87	24.51
Panama	612	738	1,490	3,022	2.7	621.35	1,027.75	0.7	58.93	75.58
Paraguay	4,578	6,293	3,083	3,905	4.2	742.89	708.71	1.2	-48.49	-61.15
Peru	10,596	9,234	9,952	12,024	0.2	461.40	449.51	-1.8	-6.47	23.20
Philippines	13,701	21,941	26,159	42,008	3.9	428.56	525.47	1.5	47.62	47.77
Poland	99,228	79,633	99,847	89,185	-1.4	2,619.36	2,332.73	-1.8	0.62	10.71
Portugal	3,393	3,643	17,746	26,392	4.7	1,793.25	2,545.53	4.7	80.88	86.20
Qatar	26,113	56,025	6,454	12,158	–	13,307.22	19,915.25	–	-304.60	-360.81
Romania	40,834	28,406	62,403	36,976	-3.1	2,688.97	1,695.91	-3.2	34.56	23.18
Russian Federation	–	1,034,519	–	617,843	–	–	4,288.47	–	–	-67.44
Saudi Arabia	372,985	462,807	65,538	126,387	5.0	4,147.19	5,774.80	1	-469.11	-266.18
Senegal	1,362	1,807	2,238	3,192	2.4	305.45	318.98	-0.3	39.14	43.39

Table L *(Continued)*
World Countries: Energy Production and Use, 1990–2002

COUNTRY	COMMERCIAL ENERGY PRODUCTION		COMMERCIAL ENERGY USE			COMMERCIAL ENERGY USE PER CAPITA			NET ENERGY IMPORTS[a]	
	Thousand Metric Tons (Kilotons) of oil equivalent		Thousand Metric Tons (Kilotons) of oil equivalent		Average Annual % Growth	Kilogram of oil equivalent		Average Annual % Growth	% of Commercial Energy Use	
	1990	2002	1990	2002	1980-2000	1990	2002	1980-2000	1990	2002
Serbia and Montenegro	–	10,876	–	16,169	–	–	1,981.50	–	–	32.74
Singapore	–	64	13,357	25,307	8.7	4,383.66	6,077.57	6	–	99.75
Slovak Republic	5,273	6,650	21,426	18,546	-1.4	4,055.65	3,447.85	-1.7	75.39	64.14
Slovenia	–	3,379	–	6,951	–	–	3,485.96	–	–	51.39
South Africa	114,534	146,506	91,229	113,458	2.1	2,591.73	2,502.09	-0.2	-25.55	-29.13
Spain	34,648	61,737	91,209	131,558	3.2	2,348.57	3,215.22	2.9	62.01	75.88
Sri Lanka	4,191	4,557	5,516	8,179	2.5	339.09	430.32	1.4	24.02	44.28
Sudan	8,775	25,013	10,627	15,850	3.1	426.32	483.37	0.7	17.43	-57.81
Sweden	29,754	32,403	46,658	51,031	1.0	5,451.34	5,718.40	0.6	36.23	36.50
Switzerland	9,831	11,942	25,106	27,139	1.4	3,740.47	3,722.77	0.7	60.84	56.00
Syrian Arab Republic	22,570	36,706	11,928	18,054	5.4	984.48	1,062.90	2.2	-89.22	-103.31
Tajikistan	–	1,330	–	3,247	–	–	518.25	–	–	59.04
Tanzania	9,063	13,286	9,808	14,339	2.0	385.08	407.57	-1	7.60	7.34
Thailand	26,496	45,303	43,860	83,339	7.4	788.92	1,352.62	6	39.59	45.64
Togo	778	1,081	1,001	1,540	3.8	289.73	323.56	0.8	22.28	29.81q
Trinidad and Tobago	12,612	21,321	5,795	9,286	3.3	4,769.55	7,121.30	2.4	-117.64	-129.60
Tunisia	6,127	6,943	5,536	8,276	3.7	678.90	846.13	1.6	-10.68	16.11
Turkey	25,857	24,432	53,005	75,418	4.6	943.92	1,083.19	2.7	51.22	67.60
Turkmenistan	–	53,645	–	16,606	–	–	3,464.71	–	–	-223.05
Ukraine	–	71,520	–	130,743	–	–	2,683.71	–	–	45.30
United Arab Emirates	109,446	142,148	17,839	36,072	8.2	10,061.48	9,608.95	2.8	-513.52	-294.07
United Kingdom	207,007	257,541	212,176	226,508	1.0	3,686.11	3,824.28	0.8	2.44	-13.70
United States	1,650,474	1,666,050	1,927,638	2,290,410	1.5	7,722.20	7,942.64	0.4	14.38	27.26
Uruguay	1,149	1,239	2,251	2,510	1.7	724.73	746.80	1	48.96	50.64
Uzbekistan	–	55,788	–	51,740	–	–	2,047.41	–	–	-7.82
Venezuela, RB	148,854	210,150	43,918	54,006	2.9	2,223.70	2,141.40	0.1	-238.94	-289.12
Vietnam	24,711	53,439	24,324	42,645	3.2	367.43	530.25	1.2	-1.59	-25.31
Yemen, Rep.	9,384	22,235	2,708	4,107	4.2	228.02	220.80	0.3	-246.53	-441.39
Zambia	4,923	6.226	5,470	6,549	1.3	702.72	639.27	-1.6	10.00	4.93
Zimbabwe	8,500	8,468	9,334	9,761	2.6	911.43	750.79	-0.3	8.94	13.25
World	**8,801,246**	**10,274,550**	**8,616,766**	**10,196,820**	**2.9**	**1,686.26**	**1,699.22**	**0.9**	**-2.31**	**-0.77**

Source: World Development Indicators 2003, (World Bank).

Table M
World Countries: Water Resources

COUNTRY	ANNUAL RENEWABLE WATER RESOURCES[a]				ANNUAL AVERAGE GROUNDWATER RESOURCES[b]	SECTORAL WITHDRAWALS (%)[c]		
	Supply Per Capita (cubic meters) 2000	Recharge	Population (2003)	Withdrawal Per Capita (cubic meters) 2000	Recharge Per Capita (cubic meters) 2000	Domestic	Industry	Agriculture
						9	20	71
WORLD	–	11,358.0	6,130	650	1,853	65	15	67
AFRICA	5,159		812	307	–	–	–	63
Algeria	460	1.7	31	181	55	34	52	14
Angola	13,203	72.0	14	54	5,143	14	10	76
Benin	3,741	1.8	6	27	300	23	10	67
Botswana	9,209	1.7	2	86	850	32	20	48
Burkina Faso	1,024	9.5	12	40	792	19	0	81
Burundi	538	2.1	7	19	300	36	64	0
Cameroon	18,378	100.0	15	38	6,667	46	19	35
Central African Republic	37,565	56.0	4	25	14,000	21	5	74
Chad	5,125	12.0	8	34	1,500	16	2	82
Congo (Zaire)	259,547	198.0	52	20	3,808	62	27	11
Congo Republic	23,639	421.0	3	10	140,333	61	16	23
Cote d'Ivoire	4,853	38.0	4	62	9,500	22	11	67
Egypt	830	1.3	65	1,055	20	11	82	86
Equatorial Guinea	53,841			30	–	81	13	6
Eritrea	1,578	–	4	–	–	–	–	–
Ethiopia	1,666	40.0	66	51	606	11	3	86
Gabon	126,789	62.0	1	70	62,000	72	22	6
Gambia	5,836	0.5	1	29	500	7	2	91
Ghana	2,637	26.0	20	35	1,300	35	13	52
Guinea	26,964	38.0	8	132	4,750	10	3	87
Guinea-Bissau	24,670	14.0	1	17	14,000	60	4	36
Kenya	947	3.0	31	87	97	20	4	76
Lesotho	1,456	0.5	2	32	250	22	22	56
Liberia	70,348	60.0	3	59	20,000	27	13	60
Libya	109	0.5	5	870	100	13	3	84
Madagascar	19,925	55.0	16	1,611	3,438	1	–	99
Malawi	1,461	1.4	11	95	127	10	3	86
Mali	8,320	20.0	24	167	833	2	1	97
Mauritania	4,029	0.3	3	923	100	2	2	92
Morocco	936	10.0	29	399	345	10	2	89
Mozambique	11,382	17.0	18	42	944	9	2	89
Namibia	9,865	2.1	2	175	1,050	29	3	68
Niger	2,891	2.5	11	69	227	16	2	82
Nigeria	2,891	2.5	130	69	19	16	2	82
Rwanda	638	3.6	9	141	400	5	2	94
Senegal	3,977	7.6	10	202	760	5	3	92
Sierra Leone	33,237	50.0	5	98	10,000	7	4	89
Somalia	1,413	3.3	9	119	367	3	0	97
South Africa	1,131	4.8	43	366	112	17	11	72

Table M *(Continued)*
World Countries: Water Resources

COUNTRY	ANNUAL RENEWABLE WATER RESOURCES[a]				ANNUAL AVERAGE GROUNDWATER RESOURCES[b]	SECTORAL WITHDRAWALS (%)[c]		
	Supply Per Capita (cubic meters) 2000	Recharge	Population (2003)	Withdrawal Per Capita (cubic meters) 2000	Recharge Per Capita (cubic meters) 2000	Domestic	Industry	Agriculture
Sudan	1,981	7.0	32	637	219	4	1	94
Tanzania	2,472	30.0	34	39	882	9	2	89
Togo	3,076	5.7	5	29	1,140	62	13	25
Tunisia	577	1.5	10	312	150	13	1	86
Uganda	2,663	29.0	23	21	1,261	32	8	60
Zambia	9,676	47.0	10	190	4,700	16	7	77
Zimbabwe	1,530	5.0	13	131	385	14	7	79
NORTH AMERICA	–	1,670.0		–	–	–	–	46
Canada	92,810	370.0	31	1,607	11,935	18	70	12
United States	10,574	1,300.0	285	1,834	4,561	13	45	42
CENTRAL AMERICA	–	359.0		–	–	–	–	–
Belize	78,763	–		485	–	12	88	0
Costa Rica	26,764	37.0	4	1,540	9,250	13	7	80
Cuba	3,382	6.5	11	475	591	49	0	51
Dominican Republic	2,430	12.0	9	1,102	1,333	11	0	89
El Salvador	3,872	6.2	6	137	1,033	34	20	46
Guatemala	9,277	34.0	12	126	2,833	9	17	74
Haiti	1,670	2.2	8	139	275	5	1	94
Honduras	14,250	39.0	7	294	5,571	4	5	91
Jamaica	3,588	3.9	3	371	1,300	15	7	77
Mexico	4,490	139.0	99	812	1,404	17	5	78
Nicaragua	36,784	59.0	5	267	11,800	14	2	84
Panama	50,299	21.0	3	685	7,000	28	2	70
Trinidad and Tobago	2,940	–	1	233	–	68	26	6
SOUTH AMERICA	–	3,693.0	349	–	10,582	–	–	–
Argentina	21,453	128.0	37	822	3,459	16	9	75
Bolivia	71,511	130.0	9	197	14,444	10	3	87
Brazil	47,125	1,874.0	172	359	10,895	21	18	61
Chile	59,143	140.0	15	1,629	9,333	5	11	84
Colombia	49,017	510.0	43	228	11,860	59	4	37
Ecuador	32,948	134.0	13	1,423	10,308	12	6	82
Guyana	314,963	–	–	1,993	–	1	1	98
Paraguay	28,148	41.0	6	112	6,833	15	7	78
Peru	72,127	303.0	26	849	11,654	7	7	86
Suriname	298,848	–	–	1,171	–	6	5	89
Uruguay	41,065	23.0	3	–	7,667	6	3	91
Venezuela	49,144	227.0	25	382	9,080	44	10	46
ASIA	2,790	–		2,007	–	1	0	99
Afghanistan	2,421		27	1,846	0	–	–	99
Armenia	2,778	4.2	4	784	1,050	30	4	66

Table M *(Continued)*
World Countries: Water Resources

COUNTRY	ANNUAL RENEWABLE WATER RESOURCES[a]				ANNUAL AVERAGE GROUNDWATER RESOURCES[b]	SECTORAL WITHDRAWALS (%)[c]		
	Supply Per Capita (cubic meters) 2000	Recharge	Population (2003)	Withdrawal Per Capita (cubic meters) 2000	Recharge Per Capita (cubic meters) 2000	Domestic	Industry	Agriculture
Azerbaijan	3,716	6.5	8	2,151	813	2	25	70
Bangladesh	8,444	21.0	133	133	158	12	2	86
Bhutan	43,214	–		13	–	36	10	54
Cambodia	34,516	18.0	12	60	1,500	5	1	94
China	2,186	829.0	1,272	439	652	5	18	78
Georgia	12,149	17.0	5	635	3,400	21	20	59
India	1,822	419.0	1,032	592	406	5	3	92
Indonesia	13,046	455.0	209	407	2,177	6	1	93
Iran	1,900	49.0	65	1,122	754	6	2	92
Iraq	3,111	1.2	24	2,478	50	3	5	92
Israel	265	0.5	6	287	83	39	7	54
Japan	3,372	27.0	127	735	213	19	17	64
Jordan	169	0.5	5	255	100	22	3	75
Kazakhstan	6,839	6.1	15	2,019	407	2	17	81
Korea, North	3,415	13.0	22	742	591	11	16	73
Korea, South	1,471	13.0	47	531	277	26	11	63
Kuwait	10	0.0	2	306	0	37	2	60
Kyrgyzstan	4,078	14.0	5	2,231	2,800	3	3	94
Laos	60,318	38.0	5	259	7,600	8	10	82
Lebanon	1,220	3.2	4	400	800	27	6	68
Malaysia	25,178	64.0	24	636	2,667	11	13	77
Mongolia	13,451	6.1	2	182	3,050	20	27	53
Myanmar (Burma)	21,358	156.0	48	103	3,250	7	3	90
Nepal	8,703	20.0	24	1,451	833	1	0	99
Oman	364	1.0	2	658	500	5	2	94
Pakistan	2,812	55.0	141	1,382	390	2	2	97
Philippines	6,093	180.0	78	811	2,308	8	4	88
Saudi Arabia	111	2.2	21	1,056	105	9	1	90
Singapore	–	–	4	–	–	45	51	4
Sri Lanka	2,592	7.8	19	574	411	2	2	96
Syria	1,541	4.2	17	844	247	8	2	90
Tajikistan	2,587	6.0	6	2,096	1,000	3	4	92
Thailand	6,371	42.0	61	605	689	5	4	91
Turkey	3,344	69.0	66	558	1,045	16	12	73
Turkmenistan	5,015	0.4	5	5,801	80	1	1	98
United Arab Emirates	56	0.1	3	896	33	24	9	67
Uzbekistan	1,968	8.8	25	2,598	352	4	2	94
Vietnam	11,109	48.0	80	822	600	4	10	87
Yemen	206	1.5	18	253	83	7	1	92
EUROPE	–	1,318.0	709	–	1,859	–	–	–
Albania	13,178	6.2	3	440	2,067	29	0	71
Austria	9,629	6.0	8	303	750	33	58	9
Belarus	5,739	18.0	10	266	1,800	22	43	35

Table M *(Continued)*
World Countries: Water Resources

COUNTRY	ANNUAL RENEWABLE WATER RESOURCES[a]				ANNUAL AVERAGE GROUNDWATER RESOURCES[b]	SECTORAL WITHDRAWALS (%)[c]		
	Supply Per Capita (cubic meters) 2000	Recharge	Population (2003)	Withdrawal Per Capita (cubic meters) 2000	Recharge Per Capita (cubic meters) 2000	Domestic	Industry	Agriculture
Belgium	1,781	0.9	10	–	90	–	–	–
Bosnia-Herzegovina	9,088	–		292	–	30	10	60
Bulgaria	2,734	6.4	8	1,573	800	22	3	75
Croatia	22,654	11.0	4	164	2,750	50	50	0
Czech Republic	1,283	1.4	10	266	140	41	57	2
Denmark	1,123	4.3	5	233	860	30	27	43
Estonia	9,413	4.0	1	106	4,000	56	39	5
Finland	21,223	2.2	5	439	440	12	85	3
France	3,414	100.0	59	547	1,695	18	72	10
Germany	1,878	46.0	82	579	561	11	69	20
Greece	6,984	10.0	11	826	909	10	3	87
Hungary	10,541	6.0	10	659	600	9	55	36
Iceland	599,944			622	–	31	63	6
Ireland	13,408	11.0	4	232	2,750	16	74	10
Italy	3,330	43.0	58	730	741	19	34	48
Latvia	14,820	2.2	2	112	1,100	55	32	13
Lithuania	6,763	1.2	3	68	400	81	16	3
Macedonia	3,121	–	2	936	–	12	15	74
Moldova	2,726	0.4	4	678	100	9	65	26
Netherlands	5,691	4.5	16	519	281	5	61	34
Norway	84,787	96.0	5	489	19,200	20	72	8
Poland	1,598	13.0	39	321	333	13	76	11
Portugal	6,837	4.0		736	–	15	37	48
Romania	9,486	8.3	22	1,141	377	8	33	59
Russian Federation	31,354	788.0	145	519	5,434	19	62	20
Serbia and Montenegro	19,815	3.0	11	1,233	273	6	86	8
Slovak Republic	9,265	1.7	5	337	340	–	–	–
Slovenia	16,070	14.0	2	642	7,000	20	80	1
Spain	2,793	30.0	41	884	732	13	19	68
Sweden	19,721	20.0	9	340	2,222	36	55	9
Switzerland	7,464	2.5	7	172	357	23	73	4
Ukraine	2,868	20.0	49	500	408	18	52	30
United Kingdom	2,464	9.8	59	204	166	20	77	3
OCEANIA	–		–	–	–	–	–	–
Australia	25,185	72.0	19	933	3,789	65	2	33
Fiji	34,330	–		–	–	20	20	60
New Zealand	85,221	–		588	–	46	10	44
Papua New Guinea	159,171	–		29	–	29	22	49
Solomon Islands	93,405	–		–	–	40	20	40

a. Annual renewable water resources usually include river flows from other countries.
b. Withdrawal data from most recent year available; varies by country from 1987 to 1995.
c. Total withdrawals may exceed 100% because of groundwater withdrawals or river inflows.

Source: World Resources 1998-99 (World Resources Institute).

Table N
World Countries: Energy Efficiency and Emissions, 1980–1999

COUNTRY	GDP PER UNIT OF ENERGY USE		TRADITIONAL FUEL USE		CARBON DIOXIDE EMISSIONS					
	PPP $ Per Kg. of Oil Equivalent		% of Total Energy Use		Total Million Metric Tons		Per Capita Metric Tons		Kg. Per PPP $ of GDP	
	1980	2000	1980	1997	1980	1999	1980	1999	1980	1999
Afghanistan	–	–	–	–	1.7	1.0	0.1	0	–	–
Albania	–	6.7	13.1	7.3	4.8	1.5	1.8	0.5	–	0.2
Algeria	5.5	6.4	1.9	1.5	66.1	90.8	3.5	3.0	1.0	0.5
Angola	–	3.6	64.9	69.7	5.3	10.3	0.8	0.8	–	0.4
Argentina	4.4	7.2	5.9	4.0	107.5	137.5	3.8	3.8	0.6	0.3
Armenia	–	4.5	–	0.0	–	3.1	–	0.8	–	0.4
Australia	2.0	4.3	3.8	4.4	202.8	344.4	13.8	18.2	1.5	0.8
Austria	3.4	7.5	1.2	4.7	52.4	61.4	6.9	7.6	0.7	0.3
Azerbaijan	–	1.9	–	0.0	–	33.6	–	4.2	–	1.8
Bangladesh	5.4	10.8	81.3	46.0	7.6	25.4	0.1	0.2	0.2	0.1
Belarus	–	3.0	–	0.8	–	57.6	–	5.7	–	0.9
Belgium	2.2	4.4	0.2	1.6	131.3	104.4	13.3	10.2	1.3	0.4
Benin	1.2	2.5	85.4	89.2	0.5	1.3	0.1	0.2	0.3	0.2
Bolivia	3.0	3.9	19.3	14.0	4.5	11.2	0.8	1.4	0.6	0.6
Bosnia-Herzegovina	–	5.2	–	10.1	–	4.8	–	1.2	–	0.2
Botswana	–	–	35.7	–	1.0	3.9	1.1	2.4	0.7	0.4
Brazil	4.2	6.7	35.5	28.7	183.4	300.7	1.5	1.8	0.4	0.3
Bulgaria	1.0	2.8	0.5	1.3	75.3	42.1	8.5	5.1	2.7	0.9
Burkina Faso	–	–	91.3	87.1	0.4	1.0	0.1	0.1	0.1	0.1
Burundi	–	–	97.0	94.2	0.1	0.2	0.0	0.0	0.1	0.1
Cambodia	–	–	100.0	89.3	0.3	0.7	0.0	0.1	–	0.0
Cameroon	2.7	3.8	51.7	69.2	3.9	4.7	0.4	0.3	0.4	0.2
Canada	1.4	3.3	0.4	4.7	420.9	438.6	17.1	14.4	1.5	0.6
Central African Republic	–	–	88.9	87.5	0.1	0.3	0.0	0.1	0.1	0.1
Chad	–	–	95.9	97.6	0.2	0.1	0.0	0.0	0.1	0.0
Chile	3.0	5.6	12.3	11.3	27.5	62.5	2.5	4.2	1.0	0.5
China	0.7	4.1	8.4	5.7	1,476.8	2,825.0	1.5	2.3	3.5	0.7
Hong Kong, China	6.2	10.9	0.9	0.7	16.3	41.2	3.2	6.2	0.5	0.3
Colombia	4.7	10.3	15.9	17.7	39.8	63.6	1.4	1.5	0.4	0.2
Democratic Republic of the Congo (formerly Zaire)	3.8	2.5	73.9	91.7	3.5	2.1	0.1	0.0	0.1	0.1
Congo Republic	0.8	3.2	77.8	53.0	0.4	2.4	0.2	0.8	0.6	0.9
Costa Rica	6.6	11.7	26.3	54.2	2.5	6.1	1.1	1.6	0.2	0.2
Côte d'Ivoire	–	–	52.8	91.5	5.3	13.3	0.6	0.9	0.5	0.6
Croatia	–	4.9	–	3.2	–	20.8	–	4.8	–	0.6
Cuba	–	–	27.9	30.2	30.8	25.4	3.2	2.3	–	–
Czech Republic	–	3.6	0.6	1.6	–	108.9	–	10.6	–	0.8
Denmark	3.0	7.9	0.4	5.9	92.9	49.7	12.3	9.3	1.1	0.3
Dominican Republic	4.1	7.4	27.5	14.3	6.4	23.3	1.1	2.8	0.4	0.4
Ecuador	2.8	4.9	26.7	17.5	13.4	23.3	1.7	1.9	0.9	0.6
Egypt	3.3	4.8	4.7	3.2	45.2	123.6	1.1	2.0	0.9	0.6
El Salvador	5.0	8.1	52.9	34.5	2.1	5.8	0.5	0.9	0.2	0.2
Eritrea	–	–	–	96.0	–	6.0	–	0.1	–	0.1
Estonia	–	2.9	–	13.8	–	16.2	–	11.7	–	1.4
Ethiopia	1.6	2.6	89.6	95.9	1.8	5.5	0.0	0.1	0.1	0.1
Finland	1.7	3.8	4.3	6.5	56.9	58.4	11.9	11.3	1.3	0.5
France	2.8	5.4	1.3	5.7	482.7	359.7	9.0	6.1	0.9	0.3

Table N *(Continued)*
World Countries: Energy Efficiency and Emissions, 1980–1999

COUNTRY	GDP PER UNIT OF ENERGY USE		TRADITIONAL FUEL USE		CARBON DIOXIDE EMISSIONS					
	PPP $ Per Kg. of Oil Equivalent		% of Total Energy Use		Total Million Metric Tons		Per Capita Metric Tons		Kg. Per PPP $ of GDP	
	1980	2000	1980	1997	1980	1999	1980	1999	1980	1999
Gabon	1.8	4.7	30.8	32.9	6.2	3.6	8.9	3.0	2.3	0.5
The Gambia	–	–	72.7	78.6	0.2	0.3	0.2	0.2	0.2	0.1
Georgia	4.6	4.5	–	1.0	–	5.4	–	1.0	–	0.5
Germany	2.2	6.1	0.3	1.3	–	792.2	–	9.7	–	0.4
Ghana	3.1	5.5	43.7	78.1	2.4	5.6	0.2	0.3	0.2	0.1
Greece	4.7	6.3	3.0	4.5	51.7	85.9	5.4	8.2	0.7	0.5
Guatemala	4.6	7.1	54.6	62.0	4.5	9.7	0.7	0.9	0.3	0.2
Guinea	–	–	71.4	74.2	0.9	1.3	0.2	0.2	–	0.1
Guinea-Bissau	–	–	80.0	57.1	0.5	0.3	0.7	0.2	1.4	0.3
Haiti	4.7	7.5	80.7	74.7	0.8	1.4	0.1	0.2	0.1	0.1
Honduras	3.2	6.0	55.3	54.8	2.1	5.0	0.6	0.8	0.3	0.3
Hungary	2.0	4.9	2.0	1.6	82.5	56.9	7.7	5.6	1.5	0.5
India	2.2	5.5	31.5	20.7	347.3	1,077.0	0.5	1.1	0.7	0.4
Indonesia	2.0	4.2	51.5	29.3	94.6	235.6	0.6	1.2	0.8	0.4
Iran	2.7	3.2	0.4	0.7	116.1	301.4	3.0	4.8	1.1	0.9
Iraq	–	–	0.3	0.1	44.0	74.2	3.4	3.3	–	–
Ireland	2.3	7.9	0.0	0.2	25.2	40.4	7.4	10.8	1.3	0.4
Israel	3.7	6.5	0.0	0.0	21.2	61.1	5.4	10.0	0.7	0.5
Italy	3.9	8.2	0.8	1.0	371.9	422.7	6.6	7.3	0.7	0.3
Jamaica	1.8	2.4	5.0	6.0	8.4	10.2	4.0	4.0	2.0	1.2
Japan	3.1	6.1	0.1	1.6	920.4	1,155.2	7.9	9.1	0.8	0.4
Jordan	3.1	3.6	0.0	0.0	4.7	14.6	2.2	3.1	0.9	0.8
Kazakhstan	–	2.2	–	0.2	–	112.8	–	7.4	–	1.5
Kenya	1.0	1.9	76.8	80.3	6.2	8.8	0.4	0.3	0.6	0.3
Korea, North	–	–	3.1	1.4	124.9	208.7	7.3	9.4	–	–
Korea, South	2.3	3.6	4.0	2.4	125.1	393.5	3.3	8.4	1.3	0.6
Kuwait	1.4	1.8	0.0	0.0	24.7	48.0	18.0	24.9	1.5	1.4
Kyrgyzstan	–	5.4	–	0.0	–	4.7	–	1.0	–	0.4
Laos	–	–	72.3	88.7	0.2	0.4	0.1	0.1	–	0.1
Latvia	19.8	4.6	–	26.2	–	6.6	–	2.8	–	0.4
Lebanon	–	3.5	2.4	2.5	6.2	16.9	2.1	4.0	–	1.0
Lesotho	–	–	–	–	–	–	–	–	–	–
Liberia	–	–	–	–	2.0	0.4	1.1	0.1	–	–
Libya	–	–	2.3	0.9	26.9	42.8	8.8	8.3	–	–
Lithuania	–	3.9	–	6.3	–	13.2	–	3.8	–	0.5
Macedonia	–	–	–	6.1	–	11.4	–	5.6	–	1.0
Madagascar	–	–	78.4	84.3	1.6	1.9	0.2	0.1	0.3	0.2
Malawi	–	–	90.6	88.6	0.7	0.8	0.1	0.1	0.3	0.1
Malaysia	2.6	4.3	15.7	5.5	28.0	123.7	2.0	5.4	0.9	0.7
Mali	–	–	86.7	88.9	0.4	0.5	0.1	0.0	0.1	0.1
Mauritania	–	–	0.0	0.0	0.6	3.0	0.4	1.2	0.3	0.6
Mauritius	–	–	59.1	36.1	0.6	2.5	0.6	2.1	0.3	0.2
Mexico	2.9	5.5	5.0	4.5	252.5	378.5	3.7	3.9	0.9	0.5
Moldova	–	3.1	–	0.5	–	6.5	–	1.5	–	0.8
Mongolia	–	–	14.4	4.3	6.8	7.5	4.1	3.2	3.7	1.9
Morocco	6.4	9.5	5.2	4.0	15.9	35.8	0.8	1.3	0.5	0.4
Mozambique	0.7	2.5	43.7	91.4	3.2	1.3	0.3	0.1	0.6	0.1
Myanmar (Burma)	–	–	69.3	60.5	4.8	9.2	0.1	0.2	–	–

Table N *(Continued)*
World Countries: Energy Efficiency and Emissions, 1980–1999

COUNTRY	GDP PER UNIT OF ENERGY USE		TRADITIONAL FUEL USE		CARBON DIOXIDE EMISSIONS					
	PPP $ Per Kg. of Oil Equivalent		% of Total Energy Use		Total Million Metric Tons		Per Capita Metric Tons		Kg. Per PPP $ of GDP	
	1980	2000	1980	1997	1980	1999	1980	1999	1980	1999
Namibia	–	12.0	–	–	–	0.1	–	0.1	–	0.0
Nepal	1.5	3.7	94.2	89.6	0.5	3.3	0.0	0.1	0.1	0.1
Netherlands	2.3	5.7	0.0	1.1	153.0	134.6	10.8	8.5	1.0	0.3
New Zealand	2.7	3.7	0.2	0.8	17.6	30.8	5.6	8.1	0.7	0.5
Nicaragua	4.0	4.6	49.2	42.2	2.0	3.8	0.7	0.8	0.3	0.3
Niger	–	–	79.5	80.6	0.6	1.1	0.1	0.1	0.1	0.1
Nigeria	0.8	1.2	66.8	67.8	68.1	40.4	1.0	0.3	1.7	0.4
Norway	2.3	5.1	0.4	1.1	38.7	38.7	9.5	8.7	0.9	0.3
Oman	4.5	3.0	0.0	–	5.9	19.9	5.3	8.5	1.3	0.7
Pakistan	2.1	4.0	24.4	29.5	31.6	98.9	0.4	0.7	0.6	0.4
Panama	4.1	6.5	26.6	14.4	3.5	8.3	1.8	2.9	0.6	0.5
Papua New Guinea	–	–	65.4	62.5	1.8	2.5	0.6	0.5	0.5	0.2
Paraguay	4.8	7.2	62.0	49.6	1.5	4.5	0.5	0.8	0.1	0.2
Peru	4.4	9.5	15.2	24.6	23.6	30.4	1.4	1.2	0.5	0.3
Philippines	5.3	6.8	37.0	26.9	36.5	73.2	0.8	1.0	0.3	0.3
Poland	–	4.0	0.4	0.8	456.2	314.4	12.8	8.1	–	0.9
Portugal	5.5	7.2	1.2	0.9	27.1	60.0	2.8	6.0	0.5	0.4
Puerto Rico	–	–	0.0	–	14.0	10.1	4.4	2.7	0.6	0.1
Romania	–	3.4	1.3	5.7	191.8	81.2	8.6	3.6	–	0.7
Russian Federation	–	1.6	–	0.8	–	1,437.3	–	9.8	–	1.6
Rwanda	–	–	89.8	88.3	0.3	0.6	0.1	0.1	0.1	0.1
Saudi Arabia	4.0	2.6	0.0	0.0	130.7	235.4	14.0	11.7	1.1	0.9
Senegal	2.2	4.5	50.8	56.2	2.8	3.7	0.5	0.4	0.7	0.3
Serbia and Montenegro	–	–	–	1.5	102.0	39.5	10.4	3.7	–	–
Sierra Leone	–	–	90.0	86.1	0.6	0.5	0.2	0.1	0.3	0.3
Singapore	2.2	3.9	0.4	0.0	30.1	54.3	12.5	13.7	2.3	0.7
Slovak Republic	–	3.6	–	0.5	–	38.6	–	7.2	–	0.7
Slovenia	–	5.0	–	1.5	–	14.4	–	7.3	–	0.5
Somalia	–	–	–	–	0.6	0.0	0.1	0.0	–	–
South Africa	3.1	4.4	4.9	43.4	211.3	334.6	7.7	7.9	1.0	0.8
Spain	3.8	6.4	0.4	1.3	200.0	273.7	5.3	6.8	0.8	0.4
Sri Lanka	3.1	7.8	53.5	46.5	3.4	8.6	0.2	0.5	0.2	0.2
Sudan	1.6	3.8	86.9	75.1	3.3	2.6	0.2	0.1	0.2	0.0
Swaziland	–	–	–	–	0.5	0.4	0.8	0.4	0.4	0.1
Sweden	2.0	4.4	7.7	17.9	71.4	46.6	8.6	5.3	0.9	0.2
Switzerland	4.4	7.5	0.9	6.0	40.9	40.6	6.5	5.7	0.4	0.2
Syria	2.6	2.9	0.0	0.0	19.3	53.4	2.2	3.4	1.4	1.1
Tajikistan	–	2.3	–	–	–	5.1	–	0.8	–	0.8
Tanzania	–	1.1	92.0	91.4	1.9	2.5	0.1	0.1	–	0.2
Thailand	2.9	5.1	40.3	24.6	40.0	199.7	0.9	3.3	0.6	0.6
Togo	4.9	4.9	35.7	71.9	0.6	1.3	0.2	0.3	0.2	0.2
Trinidad and Tobago	1.2	1.3	1.4	0.8	16.7	25.1	15.4	19.4	3.6	2.4
Tunisia	3.8	7.4	16.1	12.5	9.4	17.5	1.5	1.8	0.6	0.3
Turkey	3.2	5.3	20.5	3.1	76.3	198.5	1.7	3.1	0.8	0.5
Turkmenistan	–	1.2	–	–	–	31.0	–	6.7	–	2.5
Uganda	–	–	93.6	89.7	0.6	1.4	0.1	0.1	0.1	0.0
Ukraine	–	1.4	–	0.5	–	374.3	–	7.5	–	2.1

Table N *(Continued)*
World Countries: Energy Efficiency and Emissions, 1980–1999

COUNTRY	GDP PER UNIT OF ENERGY USE		TRADITIONAL FUEL USE		CARBON DIOXIDE EMISSIONS					
	PPP $ Per Kg. of Oil Equivalent		% of Total Energy Use		Total Million Metric Tons		Per Capita Metric Tons		Kg. Per PPP $ of GDP	
	1980	2000	1980	1997	1980	1999	1980	1999	1980	1999
United Arab Emirates	4.9	2.0	0.0	–	36.3	88.0	34.8	31.3	1.2	1.6
United Kingdom	2.5	6.0	0.0	3.3	580.3	539.3	10.3	9.2	1.2	0.4
United States	1.6	4.2	1.3	3.8	4,626.8	5,495.4	20.4	19.7	1.6	0.6
Uruguay	4.8	9.4	11.1	21.0	5.8	6.5	2.0	2.0	0.5	0.2
Uzbekistan	–	1.1	–	0.0	–	104.8	–	4.4	–	2.1
Venezuela	1.6	2.3	0.9	0.7	90.1	125.8	6.0	5.3	1.6	1.0
Vietnam	–	4.2	49.1	37.8	16.8	46.6	0.3	0.6	–	0.3
West Bank and Gaza	–	–	–	–	–	–	–	–	–	–
Yemen	–	4.0	0.0	1.4	–	18.3	–	1.1	–	1.4
Zambia	0.8	1.2	37.4	72.7	3.5	1.8	0.6	0.2	0.9	0.3
Zimbabwe	1.5	3.1	27.6	25.2	9.6	17.6	1.3	1.4	1.0	0.5
WORLD	**2.1**	**4.5**	**7.4w**	**8.2w**	**13,852.7**	**22,518.8**	**3.4**	**3.8**	**1.1**	**0.5**
Low income	2.1	4.0	46.4	29.8	774.3	2,429.2	0.5	1.0	0.6	0.5
Middle income	2.1	4.0	10.4	7.3	4,132.9	8,484.0	2.3	3.2	1.2	0.7
Lower middle income	1.6	3.7	10.7	5.7	2,682.6	6,391.3	1.8	3.0	1.6	0.7
Upper middle income	3.4	4.9	8.6	10.6	1,450.3	2,092.7	4.3	4.3	0.7	0.5
Low & middle income	2.1	4.0	18.5	12.9	4,907.1	10,613.2	1.5	2.2	1.0	0.6
East Asia & Pacific	–	–	15.1	9.7	1,833.3	3,734.4	1.3	2.1	2.2	0.6
Europe & Central Asia	–	2.3	3.2	1.3	989.0	3,144.1	–	6.6	1.3	1.2
Latin America & Carib.	3.6	6.1	18.4	16.0	848.8	1,286.7	2.4	2.5	0.6	0.4
Middle East & N. Africa	3.6	3.8	1.6	1.1	491.7	1,048.4	3.0	3.7	1.0	0.7
South Asia	2.3	5.5	34.2	23.8	392.3	1,215.1	0.4	0.9	0.6	0.4
Sub-Sarahan Africa	2.0	2.9	47.2	63.5	352.0	484.6	0.9	0.8	0.8	0.4
High Income	2.2	4.9	1.0	3.4	8,945.6	11,606.6	12.0	12.3	1.2	0.5
Europe EMU	2.8	6.2	0.7	2.5	1,565.2	2,408.4	7.5	7.9	0.8	0.4

Source: World Development Indicators 2003 (World Bank).

Part IX

Geographic Index

The geographic index contains approximately 1,500 names of cities, states, countries, rivers, lakes, mountain ranges, oceans, capes, bays, and other geographic features. The name of each geographical feature in the index is accompanied by a geographical coordinate (latitude and longitude) in degrees and by the page number of the primary map on which the geographical feature appears. Where the geographical coordinates are for specific places or points, such as a city or a mountain peak, the latitude and longitude figures give the location of the map symbol denoting that point. Thus, Los Angeles, California, is at 34N and 118W and the location of Mt. Everest is 28N and 87E.

The coordinates for political features (countries or states) or physical features (oceans, deserts) that are areas rather than points are given according to the location of the name of the feature on the map, except in those cases where the name of the feature is separated from the feature (such as a country's name appearing over an adjacent ocean area because of space requirements). In such cases, the feature's coordinates will indicate the location of the center of the feature. The coordinates for the Sahara Desert will lead the reader to the place name "Sahara Desert" on the map; the coordinates for North Carolina will show the center location of the state since the name appears over the adjacent Atlantic Ocean. Finally, the coordinates for geographical features that are lines rather than points or areas will also appear near the center of the text identifying the geographical feature.

Alphabetizing follows general conventions; the names of physical features such as lakes, rivers, mountains are given as: proper name, followed by the generic name. Thus "Mount Everest" is listed as "Everest, Mt." Where an article such as "the," "le," or "al" appears in a geographic name, the name is alphabetized according to the article. Hence, "La Paz" is found under "L" and not under "P."

Part IX *(Continued)*

Geographic Index

Geographic Index

Geographic Index

Name/Description	Latitude & Longitude	Page
Changchun, China (city)	44N 125E	115
Chari (riv., Africa)	11N 16E	123
Charleston, SC (city)	33N 80W	106
Charleston, West Virginia (city, st. cap., US)	38N 82W	106
Charlotte, NC (city)	35N 81W	106
Charlotte Waters, Aust. (city)	26S 135E	127
Charlottetown, P.E.I. (city, prov. cap., Can.)	46N 63W	106
Chelyabinsk, Russia (city)	55N 61E	107
Chengdu, China (city)	30N 104E	115
Chesapeake Bay	36N 74W	105
Chetumal, Quintana Roo (city, st. cap., Mex.)	19N 88W	106
Cheyenne, Wyoming (city, st. cap., US)	41N 105W	106
Chiapas (st., Mex.)	17N 92W	106
Chicago, IL (city)	42N 87W	106
Chiclayo, Peru (city)	7S 80W	109
Chidley, Cape	60N 65W	105
Chihuahua (st., Mex.)	30N 110W	106
Chihuahua, Chihuahua (city, st. cap., Mex.)	29N 106W	106
Chile (country)	32S 75W	109
Chiloe (island)	43S 74W	108
Chongqing, China (city)	30N 107E	115
Chimborazo, Mt. 20,702	2S 79W	108
China (country)	38N 105E	115
Chisinau, Moldova (city, nat. cap.)	47N 29E	107
Christchurch, New Zealand (city)	43S 173E	127
Chubut (st., Argentina)	44S 70W	109
Chubut, Rio (riv., S.Am.)	44S 71W	108
Cincinnati, OH (city)	39N 84W	106
Cleveland (city)	41N 82W	106
Coahuila (st., Mex.)	30N 105W	106
Coast Mountains (Can.)	55N 130W	105
Coast Ranges (US)	40N 120W	105
Coco Island	8N 88W	105
Cod, Cape	42N 70W	105
Colima (st., Mex.)	18N 104W	106
Colima, Colima (city, st. cap., Mex.)	19N 104W	106
Colombia (country)	4N 73W	109
Colombo, Sri Lanka (city, nat. cap.)	7N 80E	115
Colorado (riv., N.Am.)	36N 110W	105
Colorado (st., US)	38N 104W	106
Colorado, Rio (riv., S.Am.)	38S 70W	108
Colorado (Texas) (riv., N.Am.)	30N 100W	105
Columbia (riv., N.Am.)	45N 120W	106
Damascus, Syria (city, nat. cap.)	34N 36E	107
Danube (riv., Europe)	44N 24E	111
Dar es Salaam, Tanzania (city, nat. cap.)	7S 39E	124
Darien, Gulf of	9N 77W	108
Darling (riv., Australasia)	35S 144E	126
Darling Range	33S 116W	126
Comoros (country)	12S 44E	124
Conakry, Guinea (city, nat. cap.)	9N 14W	107
Concord, New Hampshire (city, st. cap., US)	43N 71W	106
Congo (country)	3S 15E	124
Congo (riv., Africa)	3N 22E	123
Congo Basin	4N 22E	123
Congo, Democratic Republic of (country)	5S 15E	124
Connecticut (st., US)	43N 76W	106
Connecticut (riv., N.Am.)	43N 76W	105
Cook, Mt. 12,316	44S 170E	126
Cook Strait	42S 175E	126
Copenhagen, Denmark (city, nat. cap.)	56N 12E	107
Copiapo, Chile (city)	27S 70W	109
Copiapo, Mt. 19,947	26S 70W	108
Coquimbo, Chile (city)	30S 70W	109
Coral Sea	15S 155E	126
Cordilleran Highlands	45N 118W	105
Cordoba (st., Argentina)	32S 67W	109
Cordoba, Cordoba (city, st. cap., Argen.)	32S 64W	109
Corrientes (st., Argentina)	27S 60W	109
Corrientes, Corrientes (city, st. cap., Argen.)	27S 59W	109
Corsica (island)	42N 9E	107
Cosmoledo Islands	9S 48E	123
Costa Rica (country)	15N 84W	106
Côte d'Ivoire (country)	7N 86W	124
Cotopaxi, Mt. 19,347	1S 78W	108
Crete (island)	36N 25W	111
Croatia (country)	46N 20W	107
Cuango (riv., Africa)	10S 16E	123
Cuba (country)	22N 78W	106
Cuiaba, Mato Grosso (city, st. cap., Braz.)	16S 56W	109
Cuidad Victoria, Tamaulipas (city, st. cap., Mex.)	24N 99W	106
Culiacan, Sinaloa (city, st. cap., Mex.)	25N 107W	106
Curitiba, Parana (city, st. cap., Braz.)	26S 49W	109
Cusco, Peru (city)	14S 72W	109
Cyprus (island)	36N 34E	111
Czech Republic (country)	50N 16E	107
d'Ambre, Cape	12S 50E	124
Dakar, Senegal (city, nat. cap.)	15N 17W	124
Dakhla, Western Sahara (city)	24N 16W	124
Dallas, TX (city)	33N 97W	106
Dalrymple, Mt. 4,190	22S 148E	126
Daly (riv., Australasia)	14S 132E	126
Edmonton, Alberta (city, prov. cap., Can.)	54N 114W	106
Edward, Lake	0 30E	123
Egypt (country)	23N 30E	124
El Aaiun, Western Sahara (city)	27N 13W	124
El Djouf	25N 15W	123
El Paso, TX (city)	32N 106W	106

Geographic Index

Geographic Index

Part IX *(Continued)*

Geographic Index

Geographic Index

Geographic Index

Geographic Index

Name/Description	Latitude & Longitude	Page	Name/Description	Latitude & Longitude	Page
Missouri (riv., N.Am.)	41N 96W	105	Nebraska (st., US)	42N 100W	106
Missouri (st., US)	35N 92W	106	Negro, Rio (Argentina) (riv., S.Am.)	40S 70W	108
Misti, Mt. 19,101	15S 73W	108	Negro, Rio (Brazil) (riv., S.Am.)	0 65W	108
Mitchell (riv., Australasia)	16S 143E	126	Negros (island)	10N 125E	114
Mobile, AL (city)	31N 88W	106	Nelson (riv., N.Am.)	56N 90W	105
Mocambique, Mozambique (city)	15S 40E	124	Nepal (country)	29N 85E	115
Mogadishu, Somalia (city, nat. cap.)	2N 45E	124	Netherlands (country)	54N 6E	107
Moldova (country)	49N 28E	107	Neuquen (st., Argentina)	38S 68W	109
Mombasa, Kenya (city)	4S 40E	124	Neuquen, Neuquen (city, st. cap., Argen.)	39S 68W	109
Monaco, Monaco (city)	44N 8E	107	Nevada (st., US)	37N 117W	106
Mongolia (country)	45N 100E	115	New Britain (island)	5S 152E	126
Monrovia, Liberia (city, nat. cap.)	6N 11W	124	New Brunswick (prov., Can.)	47N 67W	106
Montana (st., US)	50N 110W	106	New Caledonia (island)	21S 165E	126
Monterrey, Nuevo Leon (city, st. cap., Mex.)	26N 100W	106	New Delhi, India (city, nat. cap.)	29N 77E	115
Montevideo, Uruguay (city, nat. cap.)	35S 56W	109	New Georgia (island)	8S 157E	126
Montgomery, Alabama (city, st. cap., US)	32N 86W	106	New Guinea (island)	5S 142E	126
Montpelier, Vermont (city, st. cap., US)	44N 73W	106	New Hampshire (st., US)	45N 70W	106
Montreal, Canada (city)	45N 74W	106	New Hanover (island)	3S 153E	126
Morelin, Michoacan (city, st. cap., Mex.)	20N 100W	106	New Hebrides (island)	15S 165E	126
Morocco (country)	34N 10W	124	New Ireland (island)	4S 154E	126
Moroni, Comoros (city, nat. cap.)	12S 42E	124	New Jersey (st., US)	40N 75W	106
Moscow, Russia (city, nat. cap.)	56N 38E	107	New Mexico (st., US)	30N 108W	106
Mountain Nile (riv., Africa)	5N 30E	123	New Orleans, LA (city)	30N 90W	106
Mozambique (country)	19N 35E	124	New Siberian Islands	74N 140E	114
Mozambique Channel	19N 42E	123	New South Wales (st., Aust.)	35S 145E	127
Munich, Germany (city)	48N 12E	107	New York (city)	41N 74W	106
Murchison (riv., Australasia)	26S 115E	126	New York (st., US)	45N 75W	106
Murmansk, Russia (city)	69N 33E	107	New Zealand (country)	40S 170E	127
Murray (riv., Australasia)	36S 143E	126	Newcastle, Aust. (city)	33S 152E	127
Murrumbidgee (riv., Australasia)	35S 146E	126	Newcastle, UK (city)	55N 2W	107
Musoat, Oman (city, nat. cap.)	23N 58E	115	Newfoundland (prov., Can.)	53N 60W	106
Musgrave Ranges	28S 135E	126	Nicaragua (country)	10N 90W	106
Myanmar (Burma) (country)	20N 95E	115	Niamey, Niger (city, nat. cap.)	14N 2E	124
Nairobi, Kenya (city, nat. cap.)	1S 37E	124	Nicobar Islands	5N 93E	114
Namibe, Angola (city)	16S 13E	124	Niger (country)	10N 8E	124
Namibia (country)	20S 16E	124	Niger (riv., Africa)	12N 0	123
Namoi (riv., Australasia)	31S 150E	126	Nigeria (country)	8N 5E	124
Nan Ling Mountains	25N 110E	114	Nile (riv., Africa)	25N 31E	123
Nipigon (lake, N.Am.)	50N 87W	105	Oregon (st., US)	46N 120W	106
Nizhny-Novgorod, Russia (city)	56N 44E	107	Orinoco, Rio (riv., S.Am.)	8N 65W	108
Norfolk, VA (city)	37N 76W	106	Orizaba Peak 18,406	19N 97W	105
North Cape (NZ)	36N 174W	126	Orkney Islands	60N 0	111
North Carolina (st., US)	30N 78W	106	Osaka, Japan (city)	35N 135E	115
North Channel	56N 5W	111	Oslo, Norway (city, nat. cap.)	60N 11W	107
North Dakota (st., US)	49N 100W	106	Ossa, Mt. 5,305 (Tasm.)	43S 145E	126
North Island (NZ)	37S 175W	126	Ottawa, Canada (city, nat. cap.)	45N 76W	106
North Saskatchewan (riv., N.Am.)	55N 110W	105	Otway, Cape	40S 142W	126
North Sea	56N 3E	111	Ougadougou, Burkina Faso (city, nat. cap.)	12N 2W	124
North West Cape	22S 115W	126	Owen Stanley Range	9S 148E	126

Geographic Index

Geographic Index

Geographic Index

Geographic Index

Geographic Index

Geographic Index

Sources

Amnesty International. Online access at www.amnesty.org.

Asia-Pacific Economic Organization (APEC). Online access at www.apecsec.org.

Bercovitch, J., and R. Jackson. (1997). *International conflict: A chronological encyclopedia of conflicts and their management 1945–1995.* Washington, DC: Congressional Quarterly.

BP Statistical Review of World Energy. Online access at: http://www.bp.com/bpstats.

Canadian Forces College, Information Resources Centre. Online access at: http://www.cfcsc.dnd.ca/links/wars/index.html.

CIA. *World Factbook, 2002*. Online access at: www.cia.gov/cia/publications/factbook.

Cohen, Saul. (2002). *Geopolitics of the world system*. Lanham, MD: Rowman & Littlefield.

Commonwealth of Independent States (CIS). Online access at: www.cis.minsk.by/english.

Crabb, C. (1993, January). Soiling the planet. *Discover, 14* (1), 74–75.

DeBlij, H. J., & Muller, P. (1998). *Geography: Realms, regions and concepts* (8th ed.). New York: John Wiley.

Domke, K. (1988). *War and the changing global system.* New Haven, CT: Yale University Press.

Economic Community of West African States (ECOWAS). Online access at: www.state.gov.

European Free Trade Association (EFTA). Online access at www.efta.int.

The European Union (EU). Online access at: www.europa.eu.int.

Food and Agricultural Organization of the United Nations (FAO). Online access at www.fao.org.

Freedom House. Online access at www.freedomhouse.org.

Goode's world atlas. (1995, 19th ed.). New York: Rand McNally.

The Greater Caribbean Community (CARICOM). Online access at: www.caricom.org.

Gunnemark, Erik V., *Countries, peoples and their languages.* Gothenburg, Sweden: The Geolinguistic Handbook (n.d., early 1990s).

Hammond atlas of the world. (1993). Maplewood, NJ: Hammond.

Information please almanac, atlas, and yearbook 2002. (2002). Boston & New York: Houghton Mifflin.

International Energy Agency. (2001). *Key world energy statistics 2000.* Paris. Online access: http://www.iea.org/statist/keyworld/keystats.htm.

Johnson, D. (1977). *Population, society, and desertification.* New York: United Nations Conference on Desertification, United Nations Environment Programme.

Köppen, W., & Geiger, R. (1954). *Klima der erde* [Climate of the earth]. Darmstadt, Germany: Justus Perthes.

Lindeman, M. (1990). *The United States and the Soviet Union: Choices for the 21st century.* Guilford, CT: Dushkin Publishing Group.

Murphy, R. E. (1968). Landforms of the world [Map supplement No. 9]. *Annals of the Association of American Geographers, 58* (1), 198–200.

National Oceanic and Atmospheric Administration. (1990–1992). Unpublished data. Washington, DC: NOAA.

North Atlantic Treaty Organization (NATO). Online access at: www.nato.int.

The *New York Times.* Online access at: http://archives.nytimes.com/archives/.

The Peace Corps. Online access at www.peacecorps.gov.

Population Reference Bureau (2001). *World population data sheet.* New York: Population Reference Bureau.

Rourke, J. T. (2003). *International politics on the world stage* (9th ed.). Guilford, CT: McGraw-Hill/Dushkin.

Southern African Development Community (SADC). Online access at www.sadc.int.

Southern Cone Common Market (Mercosur). Online access at http://www.infoplease.com/ce6/history/A0846059.html.

Spector, L. S., & Smith, J. R. (1990). *Nuclear ambitions: The spread of nuclear weapons.* Boulder, CO: Westview Press.

Time atlas of world history. (1978). London.

United Nations Development Programme (UNDP, 2001). *Human development indicators, human development report 2001.* New York: Oxford University Press. Online access at: http://www.undp.org/hdr2001/back.pdf.

United Nations Food and Agriculture Organization. *FAOSTAT database.* Online access at: http://apps.fao.org/page/collections?subset=agriculture.

United Nations High Commissioner on Refugees (UNHCR), Population Data Unit, Population and Geographic Data Section. (2000). Online access at: http://www.unhcr.ch/cgibin/texis/vtx/home. *Provisional statistics on refugees and others of concern to UNHCR for the year 2000.* Geneva.

United Nations Population Division. *World population prospects: The 2002 revision*. Online access at: http://www.un.org/esa/population/unpop.htm.

United Nations Population Fund. (2000). *The state of the world's population.* New York: United Nations Population Fund.

United Nations Statistics Division, Department of Economic and Social Affairs. (2001). *Social indicators.* Online access at: http://www.un.org/depts/unsd/social/index. htm.

Uranium Institute. Online access at: http://www.uilondon.org/safetab.htm.

U.S. Arms Control and Disarmament Agency. (1993). *World military expenditures and arms transfers.* Washington, DC: U.S. Government Printing Office.

U.S. Census Bureau. (2000). *International database, United States Census Bureau*. Online access at: http://www.census.gov/ipc/www/idbnew.html.

U.S. Central Intelligence Agency, Office of Public Affairs. 2001. *The world factbook.* Washington, DC. Online access at: http://www.odci.gov/cia/publications/factbook/.

U.S. Department of State, (1999) Undersecretary for Arms Control and International Security. *World military expenditures and arms transfers*. Online access at: http://www.state.gov/www/global/arms/bureau_ac/wmeat98/wmeat98.html.

USDA Forest Service. (1989). *Ecoregions of the continents.* Washington, DC: U.S. Government Printing Office.

Watts, Ronald K. (1999). *Comparing federal systems*, 2nd ed. Queens University Press, Kingston, Ont.

The world almanac and book of facts 2001. (2000). Mahwah, NJ: World Almanac Books.

World Bank. *World development indicators, 2003*. Washington, DC.

World Bank. *World development indicators 2001*. Washington, DC. World Bank. Online access at: http://www.worldbank.org/data/.

World Bank. (2001). *World development report 2000/2001: Attacking poverty*. New York: Oxford University Press.

World Health Organization. (1998). *World health statistics annual.* Geneva: World Health Organization.

World Resources Institute. (2000). *World resources 2000–2001. People and ecosystems: The fraying web of life.* Washington, DC: World Resources Institute: Online access at: http://www.wri.org.

World Trade Organization (WTO). Online access at: www.wto.org.

Wright, John W., ed. 2003. *The New York Times almanac 2003*. New York: Penguin Reference.

Notes

Notes

Notes

Notes